INTRODUCTION TO FUZZY ARITHMETIC

Van Nostrand Reinhold
Electrical/Computer Science and Engineering Series
Sanjit Mitra, Series Editor

HANDBOOK OF ELECTRONIC DESIGN AND ANALYSIS PROCEDURES USING PROGRAMMABLE CALCULATORS, by Bruce K. Murdock

COMPILER DESIGN AND CONSTRUCTION, by Arthur B. Pyster

SINUSOIDAL ANALYSIS AND MODELING OF WEAKLY NONLINEAR CIRCUITS, by Donald D. Weiner and John F. Spina

APPLIED MULTIDIMENSIONAL SYSTEMS THEORY, by N. K. Bose

MICROWAVE SEMICONDUCTOR ENGINEERING, by Joseph F. White

INTRODUCTION TO QUARTZ CRYSTAL UNIT DESIGN, by Virgil E. Bottom

DIGITAL IMAGE PROCESSING, by William B. Green

SOFTWARE TESTING TECHNIQUES, by Boris Beizer

LIGHT TRANSMISSION OPTICS, Second edition, by Dietrich Marcuse

REAL TIME COMPUTING, edited by Duncan Mellichamp

HARDWARE AND SOFTWARE CONCEPTS IN VLSI, edited by Guy Rabbat

MODELING AND IDENTIFICATION OF DYNAMIC SYSTEMS, by N. K. Sinha and B. Kuszta

COMPUTER METHODS FOR CIRCUIT ANALYSIS AND DESIGN, by Jiri Vlach and Kishore Singhal

HANDBOOK OF SOFTWARE ENGINEERING, edited by C. R. Vick and C. V. Ramamoorthy

SWITCHED CAPACITOR CIRCUITS, by Phillip E. Allen and Edgar Sánchez-Sinencio

INTRODUCTION TO FUZZY ARITHMETIC: Theory and Applications, by Arnold Kaufmann and Madan M. Gupta

INTRODUCTION TO FUZZY ARITHMETIC

Theory and Applications

Arnold Kaufmann

University of Louvain

Madan M. Gupta

University of Saskatchewan

Van Nostrand Reinhold
Electrical/Computer Science and Engineering Series

VNR VAN NOSTRAND REINHOLD COMPANY

New York

Library of Congress Catalog Card Number: 84-15254
ISBN: 0-442-23007-9

Manufactured in the United States of America.

Published by Van Nostrand Reinhold Company Inc.
135 West 50th Street
New York, New York 10020

Van Nostrand Reinhold Company Limited
Molly Millars Lane
Wokingham, Berkshire RG11 2PY, England

Van Nostrand Reinhold
480 Latrobe Street
Melbourne, Victoria 3000, Australia

Macmillan of Canada
Division of Gage Publishing Limited
164 Commander Boulevard
Agincourt, Ontario MIS 3C7, Canada

15 14 13 12 11 10 9 8 7 6 5 4 3 2 1

Library of Congress Cataloging in Publication Data
Kaufmann, Arnold.
Introduction to fuzzy arithmetic.
(Van Nostrand Reinhold electrical/computer science and engineering series)
Bibliography: p.
Includes index.
1. Fuzzy numbers. 2. Fuzzy arithmetic. I. Gupta, Madan M.
II. Title. III. Series.
QA248.K383 1984 511.3′2 84-15254
ISBN 0-442-23007-9

To Yvette Kaufmann
A.K.

To my grandparents, parents, and teachers
M. M. G.

Contents

Foreword

Although the concept of a fuzzy number as a convex fuzzy subset of the real line was contained in the original formulation of the theory of fuzzy sets, its development began to gather momentum about a decade ago with the publication of seminal papers by A. Kaufmann, K. Tanaka and M. Mizumoto, S. Nahmias, H. Dubois and D. Prade, and others.

The concept of a fuzzy number has led to what has come to be called *fuzzy arithmetic*. Because the concept of a fuzzy number contains that of an interval as a special case, fuzzy arithmetic subsumes interval arithmetic. What matters, however, is not merely that interval arithmetic is a special case of fuzzy arithmetic. Rather, the important point is that, as a language, fuzzy arithmetic has a far greater expressive power than interval arithmetic by virtue of gradation of membership in a fuzzy set. Thus, a fuzzy number may be expressed in linguistic terms, making it possible to compute with words rather than numbers. Furthermore, the membership function of a fuzzy number may be fuzzy set valued, leading to the concept of a fuzzy number of type 2 or, equivalently, an ultrafuzzy number. In this way, the fuzziness of a fuzzy number provides additional degrees of freedom for representing various types of uncertainty as nonuniform possibility distributions over the real line.

An important property of fuzzy numbers that simplifies computation with them is their closure under linear combinations. Thus, if **A** and **B** are triangular numbers, for example, so is any linear combination of them. This property, which is analogous to the closure property of Gaussian probability distributions, makes it possible to characterize fuzzy numbers by a small number of parameters and leads to a theory of so-called L. R. numbers that was developed by Dubois and Prade.

The present volume is the first to present a comprehensive and self-contained theory of fuzzy numbers and its applications. One cannot but be greatly impressed by the wide variety of concepts that Arnold Kaufmann and Madan M. Gupta discuss so well and with so many carefully worked-out examples. Many of the concepts and techniques they describe are original and appear in the literature for the first time: in particular, the concept of a hybrid number, which is both possibilistic and probabilistic in nature; algebraic systems of fuzzy numbers discussed in chapter 3; fuzzy trigonometric functions and complex numbers; and a fuzzy generalization of catastrophy theory.

The authors' exposition of these and other techniques is thorough and well organized and points to a wide variety of applications in systems analysis, decision making under uncertainty, and other fields.

In the years ahead, fuzzy arithmetic is likely to become an important part of numerical analysis and a widely used tool for dealing with uncertainty in a computationally effective way. Professors Kaufmann and Gupta deserve to be congratulated on producing a seminal work that makes an important contribution toward the realization of this objective.

LOTFI A. ZADEH

Preface

In a world that is less and less predictable, we are more and more conscious of uncertainty. Perhaps the environment today is not more uncertain than what it was during the past, but the speed with which information is transferred to every part of the globe makes it appear so. Probably this perception is owing to the difficulties that we experience in making forecasts today. Certainly the world of yesterday appeared simpler. We are buffeted from all sides by all types of information, some of which is directly relevant to our everyday lives, but much of which is not. The difficulty that we experience is trying to decide which is which. When the man in the street states "the world is crazy," and when the scientist claims that "nothing is stationary and certain," they are both reacting to a confusing and uncertain situation. They are both expressing their frustration at not being able to predict the future as well as they would like. This does not mean, of course, that a perfectly predictable future would be good for us. As a matter of fact, we would very soon get tired of such a situation. Whether we like to admit it or not, human beings require a certain amount of uncertainty and variety in their lives. If they did not have this variety, they would soon tire of life itself. Unfortunately, too much uncertainty makes people fear the future, and this is what we must try to avoid.

As human beings we must learn to accept that uncertainty is part of our life and will continue to be part of it in the future as well. As scientists we must also accept the fact that uncertainty is an ever-increasing part of our work day. We must learn to understand it and to work with it. Every system that we deal with has some uncertainty associated with it. Some of this uncertainty appears at a very low level, but a great deal of it appears at a much higher level. In every system we encounter structure, configuration, and constraints, some of which are known and many of which are unknown. A great portion of this uncertainty actually contains useful information, albeit of a fuzzy type, but information nevertheless. This uncertainty appears in the form of *imprecision*, *vagueness*, and *ill defined*, *ill separable*, and *doubtful* data. From this uncertainty, or fuzzy information, we must learn to understand the dynamics of the systems that we are studying. We must learn to recover the information that is present in these uncertain data, we must learn to use it in the building of models of dynamic systems, and finally we must learn how to use these models that are built upon imperfect information. Above all, we must learn how much faith we can place on these models. In other words, from this fuzzy information we must be able to make good and reliable decisions. However, good and reliable decisions does not mean that after the implementation of these decisions we should not expect later regrets.

Life itself exists within a very uncertain dynamic system with a limited entropy range. Consider human beings—without material help we can survive within only a very small region of the earth. If we think about temperature, for example, the range within which we can live is very small. We die of heat at temperatures in excess of 30°C and of cold at temperatures that do not even have to fall below 0°C. Yet we have learned to cope with much hotter and much colder temperatures for long periods of time.

There is no such thing as absolute truth. At best it is a subjective criterion, but one based upon valuation. Unfortunately, too many people place their fate in the hands of the subjective without properly evaluating it.

Uncertainty and subjectivity are the opposite of certainty and objectivity. They are of considerable interest to the authors and will be discussed in this book. The theory of fuzzy sets was introduced in 1965 by Lotfi A. Zadeh in the United States, and has been studied and applied by many people in all parts of the world. More recently, the theory of fuzzy numbers has been introduced by S. Nahmais in the United States and H. Dubois and D. Prade in France. Much of what they have developed is used by us here, but we have also introduced some novel concepts. We have written this book in the hope that it will assist the reader in coming to grips with the uncertain world within which he or she lives. We hope that it will enable him or her to build better models from the fuzzy and subjective data that are available. We also introduce random data into our discussions, we show how they differ from the fuzzy material, and how random data and fuzzy material can be used together. Often, however, there is so much data that it can be gainfully employed only using a computer.

The concept of fuzziness can be approached from several different avenues. The one that we have used in this book is based on the concept of the interval of confidence. We start with this concept and extend it to a range of levels between 0 and 1. Fuzzy arithmetic, a domain of fuzzy set theory, reveals many interesting results. Some of the results described here can be extended further. Many examples are included in the hope that they will make clear the mathematical theories and descriptions that are presented.

In writing this book, it was our intent to present material that was neither too elementary nor too abstract. We will be pleased if we have succeeded, for this is one of the first books devoted entirely to fuzzy numbers, and especially to an introduction to fuzzy numbers.

We hope that this volume will encourage readers to take an active interest in this field and do additional research in it. In particular, we have shown that many practical problems can be studied by means of this theory. Whenever there is only imprecise or indefinite information, nonseparable facts, sensations, subjectivity, and intuition, combined with strong data, statistics, and correlations, fuzzy numbers and random variables, which have much in common, may be used for the solution of the problems.

A friend recently asked: "How, with very few words, would you be able to summarize the subjective construction of a fuzzy number?" We gave the following answer: "What is the smallest value given to this uncertain number? What is the highest? Further, if we were authorized to give one and only one value, what value should we give? Obviously, with these three values, different from each other, we can construct a triangular fuzzy number. Later, if we wish, we may construct with some refinements a convenient, or subjectively convenient, fuzzy number." We believe that we have addressed our friend's question simply and adequately and we hope that we have introduced the basic fuzzy arithmetic that will be found useful in solving real-life problems.

ACKNOWLEDGMENTS

The subject of approximate reasoning, information processing, and decision analysis—especially using data arising from human thought and the cognition process—has occupied a prominent place in the information processing literature since the introduction of fuzzy set theory in 1965. The authors acknowledge the work of the many researchers who have contributed to the development of new theories in the field. Some of these theories are included in this book.

Dr. M. M. Gupta acknowledges the support of the University of Saskatchewan and the encouragement that he received from many members of the College of Engineering during the preparation of this book. He also acknowledges the support that he received from his family on this project.

Finally, the authors wish to express their appreciation to E. Nikiforuk, research assistant, for the editorial assistance that she has provided. Also they acknowledge the help and patience of Bernice Pettinato, editing supervisor, and of the rest of the publisher's staff during the production phases of this book.

A. KAUFMANN
M. M. GUPTA

List of Principal Symbols

$\mathbf{A}$	Fuzzy subset, fuzzy number
A	Ordinary subset
A_α	Ordinary subset of level α, interval of confidence of level α, α – cut
$\mathcal{A}$	Fuzzy number of type 2
$\mathbf{A}_\varphi$	Φ fuzzy subset
$\mathbf{A}_f$	Random fuzzy number with density f
$\mathbf{A}^m$	Mean fuzzy number
R	Set of real numbers
R^+	Set of non-negative real numbers
R_0^+	Set of positive real numbers
R^-	Set of non-positive real numbers
R_0^-	Set of negative real numbers
Z	Set of integers, $Z = \{0, +1, -1, +2, -2, +3, -3 \ldots\}$
N	Set of natural numbers, $N = \{0, 1, 2, 3 \ldots\}$
$A \times B$	Cartesian product of two sets A and B
$\mathbf{A} \times \mathbf{B}$	Product of two fuzzy subsets **A** and **B**
$\mathbf{A}_{\rightarrow}, \mathbf{A}_{\leftarrow}$	Fuzzy numbers bi-oriented
$(+)$	Addition of fuzzy numbers by max-min convolution
$(-)$	Subtraction of fuzzy numbers by max-min convolution
$(\cdot)$	Product of fuzzy numbers by max-min convolution
$(:)$	Division of fuzzy numbers by max-min convolution

$(\wedge)$, $(\vee)$	Minimum (maximum) of fuzzy numbers by max-min convolution
$(+)_n$	Addition modulo n of two fuzzy numbers
$(\cdot)_n$	Multiplication modulo n of two fuzzy numbers
$[+]$	Addition of hybrid numbers
$[a, b]$	Interval of confidence
$\mu_{\mathbf{A}}(x)$	Membership function for the element x with respect to the fuzzy subset **A** (level of presumption), level of a fuzzy number
$(a - b)$	Interval of confidence modulo 1
$\mathbf{A}_n!$	Factorial builds with fuzzy number $\mathbf{A}_n!$
(a_1, a_2, a_3)	Representation of a triangular fuzzy number (TFN)
$\phi(\mathbf{X})$	Function of a fuzzy number **X**
$\delta(\mathbf{A}, \mathbf{B})$	Distance between two fuzzy numbers **A** and **B**
$\delta_l(\mathbf{A})$	Left deviation of fuzzy number **A**
$\delta_r(\mathbf{A})$	Right deviation of fuzzy number **A**
$\delta(\mathbf{A})$	Deviation of fuzzy number **A**
$\iota(\mathbf{A}, \mathbf{H})$	Agreement index of **A** with respect to **H**
poss $(\mathbf{A}, \mathbf{H})$	Possibility of **A** with respect to **H**
$\succ$	Partial order relation
$\prec$	Partial order relation
$<$	Strict total order relation
$>$	Strict total order relation
$\in$	Membership
$\notin$	Nonmembership
$\subset$	Inclusion (is a subset of)
$\subset\subset$	Strict inclusion
$\not\subset$	Noninclusion
$\cup$	Union
$\cap$	Intersection
$\emptyset$	Empty subset

$\Longrightarrow$	Metaimplication (one also says, usually but improperly, implication)
$\Longleftrightarrow$	Logical equivalence
$\exists_x$	Existential quantifier (there exists an x)
iff	If and only if
$\forall_x$	Universal quantifier (for all x)
(a, b)	Interval of **R** "open on the left and on the right," thus $(x \mid a < x < b)$
$(a, b]$	Interval of **R** "open on the left and closed on the right," thus, $(x \mid a < x \leq b]$
$[a, b)$	Interval of **R** "closed on the left and open on the right," thus, $[x \mid a \leq x < b)$
$[a, b]$	Interval of **R** "closed on the left and on the right," thus, $[x \mid a \leq x \leq b]$; one also says *segment*
$\max(X, Y)$ or $X \vee Y$	Maximum of X and Y
$\min(X, Y)$ or $X \wedge Y$	Minimum of X and Y

INTRODUCTION TO FUZZY ARITHMETIC

CHAPTER 1

DEFINITIONS AND MAIN PROPERTIES OF FUZZY NUMBERS

1.1 INTRODUCTION

The concept of uncertain or fuzzy numbers may be presented in many ways. In this book we consider a fuzzy number to be an extension of the concept of the interval of confidence, which is familiar to anyone who has computed using imprecise data in simple or complex systems. This extension is based on a natural and very simple idea. Instead of considering the interval of confidence at one unique level, it is considered at several levels and more generally at all levels from 0 to 1. We consider the maximum of presumption to be at level 1 and the minimum of presumption to be at level 0. The level of presumption 0 gives a restrictive hypothesis. Similarly, the level of presumption α, $\alpha \in [0, 1]$ gives an interval of confidence $A_\alpha = [a_1^{(\alpha)}, a_2^{(\alpha)}]$, which is a monotonic decreasing function of α; that is,

$$(\alpha' > \alpha) \Rightarrow (A_{\alpha'} \subset A_\alpha),$$

or

$$(\alpha' > \alpha) \Rightarrow ([a_1^{(\alpha')}, a_2^{(\alpha')}] \subset [a_1^{(\alpha)}, a_2^{(\alpha)}]),$$

for every α, $\alpha' \in [0, 1]$.

The expressions *level of presumption* or *presumption level* are well suited to the concept of fuzzy numbers. We must remember, however, not to confuse fuzzy numbers with random numbers (more usually called random variables). Uncertainty and randomness are two very different and important concepts. They can be used together, but should not be confused.

1.2 INTERVAL OF CONFIDENCE

Consider a situation in which the value is uncertain. Suppose that the information available is such that we can accept that the uncertain value belongs to the referential set R (a set of real numbers). In many situations encountered by scientists it is possible to locate the value inside a closed interval of R; that is, an *interval of confidence* of R: $[a_1, a_2]$.[1] We are thus certain that the value is greater or equal to a_1 and smaller or equal to a_2. This kind of statement often occurs in science and engineering. In this case we use the symbol

$$A = [a_1, a_2].$$

Generally the numbers a_1 and a_2 are finite, but in some cases it is useful, or even necessary, to consider that $a_1 = -\infty$ and/or $a_2 = \infty$. In other cases, instead of considering a closed interval, we consider open intervals by using the notation

$]a_1, a_2]$ or $(a_1, a_2]$: open at the left,
$[a_1, a_2[$ or $[a_1, a_2)$: open at the right,
$]a_1, a_2[$ or (a_1, a_2): open at the left and at the right; that is, open.

Note, however, that it is preferable to use brackets to indicate an open interval rather than using parentheses. Note also that the symbol $x \in [a_1, a_2]$ denotes the fact that x is an uncertain value.

Next we discuss several operations for the interval of confidence.

Addition

Assume two intervals of confidence in R:

$$A = [a_1, a_2] \tag{1.1}$$

and

$$B = [b_1, b_2]. \tag{1.2}$$

Hence, if

$$x \in [a_1, a_2] \tag{1.3}$$

and

$$y \in [b_1, b_2], \tag{1.4}$$

then

$$x + y \in [a_1 + b_1, a_2 + b_2]. \tag{1.5}$$

Symbolically, we write

$$\begin{aligned} A\,(+)\,B &= [a_1, a_2]\,(+)\,[b_1, b_2] \\ &= [a_1 + b_1, a_2 + b_2]. \end{aligned} \tag{1.6}$$

The proof of (1.5) is trivial, because if $x \geq a_1$ and $y \geq b_1$, $(x + y) \geq (a_1 + b_1)$, and because $x \leq a_2$ and $y \leq b_2$, then $(x + y) \leq (a_2 + b_2)$.

Subtraction

If $x \in [a_1, a_2]$ and $y \in [b_1, b_2]$, then $(x - y) \in [a_1 - b_2, a_2 - b_1]$. In fact, we must subtract the largest value in $[b_1, b_2]$ from a_1 and the smallest in $[b_1, b_2]$ from a_2. Again, we can write this symbolically as follows:

$$\begin{aligned} A\,(-)\,B &= [a_1, a_2]\,(-)\,[b_1, b_2] \\ &= [a_1 - b_2, a_2 - b_1]. \end{aligned}$$

Reduction to a Certain Number

If a certain number is a singleton in R, we write $\mathbf{L} = [\ell, \ell]$, or, using the relation between the singleton and its unique element, $\ell = [\ell, \ell]$. We may then write

$$0 = [0, 0] \tag{1.7}$$

and

$$1 = [1, 1]. \tag{1.8}$$

To add an ordinary positive number to an interval of confidence we must move the interval to the right, and to subtract we must move it to the left by adding the corresponding negative number.

Image

If $x \in [a_1, a_2]$, *then* $-x \in [-a_2, -a_1]$. Hence, if A is an interval of confidence, its *image* is defined as

$$A^- = [-a_2, -a_1],$$

and, note that

$$\begin{aligned} A\,(+)\,A^- &= [a_1, a_2]\,(+)\,[-a_2, -a_1] \\ &= [a_1 - a_2, a_2 - a_1] \neq 0. \end{aligned} \tag{1.9}$$

Properties and Structure

Using (1.1–1.6) and (1.9), the following properties can be proved:

$\forall A, B, C \subset R$:

$$A\ (+)\ B = B\ (+)\ A, \qquad \text{(commutative)}, \tag{1.10}$$

$$A\ (+)\ (B\ (+)\ C) = (A\ (+)\ B)\ (+)\ C, \qquad \text{(associative)}. \tag{1.11}$$

If a neutral, a number 0, exits at the left and another neutral at the right to (1.7), then

$$A\ (+)\ 0 = 0\ (+)\ A = A. \tag{1.12}$$

The image, however, is not symmetric in the sense of the set theory; that is,

$$A\ (+)\ A^- = A^-\ (+)\ A \neq 0 \tag{1.13}$$

(see 1.9). This property shows that the set of the interval of confidence in R has a monoidal structure (semigroup) for addition, but does not have a group structure. This monoid is commutative.

Subtraction is not, however, monoidal or associative, which may be proved easily.

EXAMPLE 1.1

Consider the following numerical example:

$A = [3.52, 5.83]$, $B = [-2.24, 6.04]$, and $C = [-4, -1.17]$. Then

$$\begin{aligned}
A\ (+)\ B &= [3.52, 5.83]\ (+)\ [-2.24, 6.04] \\
&= [3.52 - 2.24, 5.83 + 6.04] \\
&= [1.28, 11.87]. \\
A\ (-)\ B &= [3.52, 5.83]\ (-)\ [-2.24, 6.04] \\
&= [3.52 - 6.04, 5.83 + 2.24] \\
&= [-2.52, 8.07]. \\
(A\ (+)\ B) - C &= ([3.52, 5.83]\ (+)\ [-2.24, 6.04])\ (-)\ [-4, -1.17] \\
&= [1.28, 11.87]\ (-)\ [-4, -1.17] \\
&= [1.28 + 1.17, 11.87 + 4] \\
&= [2.45, 15.87]. \\
A &= [-5.83, -3.52]. \\
A\ (+)\ A &= [3.52, 5.83]\ (+)\ [-5.83, -3.52] \\
&= [-2.31, 2.31].
\end{aligned}$$

Multiplication

If the intervals of confidence belong to R^+, and if $x \in [a_1, a_2]$ and $y \in [b_1, b_2]$, then $x \cdot y \in [a_1 \cdot b_1, a_2 \cdot b_2]$. We can therefore write

$$\begin{aligned} A\,(\cdot)\,B &= [a_1, a_2]\,(\cdot)\,[b_1, b_2] \\ &= [a_1 \cdot b_1, a_2 \cdot b_2]. \end{aligned} \tag{1.14}$$

When A and B belong to R instead R^+, the result is more complicated because a_1, a_2, b_1, and b_2 can be negative, and nine combinations are therefore possible. In Appendix A we present a general case in which the reference set is R.

Division

Division is defined here only in R^+:

$$\begin{aligned} A\,(:)\,B &= [a_1, a_2]\,(:)\,[b_1, b_2] \\ &= [a_1/b_2, a_2/b_1]. \end{aligned}$$

If $b_1 = 0$, the upper bound increases to $+\infty$. If $b_1 = b_2 = 0$, the interval of confidence is extended to $+\infty$.

Inverse

If $x \in [a_1, a_2] \subset R_0^+$, then $1/x \in [1/a_2, 1/a_1]$, *and*

$$\begin{aligned} A^{-1} &= [a_1, a_2]^{-1} \\ &= [1/a_2, 1/a_1]. \end{aligned}$$

Properties and Structure

$\forall A, B \subset R^+$:

$$A\,(\cdot)\,B = B\,(\cdot)\,A \quad \text{(commutative)}$$

This property is obvious from (1.14). Associativity must be proved, however, since it is not as obvious.

$\forall A, B, C \subset R^+$:

$$(A\,(\cdot)\,B)\,(\cdot)\,C = A\,(\cdot)\,(B\,(\cdot)\,C).$$

Indeed,

$$\begin{aligned} &([a_1, a_2]\,(\cdot)\,[b_1, b_2])\,(\cdot)\,[c_1, c_2] \\ &\qquad = [a_1 \cdot b_1, a_2 \cdot b_2]\,(\cdot)\,[c_1, c_2] \\ &\qquad = [a_1 \cdot b_1 \cdot c_1, a_2 \cdot b_2 \cdot c_2]. \end{aligned}$$

Also,

$$\begin{aligned} &[a_1, a_2]\,(\cdot)\,([b_1, b_2]\,(\cdot)\,[c_1, c_2]) \\ &\qquad = [a_1, a_2]\,(\cdot)\,[b_1 \cdot c_1, b_2 \cdot c_2] \\ &\qquad = [a_1 \cdot b_1 \cdot c_1, a_2 \cdot b_2 \cdot c_2]. \end{aligned}$$

Therefore,

$$([a_1, a_2] (\cdot) [b_1, b_2]) (\cdot) [c_1, c_2] = [a_1, a_2] (\cdot) ([b_1, b_2] (\cdot) [c_1, c_2]).$$

If a neutral exists at the left and a second neutral exists at the right for multiplication, this number is 1, (1.8), that is,

$$A (\cdot) 1 = 1 (\cdot) A = A.$$

The inverse is not symmetric in the sense of set theory; that is,

$$\begin{aligned} A (\cdot) A^{-1} &= [a_1, a_2] (\cdot) [1/a_2, 1/a_1] \\ &= [a_1/a_2, a_2/a_1] \neq 1. \end{aligned}$$

The set of the interval of confidence in R^+ has, therefore, a monoidal structure (semigroup) for multiplication but not group structure. This monoid is commutative.

EXAMPLE 1.2

Look at the following numerical example:
$A = [2.62, 3.55]$, $B = [0.77, 4.50]$, and $C = [3.11, 4.86]$. Then

$$\begin{aligned} A (\cdot) B &= [2.62, 3.55] (\cdot) [0.77, 4.50] \\ &= [2.0174, 15.9750]. \\ A (\cdot) B (\cdot) C &= [2.0174, 15.9750] (\cdot) [3.11, 4.86] \\ &= [6.274114, 77.638500]. \\ A (:) B &= [2.62, 3.55] (:) [0.77, 4.50] \\ &= [2.62/4.50, 3.55/0.77] \\ &= [0.58222, 4.61038]. \\ A^{-1} &= [1/3.55, 1/2.62] = [0.28169, 0.38167]. \end{aligned}$$

Multiplication by a Nonnegative Number

Suppose that $k \in R^+$. We may then write

$$k = [k, k],$$

and then

$\forall A \in R^+$:

$$\begin{aligned} k \cdot A &= k \cdot [a_1, a_2] \\ &= [k, k] (\cdot) [a_1, a_2] \\ &= [ka_1, ka_2]. \end{aligned}$$

Division by $k > 0$ is equivalent to multiplication by $1/k$.

Intervals of Confidence in *Z* (Integers)

We recall that

$$Z = \{\ldots, -3, -2, -1, 0, 1, 2, 3, \ldots\}, \tag{1.15}$$

and

$$N = \{0, 1, 2, 3, \ldots\}. \qquad \textbf{(1.16)}$$

Although we are actually misusing one of our definitions, we employ the term *interval of confidence* for **A** when a_1 and a_2 belong to Z and are the lower and upper bounds of a sequence of integers ranging from a_1 to a_2. For instance,

$$A = [-3, 2] = \{-3, -2, -1, 0, 1, 2\}. \qquad \textbf{(1.17)}$$

As far as the addition of two intervals of confidence in Z is concerned, by definition, if $x \in [a_1, a_2]$ and $y \in [b_1, b_2]$ then $(x + y) \in [a_1 + b_1, a_2 + b_2]$. It is easy to check that this structure is still a commutative monoid in Z. We note that the opposite does not exist in N.

For the operation of multiplication we consider at this point only the referential N, for which we have a commutative monoid. By definition, if $x \in [a_1, a_2]$ and $y \in [b_1, b_2]$, *then* $x \cdot y \in [a_1 \cdot b_1, a_2 \cdot b_2]$, where

$$[a_1 \cdot b_1, a_2 \cdot b_2] = \{a_1 \cdot b_1, a_1 \cdot b_1 + 1, a_1 \cdot b_1 + 2, \ldots, a_2 \cdot b_2 - 2, a_2 \cdot b_2 - 1, a_2 \cdot b_2\}.$$

The inverse is not defined for multiplication in N because we would have to pass from N to R^+.

The multiplication of an interval of confidence in N by a number $k \in N$ does not cause any problems since

$$k = [k, k]. \qquad \textbf{(1.18)}$$

If $k \in Z$, $k < 0$, we use Z instead of N.

Except for trivial cases, the process of dividing an interval of confidence belonging to Z by another leads to a subset of R.

EXAMPLE 1.3

Consider the following numerical example:

$$\begin{aligned}
A &= [-3, 2] = \{-3, -2, -1, 0, 1, 2\}, \\
B &= [1, 4] = \{1, 2, 3, 4,\}. \\
A\,(+)\,B &= [-3, 2]\,(+)\,[1, 4] \\
&= [-2, 6] = \{-2, -1, 0, 1, 2, 3, 4, 5, 6\}. \\
C &= [3, 5]. \\
B\,(\cdot)\,C &= [1, 4]\,(\cdot)\,[3, 5] \\
&= [3, 20] = \{3, 4, 5, \ldots, 18, 19, 20\}. \\
3 &= [3, 3], \\
3\,(\cdot)\,B &= [3, 3]\,(\cdot)\,[1, 4] \\
&= [3, 12] = \{3, 4, 5, \ldots, 10, 11, 12\}.
\end{aligned}$$

Let us now discuss the *maximum* and *minimum* of two intervals of confidence. We consider only R, but our conclusions also apply to Z.

Let us recall the meaning of the symbols $\wedge$ and $\vee$. Consider two numbers in R, such that, by definition,

$$\begin{aligned} a \wedge b = \text{Min}(a, b) &= a, \quad \text{if } a \leq b, \\ &= b, \quad \text{if } b \leq a. \\ a \vee b = \text{Max}(a, b) &= b, \quad \text{if } a \leq b, \\ &= a, \quad \text{if } b \leq a. \end{aligned}$$

Now let us consider two intervals of confidence in R, $A = [a_1, a_2]$, and $B = [b_1, b_2]$. We introduce two operators $(\wedge)$ and $(\vee)$ as follows:

$$\begin{aligned} A\,(\wedge)\,B &= [a_1, a_2]\,(\wedge)\,[b_1, b_2] \\ &= [a_1 \wedge b_1, a_2 \wedge b_2]. \\ A\,(\vee)\,B &= [a_1, a_2]\,(\vee)\,[b_1, b_2] \\ &= [a_1 \vee b_1, a_2 \vee b_2]. \end{aligned}$$

Misusing the term (because intervals of confidence are not totally ordered but only partially ordered), we call the operation $(\wedge)$ the *minimum of A and B* and the operation $(\vee)$ the *maximum of A and B*. We note that $(\wedge)$ and $(\vee)$ satisfy the commutative and associative properties, but a finite neutral does not exist.

EXAMPLE 1.4

$$\begin{aligned} &A = [-3, 5],\ B = [2, 6],\ C = [-4, 7]. \\ A\,(\wedge)\,B &= [-3, 5]\,(\wedge)\,[2, 6] \\ &= [-3 \wedge 2, 5 \wedge 6] = [-3, 5]. \\ A\,(\vee)\,B &= [-3, 5]\,(\vee)\,[2, 6] \\ &= [-3 \vee 2, 5 \vee 6] = [2, 6]. \\ A\,(\wedge)\,C &= [-3, 5]\,(\wedge)\,[-4, 7] \\ &= [-3 \wedge -4, 5 \wedge 7] = [-4, 5]. \\ A\,(\vee)\,C &= [-3, 5]\,(\vee)\,[-4, 7] \\ &= [-3 \vee -4, 5 \vee 7] = [-3, 7]. \\ B\,(\wedge)\,C &= [2, 6]\,(\wedge)\,[-4, 7] \\ &= [2 \wedge -4, 6 \wedge 7] = [-4, 6]. \\ B\,(\vee)\,C &= [2, 6]\,(\vee)\,[-4, 7] \\ &= [2 \vee -4, 6 \vee 7] = [2, 7]. \end{aligned}$$

Since theory of the intervals of confidence has many interesting applications, we next consider a generalization of it.

1.3. UNCERTAIN NUMBERS OR FUZZY NUMBERS

An interval of confidence is one way of reducing the uncertainty of using lower and upper bounds. It is a practical and logical process for treating uncertainty with whatever information is available. This information can be objective (we are certain that the dimension has a position between the two measured data), or subjective (the information is obtained from experience or from the opinion of the experts).

Let us relate the concept of the interval of confidence to another called the *level of presumption*.[2] Let us assume, for example, that a certain job is to be completed between two dates, say May 15 and May 31. This is an interval of confidence. On the other hand, let us assume that this same job is to be completed on May 22, a possible date. The interval of confidence in the first case is [May 15, May 31] while in the second case it is [May 22, May 22]. If we wish, we may assign two levels of confidence to these two situations, 0 for [May 15, May 31] and 1 for [May 22, May 22]. These two levels of confidence are in fact levels of presumption, and we can represent them by [0, 1]. Of course, there is no reason why we should limit ourselves to only the two values 0 and 1. We could have selected a much larger set of values such as

$$\forall \alpha_1, \alpha_2 \in [0, 1]: \quad (\alpha_1 < \alpha_2) \Rightarrow ([a_1^{(\alpha_2)}, a_2^{(\alpha_2)}] \subset [a_1^{(\alpha_1)}, a_2^{(\alpha_1)}]).$$

This means that if α increases, the interval of confidence never increases. Such a situation is shown in Figure 1.1.

Two concepts have been associated in this discussion, the interval of confidence and the level of presumption. The coupling between the level α of presumption and the interval of confidence at level α will be a way defining the concept of an *uncertain number* or a *fuzzy number*. This association corresponds to the natural, often implicit, mechanism of human thinking in the subjective estimation of a value for a dimension.

The curve that represents the modification of the interval of confidence from 0 to 1 can be one of two types. It may be a smooth curve as shown in Figure 1.1 or it may have a deflection in its slope, resulting in a flat region as is shown in Figure 1.2.

The concept of an uncertain number or a fuzzy number can be defined in R or Z, R^+ or N, in every referential set that is totally ordered (lineraly ordered). In this book we study the theory of such numbers. Many of their properties are novel and have interesting applications.

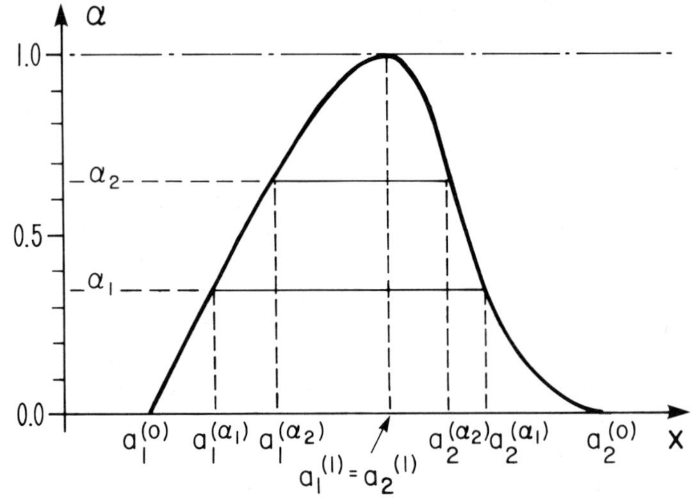

Figure 1.1 Definition of fuzzy numbers.

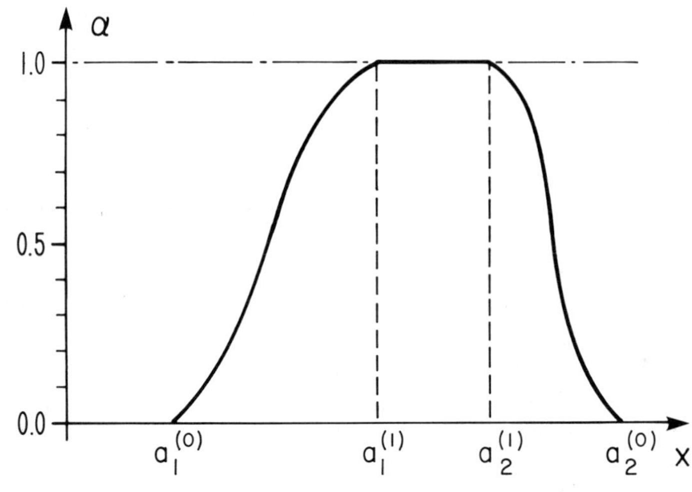

Figure 1.2 Fuzzy number with a flat region.

The word *fuzzy* was first introduced by Zadeh [1] in his famous paper "Fuzzy Sets." He used this word to generalize the mathematical concept of *set* to one of *fuzzy set* or *fuzzy subset*, where in a fuzzy set, a *membership function* is defined for each element of the referential set.[3] (In Boolean algebra this concept is usually called the *characteristic function*.) The membership function takes its values in the interval $[0, 1] \subset R^{+}$ instead of $\{0, 1\}$ as in Boolean algebra. Since the appearance of Zadeh's paper, more than 4000 papers and books have appeared on the theory of this topic and its applications. The field is growing exponentially. More recently, additional important works on the concept of fuzzy number have been written by Nahmias [2-3] in the United States and by Dubois and Prade [4-5] France and by several other authors.

In the present work, the concept of a fuzzy number is presented using the couple: level of presumption and interval of confidence. This is not, however, the classical way of introducing a fuzzy number, and we will introduce this concept from the viewpoint of fuzzy set theory. With this in mind, let us first explain how to distinguish the concept of fuzzy number from the concept of fuzzy set.

Let E be a referential set (for example, R or Z). An ordinary subset, A, of this referential set is defined by its characteristic function.

$\forall\, x \in E$:

$$\mu_{\mathbf{A}}(x) \in \{0, 1\},$$

which shows that an element of E belongs to, or does not belong to A according to the value of the characteristic function (1 or 0).

For the same referential set E a fuzzy subset **A** will be defined by its characteristic function, called the membership function, which takes its values in the interval $[0, 1]$ instead of in the binary set $\{0, 1\}$.

$\forall\, x \in E$:

$$\mu_{\mathbf{A}}(x) \in [0, 1],$$

that is, that the elements of E belong to **A** with a level located in $[0, 1]$. This extension from ordinary subsets to fuzzy subsets is specified by the use of a boldface letter.

Figure 1.3 shows an ordinary subset in R, and Figure 1.4 shows a fuzzy subset in R.

We recall the two properties previously discussed so that we may define the concept of uncertain numbers or fuzzy numbers from the concept of fuzzy subset. These properties are the convexity and the normality of a fuzzy subset.

A fuzzy subset $\mathbf{A} \subset R$ is *convex* if and only if every ordinary

$$A_\alpha = \{x \mid \mu_{\mathbf{A}}(x) \geq \alpha\}, \qquad \alpha \in [0, 1],$$

subset is convex; that is, if it is a closed interval of R. Figure 1.5 shows a convex fuzzy subset.

Another definition that we may use is

$\forall\, x^{(1)}, x^{(2)} \in R$:

$$\mu_{\mathbf{A}}[\lambda x^{(1)} + (1 - \lambda)x^{(2)}] \geq \mu_{\mathbf{A}}(x^{(1)}) \wedge \mu_{\mathbf{A}}(x^{(2)}),$$

$\forall\, \lambda \in [0, 1]$.

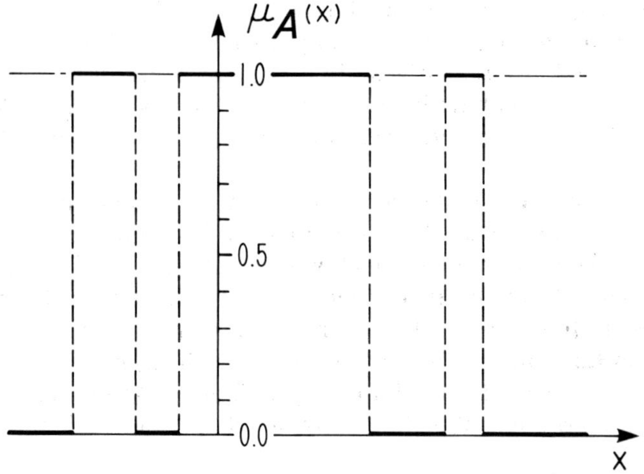

Figure 1.3 An ordinary subset in R.

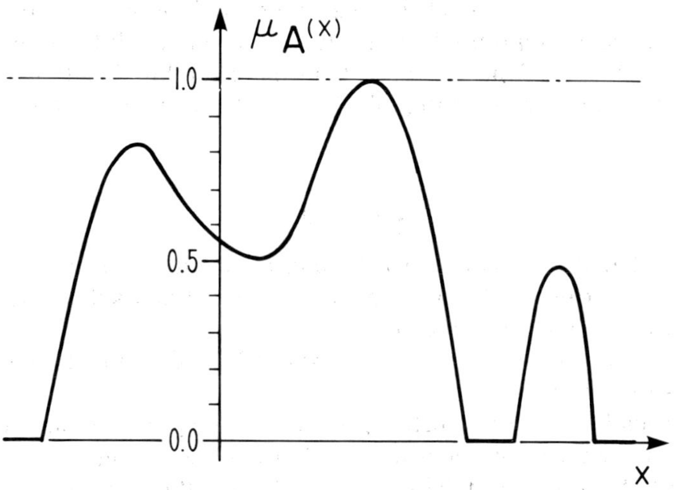

Figure 1.4 A fuzzy subset in R.

We now define normality. A fuzzy subset $\mathbf{A} \subset R$ is *normal* if and only if

$\forall x \in R$:

$$\bigvee_{x} \mu_{\mathbf{A}}(x) = 1.$$

This means that the highest value of $\mu_{\mathbf{A}}(x)$ is equal to 1. This maximum may or may not be unique. Figure 1.5 shows a nonnormal fuzzy subset, and Figure 1.6 shows a normal fuzzy subset.

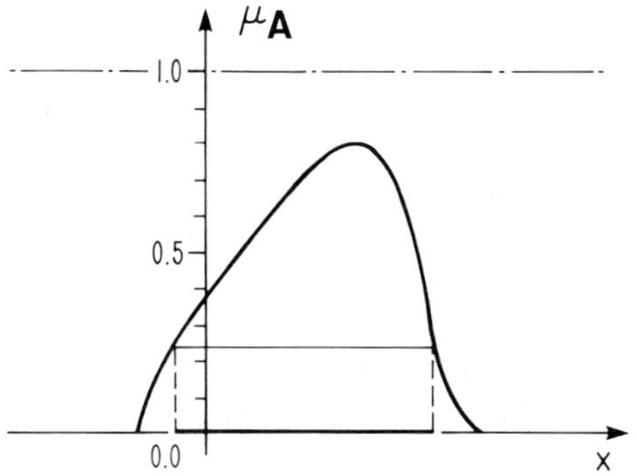

Figure 1.5 A convex fuzzy subset (nonnormal).

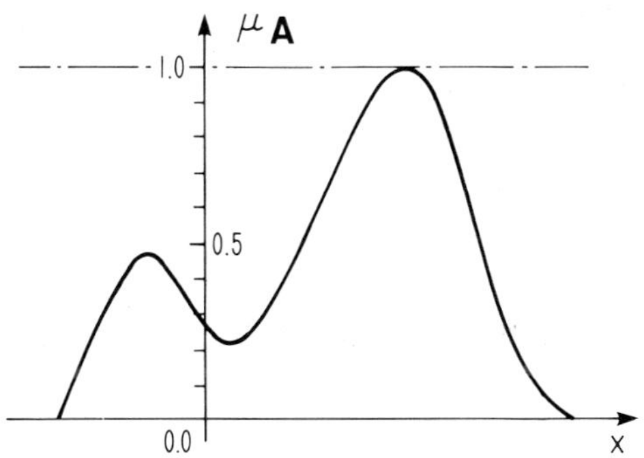

Figure 1.6 A fuzzy subset (normal).

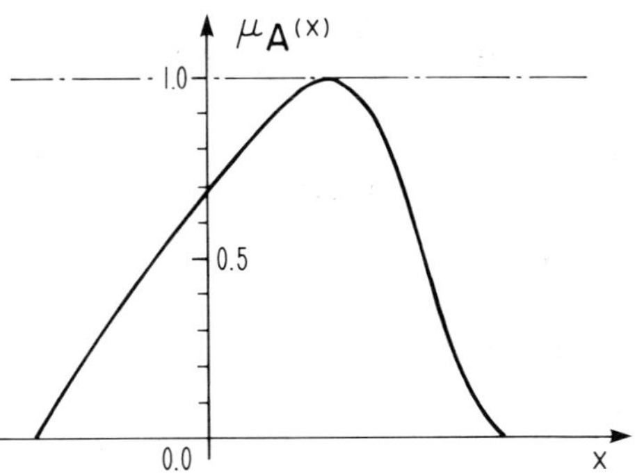

Figure 1.7 A convex and normal fuzzy set.

We are now able to give a definition for uncertain or fuzzy numbers. A fuzzy number in R is a fuzzy subset of R that is convex and normal; this is shown in Figure 1.7. Thus a fuzzy number can be considered a generalization of the interval of confidence. It is not, however, a random variable. A random variable is defined in terms of the theory of probability, which has evolved from theory of measurement. A random variable is an objective datum, whereas a fuzzy number is a subjective datum. It is a valuation, not a measure. However, we describe an association between random variables and fuzzy numbers that may be very useful for many practical problems.

1.4. ADDITION OF FUZZY NUMBERS

In section 1.2 we showed how to add two intervals of confidence. The addition of fuzzy numbers follows the same process, but level by level. For example, let **A** and **B** be two fuzzy numbers and A_α and B_α their intervals of confidence for the level of presumption α, $\alpha \in [0, 1]$. We can then write

$$\begin{aligned} A_\alpha (+) B_\alpha &= [a_1^{(\alpha)}, a_2^{(\alpha)}] (+) [b_1^{(\alpha)}, b_2^{(\alpha)}] \\ &= [a_1^{(\alpha)} + b_1^{(\alpha)}, a_2^{(\alpha)} + b_2^{(\alpha)}]. \end{aligned} \quad \mathbf{(1.19)}$$

If **A**, **B** $\subset$ R, then for the intervals of confidence at the level α, we may define the following ordinary subsets A_α and B_α:

$$A_\alpha = \{x \mid \mu_{\mathbf{A}}(x) \geq \alpha\}, \quad \mathbf{(1.20)}$$

$$B_\alpha = \{x \mid \mu_{\mathbf{B}}(x) \geq \alpha\}. \quad \mathbf{(1.21)}$$

Let us now consider another method for the addition of fuzzy numbers. Let

$\mathbf{A}, \mathbf{B} \subset R, \qquad \forall\, x, y, z \in R:$

$$\mu_{\mathbf{A}(+)\mathbf{B}}(z) = \bigvee_{z = x + y} (\mu_{\mathbf{A}}(x) \wedge \mu_{\mathbf{B}}(y)). \quad \mathbf{(1.22)}$$

We will now prove that (1.19) and (1.22) describe the same operation. From (1.20) and (1.21) we can write, by nesting, that[4]

$$\mathbf{A} = \bigcup_\alpha \alpha \cdot A_\alpha = \bigcup_\alpha \alpha \cdot [a_1^{(\alpha)}, a_2^{(\alpha)}], \quad \mathbf{(1.23)}$$

$$\mathbf{B} = \bigcup_\alpha \alpha \cdot B_\alpha = \bigcup_\alpha \alpha \cdot [b_1^{(\alpha)}, b_2^{(\alpha)}]. \quad \mathbf{(1.24)}$$

It is easy to prove that equations (1.19) and (1.22) are as valid for numbers in Z as in N, as has been described in (1.15–1.18). Two examples will clarify this situation.

$$\mu_{\mathbf{A}}(x)=1, \quad \text{if } x \in [a_1^{(\alpha)}, a_2^{(\alpha)}],$$
$$=0, \quad \text{if } x \notin [a_1^{(\alpha)}, a_2^{(\alpha)}].$$
$$\mu_{\mathbf{B}}(x)=1, \quad \text{if } x \in [b_1^{(\alpha)}, b_2^{(\alpha)}],$$
$$=0, \quad \text{if } x \notin [b_1^{(\alpha)}, b_2^{(\alpha)}]. \qquad \textbf{(1.25)}$$

Note that we have used symbolic writing in (1.23), (1.24) and (1.25). We can also write that

$$\mu_{\mathbf{A}(+)\mathbf{B}}(z)=1 \quad \text{if} \quad z \in [a_1^{(\alpha)}+b_1^{(\alpha)}, a_2^{(\alpha)}+b_2^{(\alpha)}],$$
$$=0 \quad \text{if} \quad z \notin [a_1^{(\alpha)}+b_1^{(\alpha)}, a_2^{(\alpha)}+b_2^{(\alpha)}].$$

Now let us return to (1.22) and apply it to every level α. We find that

$\forall\ x, y, z \in R$:

$$\mu_{A_\alpha(+)B_\alpha}(z)=\bigvee_{z=x+y}(\mu_{A_\alpha}(x) \wedge \mu_{B_\alpha}(y)). \qquad \textbf{(1.26)}$$

For all values of x and y such that

$$\mu_{A_\alpha}(x) = 1 \text{ and } \mu_{B_\alpha}(x) = 1,$$

the right side of (1.26) gives 1. If this is not true, it gives 0, and since $z=x+y$, we write

$$z \in [a_1^{(\alpha)} + b_1^{(\alpha)}, a_2^{(\alpha)} + b_2^{(\alpha)}].$$

Referring to (1.23) and (1.24) we write

$$A_\alpha\ (+)\ B_\alpha = \bigcup_\alpha \alpha \cdot [a_1^{(\alpha)} + b_1^{(\alpha)}, a_2^{(\alpha)} + b_2^{(\alpha)}]. \qquad \textbf{(1.27)}$$

Figure 1.8 shows the addition of two fuzzy numbers.

It is easy to prove that equations (1.19) and (1.25) are valid for numbers in Z as in N reporting to (1.15-1.18). Two examples will clarify this.

EXAMPLE 1.5

Assume the two triangular fuzzy numbers shown in Figure 1.9 and compute their sum, where $\mu_{\mathbf{A}}(x)$ and $\mu_{\mathbf{B}}(x)$ are defined as follows:

$\forall\ x \in R$:

$$\mu_{\mathbf{A}}(x) = 0, \qquad x \le -5,$$
$$= x/3 + 5/3, \qquad -5 \le x \le -2,$$
$$= -x/3 + 1/3, \qquad -2 \le x \le 1,$$
$$= 0, \qquad x \ge 1, \qquad \textbf{(1.28)}$$

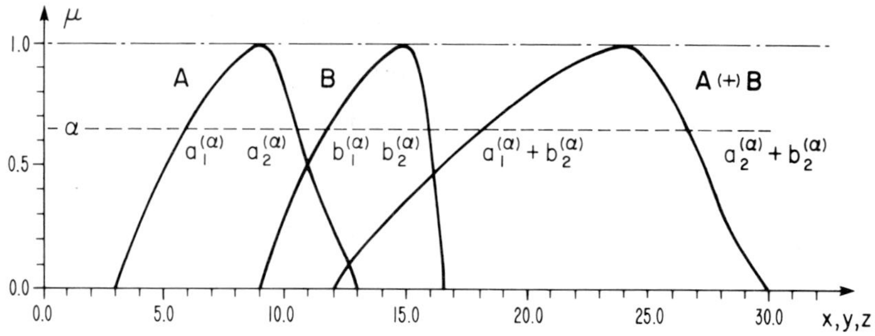

Figure 1.8 Addition of two fuzzy numbers.

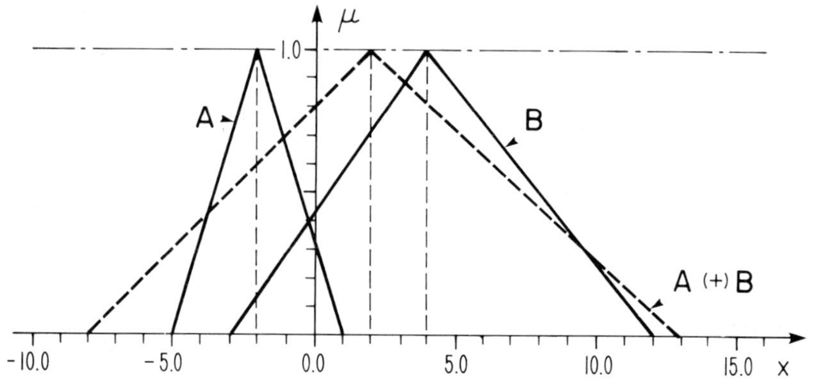

Figure 1.9 Addition of two triangular fuzzy numbers (Example 1.5).

and

$$\begin{aligned}\mu_{\mathbf{B}}(x) &= 0, & x &\le -3,\\ &= x/7 + 3/7; & -3 \le x &\le 4,\\ &= -x/8 + 12/8, & 4 \le x &\le 12,\\ &= 0, & x &\ge 12.\end{aligned} \qquad \textbf{(1.29)}$$

To compute the intervals of confidence for each level α the triangular shapes will be described by functions of α in the following manner:

From (1.28),

$$\alpha = a_1^{(\alpha)}/3 + 5/3$$

and

$$\alpha = -a_2^{(\alpha)}/3 + 1/3.$$

Hence, the interval of confidence at the level α is given by

$$\begin{aligned}A_\alpha &= [a_1^{(\alpha)}, a_2^{(\alpha)}]\\ &= [3\alpha - 5, -3\alpha + 1].\end{aligned} \qquad \textbf{(1.30)}$$

From (1.29),

$$\alpha = b_1^{(\alpha)}/7 + 3/7$$

and

$$\alpha = -b_2^{(\alpha)}/8 + 12/8.$$

Therefore

$$\begin{aligned} B_\alpha &= [b_1^{(\alpha)}, b_2^{(\alpha)}] \\ &= [7\alpha - 3, -8\alpha + 12], \end{aligned} \qquad \textbf{(1.31)}$$

Adding (1.30) and (1.31) gives

$$\begin{aligned} A_\alpha (+) b_\alpha &= [a_1^{(\alpha)} + b_1^{(\alpha)}, a_2^{(\alpha)} + b_2^{(\alpha)}] \\ &= [3\alpha - 5, -3\alpha + 1] (+) [7\alpha - 3, -8\alpha + 12] \\ &= [10\alpha - 8, -11\alpha + 13]. \end{aligned} \qquad \textbf{(1.32)}$$

From

$$a_1^{(\alpha)} + b_1^{(\alpha)} = 10\alpha - 8$$

and

$$a_1^{(\alpha)} + b_2^{(\alpha)} = -11\alpha + 13,$$

We obtain

$$\begin{aligned} \mu_{\mathbf{A}\,(+)\,\mathbf{B}}(x) &= 0, && x \leq -8, \\ &= x/10 + 8/10, && -8 \leq x \leq 2, \\ &= -x/11 + 13/11, && 2 \leq x \leq 13, \\ &= 0, && x \geq 13. \end{aligned}$$

These computations were very simple because of the triangular shapes that were used. A similar procedure can be used, however, for any other shape.

EXAMPLE 1.6

Consider the following numerical example, with a fuzzy number in N:

	0	1	2	3	4	5	6	7	8	
A =	0	0.1	0.3	0.8	1	0.7	0.3	0	0	. . .

	0	1	2	3	4	5	6	7	8	
B =	0	0.3	0.6	1	0.7	0.2	0.1	0	0	. . .

Table 1.1

α	0	1	2	3	4	5	6	7
1					1			
.9					1			
.8				1	1			
.7				1	1	1		
.6				1	1	1		
.5				1	1	1		
.4				1	1	1		
.3			1	1	1	1	1	
.2			1	1	1	1	1	
.1		1	1	1	1	1	1	
0	1	1	1	1	1	1	1	1
	0	0.1	0.3	0.8	1	0.7	0.3	0

(+)

0	1	2	3	4	5	6	7
			1				
			1				
			1				
			1	1			
		1	1	1			
		1	1	1			
		1	1	1			
	1	1	1	1			
	1	1	1	1	1		
	1	1	1	1	1	1	
1	1	1	1	1	1	1	1
0	0.3	0.6	1.0	0.7	0.2	0.1	0

=

1	2	3	4	5	6	7	8	9	10	11	12	13
						1						
						1						
					1	1						
					1	1	1	1				
				1	1	1	1	1				
				1	1	1	1	1				
				1	1	1	1	1				
		1	1	1	1	1	1	1	1			
		1	1	1	1	1	1	1	1	1		
	1	1	1	1	1	1	1	1	1	1	1	
1	1	1	1	1	1	1	1	1	1	1	1	1
0	0.1	0.3	0.3	0.6	0.8	1	0.7	0.7	0.3	0.2	0.1	0

To compute the sum of the interval of confidence at level α, we shall use (1.19) to obtain Table 1.1. Hence

$$\mathbf{C} = \mathbf{A} \ (+) \ \mathbf{B},$$

or, alternatively,

	1	2	3	4	5	6	7	8	9	10	11	12	13	
C =	0	0.1	0.3	0.3	0.6	0.8	1	0.7	0.7	0.3	0.2	0.1	0	**(1.33)**

Note that we could have computed the sum given in Table 1.1 by using (1.22). Using these computations, we obtain the following set of equations:

$$\mu_{\mathbf{C}}(1) = (0 \wedge 0.3) \vee (0 \wedge 0.1) = 0,$$
$$\mu_{\mathbf{C}}(2) = (0 \wedge 0.6) \vee (0.1 \wedge 0.3) \vee (0.3 \wedge 0) = 0.1,$$
$$\mu_{\mathbf{C}}(3) = (0 \wedge 1) \vee (0.1 \wedge 0.6) \vee (0.3 \wedge 0.3) \vee (0.8 \wedge 0) = 0.3,$$

$$\begin{aligned}
\mu_{\mathbf{C}}(4) &= (0 \wedge 0.7) \vee (0.1 \wedge 1) \vee (0.3 \wedge 0.6) \vee (0.8 \wedge 0.3) \vee \\
&\quad (1 \wedge 0) = 0.3, \\
\mu_{\mathbf{C}}(5) &= (0 \wedge 0.2) \vee (0.1 \wedge 0.7) \vee (0.3 \wedge 1) \vee (0.8 \wedge 0.6) \vee \\
&\quad (1 \wedge 0.3) \vee (0.7 \wedge 0) = 0.6, \\
\mu_{\mathbf{C}}(6) &= (0 \wedge 0.1) \vee (0.1 \wedge 0.2) \vee (0.3 \wedge 0.7) \vee (0.8 \wedge 1) \vee \\
&\quad (1 \wedge 0.6) \vee (0.7 \wedge 0.3) \vee (0.3 \wedge 0) = 0.8, \\
\mu_{\mathbf{C}}(7) &= (0 \wedge 0) \vee (0.3 \wedge 0.3) \vee (0.7 \wedge 0.6) \vee (1 \wedge 1) \\
&\quad (0.8 \wedge 0.7) \vee (0.3 \wedge 0.2) \vee (0.1 \wedge 0.1) \vee (0 \wedge 0) = 1, \\
\mu_{\mathbf{C}}(8) &= (0.1 \wedge 0) \vee (0.3 \wedge 0.1) \vee (0.8 \wedge 0.2) \vee (1 \wedge 0.7) \vee \\
&\quad (0.7 \wedge 1) \vee (0.3 \wedge 0.6) \vee (0 \wedge 0.3) = 0.7, \\
\mu_{\mathbf{C}}(9) &= (0.3 \wedge 0) \vee (0.8 \wedge 0.1) \vee (1 \wedge 0.2) \vee (0.7 \wedge 0.7) \vee \\
&\quad (0.3 \wedge 1) \vee (0 \wedge 0.6) = 0.7, \\
\mu_{\mathbf{C}}(10) &= (0.8 \wedge 0) \vee (1 \wedge 0.1) \vee (0.7 \wedge 0.2) \vee (0.3 \wedge 0.7) \vee \\
&\quad (0 \wedge 1) \vee = 0.3, \\
\mu_{\mathbf{C}}(11) &= (1 \wedge 0) \vee (0.7 \wedge 0.1) \vee (0.3 \wedge 0.2) \vee (0 \wedge 0.7) = 0.2, \\
\mu_{\mathbf{C}}(12) &= (0.7 \wedge 0) \vee (0.3 \wedge 0.1) \vee (0 \wedge 0.2) = 0.1, \\
\mu_{\mathbf{C}}(13) &= (0.3 \wedge 0) \vee (0 \wedge 0.1) = 0.
\end{aligned}$$

As can be seen, this is the same result as in (1.33).

A fuzzy number in R may be added to a fuzzy number in Z by transforming the integer fuzzy number to one that is continuous by interval. Let us consider the number

−7	−6	−5	−4	−3	−2	−1	0	1	2	3	4
0	0.1	0.1	0.2	0.3	0.3	0.5	0.5	0.6	0.8	0.9	1

5	6	7	8	9	10	11	12	13	
0.9	0.7	0.4	0.4	0.3	0.2	0.2	0.1	0	. . .

(1.34)

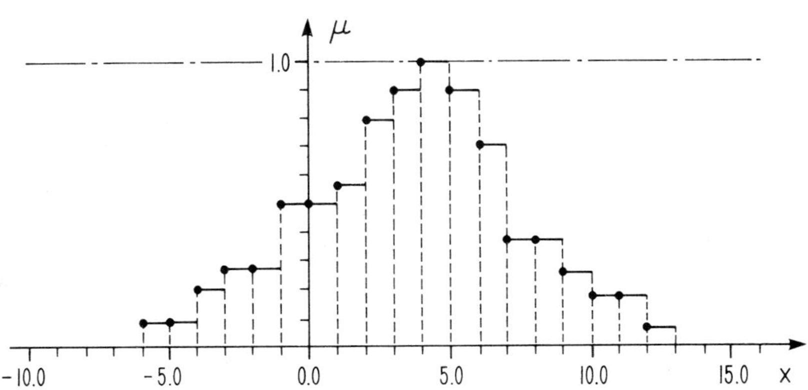

Figure 1.10 A fuzzy number in Z.

This number is represented in Figure 1.10. The number 1.34 becomes then a fuzzy number in R.

A problem of concern at this point is this: If **A** and **B** are fuzzy numbers, is **A** (+) **B** also a fuzzy number? In other words, is **A** (+) **B** convex and normal? What we have done is have reversed the process and explained addition before we have proved it. What we will have to do now is to give two simple, easy-to-prove theorems.

THEOREM 1.1

If **A** and **B** are fuzzy numbers in R, then **A** (+) **B** is also a fuzzy subset in R that is convex.

Proof:

Consider two levels α and α', where $\alpha' > \alpha$. We may then write

$$\begin{aligned} A_\alpha &= [a_1^{(\alpha)}, a_2^{(\alpha)}],\\ B_\alpha &= [b_1^{(\alpha)}, b_2^{(\alpha)}],\\ A_{\alpha'} &= [a_1^{(\alpha')}, a_2^{(\alpha')}],\\ B_{\alpha'} &= [b_1^{(\alpha')}, b_2^{(\alpha')}],\\ (\alpha' > \alpha) &\Rightarrow ([a_1^{(\alpha')}, a_2^{(\alpha')}] \subset [a_1^{(\alpha)}, a_2^{(\alpha)}]),\\ (\alpha' > \alpha) &\Rightarrow ([b_1^{(\alpha')}, b_2^{(\alpha')}] \subset [b_1^{(\alpha)}, b_2^{(\alpha)}]). \end{aligned} \tag{1.35}$$

By construction,

$$A_\alpha \,(+)\, B_\alpha = [a_1^{(\alpha)} + b_1^{(\alpha)},\ a_2^{(\alpha)} + b_2^{(\alpha)}]$$

and

$$A_{\alpha'} \,(+)\, B_{\alpha'} = [a_1^{(\alpha')} + b_1^{(\alpha')},\ a_2^{(\alpha')} + b_2^{(\alpha')}].$$

Referring to (1.35), we have

$$\begin{aligned} (\alpha' > \alpha) \Rightarrow ([a_1^{(\alpha')} + b_1^{(\alpha')},\ a_2^{(\alpha')} + b_2^{(\alpha')}]\\ \subset [a_1^{(\alpha)} + b_1^{(\alpha)},\ a_2^{(\alpha)} + b_2^{(\alpha)}]). \end{aligned}$$

Monotonicity and convexity are therefore preserved by addition.

THEOREM 1.2

If **A** and **B** are fuzzy numbers in R, then **A** (+) **B** is a fuzzy subset in R, which is normal.

Proof:

At the level $\alpha = 1$, we have

$$A_{\alpha=1} = [a_1^{(1)}, a_2^{(1)}]$$

and

$$B_{\alpha=1} = [b_1^{(1)}, b_2^{(1)}].$$

Then

$$A_{\alpha=1} (+) B_{\alpha=1} = [a_1^{(1)} + b_1^{(1)}, a_2^{(1)} + b_2^{(1)}] \neq \emptyset.$$

Hence **A** (+) **B** is normal.

This proves that if **A** and **B** are fuzzy numbers in R, **A** (+) **B** is also a fuzzy number.

Let us now examine the algebraic properties of addition for fuzzy numbers. It is sufficient to return to (1.10–1.13) and to consider intervals of confidence for each level α. We have, therefore, for all **A**, **B**, and **C** fuzzy numbers in R

$$\mathbf{A} (+) \mathbf{B} = \mathbf{B} (+) \mathbf{A}, \qquad \text{(commutative)},$$
$$(\mathbf{A} (+) \mathbf{B}) (+) \mathbf{C} = \mathbf{A} (+) (\mathbf{B} (+) \mathbf{C}), \qquad \text{(associative)}.$$

If a neutral exists at the left and at the right, it is the oridinary number 0. Thus

$$\mathbf{A} (+) 0 = 0 (+) \mathbf{A} = \mathbf{A}.$$

The image is not symmetric, however.

$$\mathbf{A} (+) \mathbf{A}^- = \mathbf{A}^- (+) \mathbf{A} \neq 0$$

where

$$\mu_{\mathbf{A}^-_\alpha}(x) = [-a_2^{(\alpha)}, -a_1^{(\alpha)}].$$

This shows that fuzzy numbers have a monoidal structure for addition that is commutative. It can be verified easily that it is true not only in R, but also in Z, R^+, and N.

Subtraction

The definition of addition can also be extended to the definition of subtraction. Consider the following definitions and symbols.

$\forall\ \alpha \in [0, 1]$:

$$\begin{aligned}\mathbf{A} (-) \mathbf{B} &= [a_1^{(\alpha)}, a_2^{(\alpha)}] (-) [b_1^{(\alpha)}, b_2^{(\alpha)}] \\ &= [a_1^{(\alpha)} - b_2^{(\alpha)}, a_2^{(\alpha)} - b_1^{(\alpha)}],\end{aligned}$$

or

$\forall\ x, y, z \in R$:

$$\mu_{\mathbf{A}(-)\mathbf{B}}(z) = \bigvee_{z=x-y} (\mu_{\mathbf{A}}(x) \wedge \mu_{\mathbf{B}}(y)). \qquad \textbf{(1.36)}$$

Subtraction is, in fact, the addition of the image of $\mathbf{B}^-$ to $\mathbf{A}$, where $\forall\ \alpha \in [0, 1]$:

$$B^-_\alpha = [-b_2^{(\alpha)}, -b_1^{(\alpha)}].$$

Subtraction is neither commutative nor associative. It is defined in Z as in R, but not in R^+ or N because negative numbers could appear. We now give two examples involving subtraction.

EXAMPLE 1.7

Let us consider two fuzzy numbers with a triangular shape as shown in Figure 1.11.

$\forall\ x \in R$:

$$\begin{aligned}
\mu_{\mathbf{A}}(x) &= 0, & x &\le 7,\\
&= x/7 - 1, & 7 \le x &\le 14,\\
&= -x/5 + 19/5, & 14 \le x &\le 19,\\
&= 0, & x &\ge 19. \qquad \mathbf{(1.37)}\\
\mu_{\mathbf{B}}(x) &= 0, & x &\le 3,\\
&= x/2 - 3/2, & 3 \le x &\le 5,\\
&= -x/5 + 10/5, & 5 \le x &\le 10,\\
&= 0, & x &\ge 10. \qquad \mathbf{(1.38)}
\end{aligned}$$

Now using (1.37), let

$$\alpha = a_1^{(\alpha)}/7 - 1,$$
$$\alpha = -\ a_2^{(\alpha)}/5 + 19/5.$$

from which

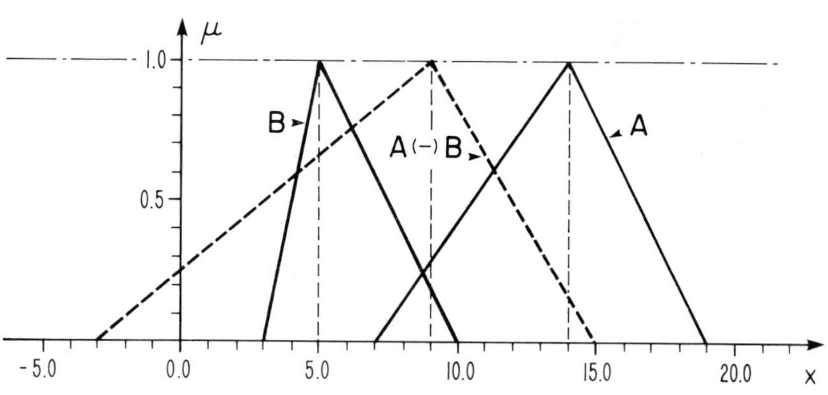

Figure 1.11 Subtraction of two fuzzy numbers (Example 1.7).

$$A_\alpha = [a_1^{(\alpha)}, a_2^{(\alpha)}] = [7\alpha + 7, -5\alpha + 19]$$

Now using (1.38), we obtain

$$\begin{aligned} \alpha &= b_1^{(\alpha)}/2 - 3/2, \\ \alpha &= -b_2^{(\alpha)}/5 + 10/5, \end{aligned} \quad \textbf{(1.39)}$$

and

$$B_\alpha = [b_1^{(\alpha)}, b_2^{(\alpha)}] = [2\alpha + 3, -5\alpha + 10]. \quad \textbf{(1.40)}$$

Subtracting (1.40) from 1.39) gives

$$\begin{aligned} A_\alpha (-) B_\alpha &= [7\alpha + 7, -5\alpha + 19] (-) [2\alpha + 3, -5\alpha + 10], \\ &= [7\alpha + 7 - (-5\alpha + 10), -5\alpha + 19 - (2\alpha + 3)], \\ &= [12\alpha - 3, -7\alpha + 16]. \end{aligned}$$

Thus if we define

$$\begin{aligned} a_1^{(\alpha)} - b_2^{(\alpha)} &= 12\alpha - 3, \\ a_2^{(\alpha)} - b_1^{(\alpha)} &= -7\alpha + 16, \end{aligned}$$

we obtain

$$\begin{aligned} \mu_{\mathbf{A}(-)\mathbf{B}}(x) &= 0, & x &\le -3, \\ &= x/12 + 3/12, & -3 \le x &\le 9, \\ &= -x/7 + 16/7, & 9 \le x &\le 16, \\ &= 0, & x &\ge 16. \end{aligned}$$

EXAMPLE 1.8

To see more clearly what it is that we have been doing, we now consider a numerical example in Z, which is given as follows:

Let

	−3	−2	−1	0	1	2	3	4	
A =	0	0.1	0.3	0.7	0.9	1.0	0.5	0	. . .

	−2	−1	0	1	2	3	4	
B =	0	0.1	0.6	1.0	0.8	0.3	0	. . .

Then we have **A** (−) **B** as shown in Table 1.2.

Table 1.2

	−3	−2	−1	0	1	2	3	4
1						1		
0.9					1	1		
0.8					1	1		
0.7				1	1	1		
0.6				1	1	1		
0.5				1	1	1	1	
0.4				1	1	1	1	
0.3			1	1	1	1	1	
0.2			1	1	1	1	1	
0.1		1	1	1	1	1	1	
0	1	1	1	1	1	1	1	1
	0	0.1	0.3	0.7	0.9	1	0.5	0

(−)

−2	−1	0	1	2	3	4
			1			
			1			
			1	1		
			1	1		
		1	1	1		
		1	1	1		
		1	1	1		
		1	1	1	1	
		1	1	1	1	
	1	1	1	1	1	
1	1	1	1	1	1	1
0	0.1	0.6	1	0.8	0.3	0

=

−7	−6	−5	−4	−3	−2	−1	0	1	2	3	4	5	6
								1					
							1	1					
						1	1	1					
					1	1	1	1					
					1	1	1	1	1				
					1	1	1	1	1	1			
					1	1	1	1	1	1			
			1	1	1	1	1	1	1	1			
			1	1	1	1	1	1	1	1			
		1	1	1	1	1	1	1	1	1	1		
1	1	1	1	1	1	1	1	1	1	1	1	1	1
0	0	0.1	0.3	0.3	0.7	0.8	0.9	1	0.6	0.5	0.1	0	0

We now use (1.36) to give

$\mu_{\mathbf{A}(-)\mathbf{B}}(-7) = 0 \wedge 0 = 0,$

$\mu_{\mathbf{A}(-)\mathbf{B}}(-6) = (0 \wedge 0.3) \vee (0.1 \wedge 0) = 0,$

$\mu_{\mathbf{A}(-)\mathbf{B}}(-5) = (0 \wedge 0.8) \vee (0.1 \wedge 0.3) \vee (0.3 \wedge 0) = 0.1,$

$\mu_{\mathbf{A}(-)\mathbf{B}}(-4) = (0 \wedge 1) \vee (0.1 \wedge 0.8) \vee (0.3 \wedge 0.3) \vee (0.7 \wedge 0) = 0.3,$

$$\mu_{\mathbf{A}(-)\mathbf{B}}(-3) = (0 \wedge 0.6) \vee (0.1 \wedge 1) \vee (0.3 \wedge 0.8) \vee (0.7 \wedge 0.3) \vee (0.9 \wedge 0) = 0.3,$$

$$\mu_{\mathbf{A}(-)\mathbf{B}}(-2) = (0 \wedge 0.1) \vee (0.1 \wedge 0.6) \vee (0.3 \wedge 1) \vee (0.7 \wedge 8) \vee (0.9 \wedge 0.3) \vee (1 \wedge 0) = 0.7,$$

$$\mu_{\mathbf{A}(-)\mathbf{B}}(-1) = (0 \wedge 0) \vee (0.1 \wedge 0.1) \vee (0.3 \wedge 0.6) \vee (0.7 \wedge 1) \vee (0.9 \wedge 0.8) \vee (1 \wedge 0.3) \vee (0.5 \wedge 0) = 0.8,$$

$$\mu_{\mathbf{A}(-)\mathbf{B}}(0) = (0.1 \wedge 0) \vee (0.3 \wedge 0.1) \vee (0.7 \wedge 0.6) \vee (0.9 \wedge 1) \; (1 \wedge 0.8) \vee (0.5 \wedge 0.3) \vee (0 \wedge 0) = 0.9,$$

$$\mu_{\mathbf{A}(-)\mathbf{B}}(1) = (0.3 \wedge 0) \vee (0.7 \wedge 0.1) \vee (0.9 \wedge 0.6) \vee (1 \wedge 1) \vee (0.5 \wedge 0.8) = 1,$$

$$\mu_{\mathbf{A}(-)\mathbf{B}}(2) = (0.7 \wedge 0) \vee (0.9 \wedge 0.1) \vee (1 \wedge 0.6) \vee (0.5 \wedge 0.8) = 0.6,$$

$$\mu_{\mathbf{A}(-)\mathbf{B}}(3) = (0.9 \wedge 0) \vee (1 \wedge 0.1) \vee (0.5 \wedge 0.6) = 0.5,$$

$$\mu_{\mathbf{A}(-)\mathbf{B}}(4) = (1 \wedge 0) \vee (0.5 \wedge 0.1) = 0.1,$$

$$\mu_{\mathbf{A}(-)\mathbf{B}}(5) = (0.5 \wedge 0) = 0.$$

1.5 MULTIPLICATION OF FUZZY NUMBERS

At this point we consider multiplication[5] in R^+ and N. Let us consider two fuzzy numbers **A** and **B** in R^+. From the level α of presumption, we can write

$$\begin{aligned} A_\alpha \,(\cdot)\, B_\alpha &= [a_1^{(\alpha)}, a_2^{(\alpha)}] \,(\cdot)\, [b_1^{(\alpha)}, b_2^{(\alpha)}] \\ &= [a_1^{(\alpha)} \cdot b_1^{(\alpha)}, a_2^{(\alpha)} \cdot b_2^{(\alpha)}]. \end{aligned} \quad \textbf{(1.41)}$$

Multiplication can also be given by

$\forall\, x, y, z \in R^+$:

$$\mu_{\mathbf{A}(\cdot)\mathbf{B}}(z) = \bigvee_{z = x \cdot y} (\mu_{\mathbf{A}}(x) \wedge \mu_{\mathbf{B}}(y)). \quad \textbf{(1.42)}$$

Equations (1.41) and (1.42) are equivalent, and this may be proven in the same way as was given for addition in the previous section.

EXAMPLE 1.9

For this example we again use triangular fuzzy numbers because they are so easy to work with.

$\forall\, x \in R^+$:

$$\begin{aligned} \mu_{\mathbf{A}}(x) &= 0, & x &\le 2, \\ &= x - 2, & 2 \le x &\le 3, \\ &= -x/2 + 5/2, & 3 \le x &\le 5, \\ &= 0, & x &\ge 5. \end{aligned} \quad \textbf{(1.43)}$$

$$\begin{aligned} \mu_{\mathbf{B}}(x) &= 0, & x &\le 3, \\ &= x/2 - 3/2, & 3 \le x &\le 5, \\ &= -x + 6, & 5 \le x &\le 6, \\ &= 0, & x &\ge 6. \end{aligned} \quad \textbf{(1.44)}$$

For the level α in Figure 1.12 and using (1.43) we have

$$\alpha = a_1^{(\alpha)} - 2, \quad \textbf{(1.45)}$$

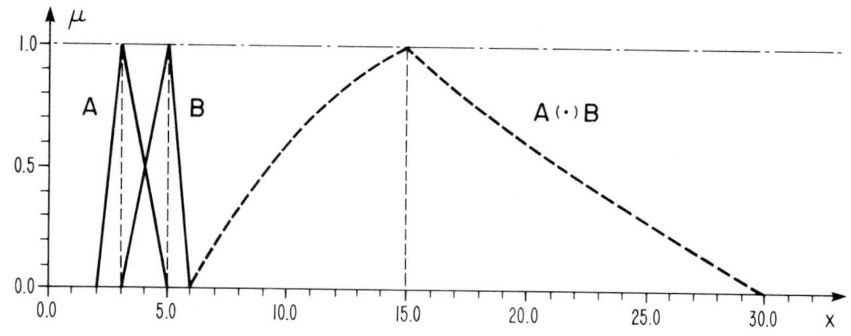

Figure 1.12 Multiplication of two fuzzy numbers (Example 1.9).

and

$$\alpha = -a_2^{(\alpha)}/2 + 5/2.$$

Hence

$$A_\alpha = [\alpha + 2, -2\alpha + 5].$$

Using (1.44) we also have

$$\alpha = b_1^{(\alpha)}/2 - 3/2 \qquad \text{and} \qquad \alpha = -b_2^{(\alpha)} + 6. \text{ Hence}$$

$$B_\alpha = [2\alpha + 3, -\alpha + 6].$$

Thus we obtain the multiplication

$$\begin{aligned} A_\alpha (\cdot) B_\alpha &= [\alpha + 2)(2\alpha + 3), (-2\alpha + 5)(-\alpha + 6)] \\ &= [2\alpha^2 + 7\alpha + 6, 2\alpha^2 - 17\alpha + 30]. \end{aligned}$$

We now have two equations to solve, namely,

$$2\alpha^2 + 7\alpha + 6 - x = 0 \tag{1.46}$$

and

$$2\alpha^2 - 17\alpha + 30 - x = 0. \tag{1.47}$$

We will retain only two roots in [0, 1]. For (1.46)

$$\alpha = (-7 + \sqrt{1 + 8x})/4,$$

and for (1.47)

$$\alpha = (17 - \sqrt{49 + 8x})/4.$$

Finally,

$\forall\ x \in R^+$:

$$\begin{aligned}\mu_{\mathbf{A}(\cdot)\mathbf{B}}(x) &= 0 & x &\le 6,\\ &= (-7+\sqrt{1+8x})/4, & 6 \le x &\le 15,\\ &= (17-\sqrt{49+8x})/4, & 15 \le x &\le 30,\\ &= 0, & x &\ge 30.\end{aligned}$$

The resulting multiplication curve is shown in Figure 1.12. Note that **A** (·) **B** does not yield a triangular shape.

EXAMPLE 1.10

In N let us use

$$\mathbf{A} = \begin{array}{|c|c|c|c|c|} \hline 2 & 3 & 4 & 5 & 6 \\ \hline 0 & 0.4 & 1 & 0.7 & 0 \\ \hline \end{array} \qquad \textbf{(1.48)}$$

$$\mathbf{B} = \begin{array}{|c|c|c|c|c|c|} \hline 2 & 3 & 4 & 5 & 6 & 7 \\ \hline 0 & 0.1 & 0.8 & 1 & 0.3 & 0 \\ \hline \end{array} \qquad \textbf{(1.49)}$$

Using (1.41) we obtain the result shown in Table 1.3 (see page **28**). We thus find that

A (·) **B** =

8	9	10	11	12	13	14	15	16	17	18	19
0	0.1	0.1	0.1	0.4	0.4	0.4	0.4	0.8	0.8	0.8	0.8

20	21	22	23	24	25	26	27	28	29	30	31
1	0.7	0.7	0.7	0.7	0.7	0.3	0.3	0.3	0.3	0.3	0

We now examine how (1.42) may be used. Consider Figure 1.10, where we showed how a fuzzy number in Z can be transformed to a fuzzy number in R. The fuzzy numbers increase monotonically to the left of the normal values ($\mu = 1$) and decrease monotonically to the right of the normal values. Let us now do the following:

1. At the left let us take into account all couplets in the parenthesis in (1.42), where $x \cdot y \le z$.
2. At the right let us take into account all couplets where $x \cdot y \ge z$. To simplify the process, omit from consideration any couplets

Table 1.3

	3	4	5
1		1	
0.9		1	
0.8		1	
0.7		1	1
0.6		1	1
0.5		1	1
0.4	1	1	1
0.3	1	1	1
0.2	1	1	1
0.1	1	1	1
0	1	1	1
	0.4	1	0.2

(·)

3	4	5	6
		1	
		1	
	1	1	
	1	1	
	1	1	
	1	1	
	1	1	
	1	1	1
	1	1	1
1	1	1	1
1	1	1	1
0.1	0.8	1	0.3

=

9	10	11	12	13	14	15	16	17	18	19	20	21	22	23	24	25	26	27	28	29	30
											1										
											1										
							1	1	1	1	1										
							1	1	1	1	1	1	1	1	1	1					
							1	1	1	1	1	1	1	1	1	1					
							1	1	1	1	1	1	1	1	1	1					
			1	1	1	1	1	1	1	1	1	1	1	1	1	1					
			1	1	1	1	1	1	1	1	1	1	1	1	1	1	1	1	1	1	1
			1	1	1	1	1	1	1	1	1	1	1	1	1	1	1	1	1	1	1
1	1	1	1	1	1	1	1	1	1	1	1	1	1	1	1	1	1	1	1	1	1
1	1	1	1	1	1	1	1	1	1	1	1	1	1	1	1	1	1	1	1	1	1
0.1	0.1	0.1	0.4	0.4	0.4	0.4	0.8	0.8	0.8	0.8	1	0.7	0.7	0.7	0.7	0.7	0.3	0.3	0.3	0.3	0.3

where at least one zero exists. Conversely, we compute values of z for which $\mu = 1$. This will show what value of z occurs when we pass from the left to the right of the maximum. For this purpose, let us use (1.48) and (1.49).

The value $\alpha = 1$ exists only for the couplet (4, 5). Hence

$$\mu_{\mathbf{A}(\cdot)\mathbf{B}}(z) = \mu(x \cdot y = 20) = \mu(4 \cdot 5) = 1.$$

Since 20 has not been included up to this point, we shall apply rule 1 to the left of 20 and rule 2 beyond 20.[6]

$$\mu(9) = \overbrace{(0.4 \wedge 0.1)}^{3 \times 3} = 0.1,$$

$$\mu(10) = \overbrace{(0.4 \wedge 0.1)}^{3 \times 3} = 0.1,$$

$$\mu(11) = \overbrace{(0.4 \wedge 0.1)}^{3 \times 3} = 0.1,$$

$$\mu(12) = \overbrace{(0.4 \wedge 0.1)}^{3 \times 3} \vee \overbrace{(0.4 \wedge 0.8)}^{3 \times 4} \vee \overbrace{(1 \wedge 0.1)}^{4 \times 3} = 0.4,$$

$$\mu(13) = \overbrace{(0.4 \wedge 0.1)}^{3 \times 3} \vee \overbrace{(0.4 \wedge 0.8)}^{3 \times 4} \vee \overbrace{(1 \wedge 0.1)}^{4 \times 3} = 0.4,$$

$$\mu(14) = \overbrace{(0.4 \wedge 0.1)}^{3 \times 3} \vee \overbrace{(0.4 \wedge 0.8)}^{3 \times 4} \vee \overbrace{(1 \wedge 0.1)}^{4 \times 3} = 0.4,$$

$$\mu(15) = \overbrace{(0.4 \wedge 0.1)}^{3 \times 3} \vee \overbrace{(0.4 \wedge 0.8)}^{3 \times 4} \vee \overbrace{(1 \wedge 0.3)}^{4 \times 3} \vee \overbrace{(0.4 \wedge 1)}^{3 \times 5} \vee \overbrace{(0.7 \wedge 0.1)}^{5 \times 3} = 0.4,$$

$$\mu(16) = \overbrace{(0.4 \wedge 0.1)}^{3 \times 3} \vee \overbrace{(0.4 \wedge 0.8)}^{3 \times 4} \vee \overbrace{(1 \wedge 0.3)}^{4 \times 3} \vee \overbrace{(0.4 \wedge 1)}^{3 \times 5} \vee \overbrace{(0.7 \wedge 0.1)}^{5 \times 3} \vee \overbrace{(1 \wedge 0.8)}^{4 \times 4} = 0.8,$$

$$\mu(17) = \overbrace{(0.4 \wedge 0.1)}^{3 \times 3} \vee \overbrace{(0.4 \wedge 0.8)}^{3 \times 4} \vee \overbrace{(1 \wedge 0.3)}^{4 \times 3} \vee \overbrace{(0.4 \wedge 1)}^{3 \times 5} \vee \overbrace{(0.7 \wedge 0.1)}^{5 \times 3} \vee \overbrace{(1 \wedge 0.8)}^{4 \times 4} = 0.8,$$

$$\mu(18) = \overbrace{(0.4 \wedge 0.1)}^{3 \times 3} \vee \overbrace{(0.4 \wedge 0.8)}^{3 \times 4} \vee \overbrace{(1 \wedge 0.3)}^{4 \times 3} \vee \overbrace{(0.4 \wedge 1)}^{3 \times 5} \vee$$

$$\overbrace{(0.7 \wedge 0.1)}^{5\times 3} \vee \overbrace{(1 \wedge 0.8)}^{4\times 4} \vee \overbrace{(0.4 \wedge 0.3)}^{3\times 6} = 0.8,$$

$$\mu(19) = \overbrace{(0.4 \wedge 0.1)}^{3\times 3} \vee \overbrace{(0.4 \wedge 0.8)}^{3\times 4} \vee \overbrace{(1 \wedge 0.3)}^{4\times 3} \vee \overbrace{(0.4 \wedge 0.1)}^{3\times 5} \vee$$
$$\overbrace{(0.7 \wedge 0.1)}^{5\times 3} \vee \overbrace{(1 \wedge 0.8)}^{4\times 4} \vee \overbrace{(0.4 \wedge 0.3)}^{3\times 6} = 0.8,$$

Beyond $z = 20$ we find

$$\mu(21) = \overbrace{(1 \wedge 0.3)}^{4\times 6} \vee \overbrace{(0.7 \wedge 1)}^{5\times 5} \vee \overbrace{(0.7 \wedge 0.3)}^{5\times 6} = 0.7,$$

$$\mu(22) = \overbrace{(1 \wedge 0.3)}^{4\times 6} \vee \overbrace{(0.7 \wedge 1)}^{5\times 5} \vee \overbrace{(0.7 \wedge 0.3)}^{5\times 6} = 0.7,$$

$$\mu(23) = \overbrace{(1 \wedge 0.3)}^{4\times 6} \vee \overbrace{(0.7 \wedge 1)}^{5\times 5} \vee \overbrace{(0.7 \wedge 0.3)}^{5\times 6} = 0.7,$$

$$\mu(24) = \overbrace{(1 \wedge 0.3)}^{4\times 6} \vee \overbrace{(0.7 \wedge 1)}^{5\times 5} \vee \overbrace{(0.7 \wedge 0.3)}^{5\times 6} = 0.7,$$

$$\mu(25) = \overbrace{(0.7 \wedge 1)}^{5\times 5} \vee \overbrace{(0.7 \wedge 0.3)}^{5\times 6} = 0.7,$$

$$\mu(26) = \overbrace{(0.7 \wedge 0.3)}^{5\times 6} = 0.3,$$

$$\mu(27) = \overbrace{(0.7 \wedge 0.3)}^{5\times 6} = 0.3,$$

$$\mu(28) = \overbrace{(0.7 \wedge 0.3)}^{5\times 6} = 0.3,$$

$$\mu(29) = \overbrace{(0.7 \wedge 0.3)}^{5\times 6} = 0.3,$$

$$\mu(30) = \overbrace{(0.7 \wedge 0.3)}^{5\times 6} = 0.3,$$

$$\mu(z > 30) = 0.$$

This computation could have been presented more simply, but all of the details were given here to show you how it is done.

We will now present two theorems, as we did for addition.

THEOREM 1.3

If **A** and **B** are fuzzy numbers in R^+, then **A** $(\cdot)$ **B** is a fuzzy subset that is convex. The proof here is limited to R^+, but in Appendix B we show that this property is also true in R.

Proof:

This proof is very similar to that for the addition of numbers in R^+

THEOREM 1.4

If **A** and **B** are fuzzy numbers in R^+, then **A** $(\cdot)$ **B** is a fuzzy subset that is normal.

Proof:

This proof is also very similar to that for the addition of numbers in R^+.

We now give some algebraic properties.

$\forall$ **A**, **B**, **C** $\subset R^+$:

$$\mathbf{A}\ (\cdot)\ \mathbf{B} = \mathbf{B}\ (\cdot)\ \mathbf{A} \qquad \text{(commutative)},$$
$$(\mathbf{A}\ (\cdot)\ \mathbf{B})\ (\cdot)\ \mathbf{C} = \mathbf{A}\ (\cdot)\ (\mathbf{B}\ (\cdot)\ \mathbf{C}) \qquad \text{(associative)}.$$

If a neutral exists at the left and at the right, it is the ordinary number 1.

$$\mathbf{A}\ (\cdot)\ 1 = 1\ (\cdot)\ \mathbf{A} = \mathbf{A}.$$

If we represent the inverse of **A** by $\mathbf{A}^{-1}$, the inverse is not symmetric. That is,

$$\mathbf{A}\ (\cdot)\ \mathbf{A}^{-1} = \mathbf{A}^{-1}\ (\cdot)\ \mathbf{A} \neq 1,$$

where

$$A_\alpha^{-1} = [1/a_2^{(\alpha)},\ 1/a_1^{(\alpha)}].$$

This calculation shows that fuzzy numbers have a monoidal structure for multiplication and that this monoidal structure is commutative. In Appendix B we show that this monoidal structure property holds both in R^+ and in R.

Division

Division of two fuzzy numbers is defined in R^+ by

$$\begin{aligned}\mathbf{A}\ (:)\ \mathbf{B} &= [a_1^{(\alpha)},\ a_2^{(\alpha)}]\ (:)\ [b_1^{(\alpha)},\ b_2^{(\alpha)}]\\ &= [a_1^{(\alpha)}/b_2^{(\alpha)},\ a_2^{(\alpha)}/b_1^{(\alpha)}], \qquad b_2^{(\alpha)} > 0, \qquad \forall\ \alpha \in [0, 1],\end{aligned}$$

or

$\forall$ x, y, z $\in R^+$:

$$\mu_{\mathbf{A}(:)\mathbf{B}}(z) = \bigvee_{z=x/y} (\mu_{\mathbf{A}}(x) \wedge \mu_{\mathbf{B}}(y)).$$

Division, as has been mentioned previously, is a multiplication by the inverse; that is, by

$$B_\alpha^{-1} = [1/b_2^{(\alpha)}, 1/b_1^{(\alpha)}], \qquad b_2^{(\alpha)} > 0, \qquad \forall\ \alpha \in [0, 1]. \quad \textbf{(1.50)}$$

Division is not, however, associative or commutative. In connection with this let us study an example in R^+.

EXAMPLE 1.11

We now consider a numerical example. Let us use the triangular shape shown in Figure 1.13, and let

$\forall\ x \in R^+$:

$$\begin{aligned} \mu_{\mathbf{A}}(x) &= 0, & x &\le 18,\\ &= x/4 - 18/4, & 18 \le x &\le 22,\\ &= -x/11 + 3, & 22 \le x &\le 33,\\ &= 0, & x &\ge 33. \end{aligned} \quad \textbf{(1.51)}$$

$$\begin{aligned} \mu_{\mathbf{B}}(x) &= 0, & x &\le 5,\\ &= x - 5, & 5 \le x &\le 6,\\ &= -x/2 + 4, & 6 \le x &\le 8,\\ &= 0, & x &\ge 8. \end{aligned} \quad \textbf{(1.52)}$$

In (1.51), let $\alpha = a_1^{(\alpha)}/4 - 18/4$ and $\alpha = -a_2^{(\alpha)}/11 + 3$, from which

$$A_\alpha = [4\alpha + 18,\ -11\alpha + 33].$$

In (1.52), let $\alpha = b_1^{(\alpha)} - 5$ and $\alpha = -b_2^{(\alpha)}/2 + 4$, from which

$$B_\alpha = [\alpha + 5,\ -2\alpha + 8].$$

Thus

$$A_\alpha\ (:)\ B_\alpha = [4\alpha + 18, -11\alpha + 33]\ (:)\ [\alpha + 5, -2\alpha + 8]$$

$$= \left(\frac{4\alpha + 18}{-2\alpha + 8},\ \frac{-11\alpha + 33}{\alpha + 5}\right)$$

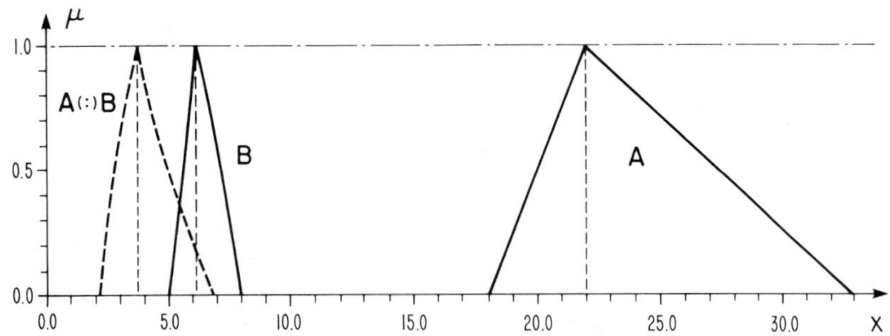

Figure 1.13 Division of two fuzzy numbers (Example 1.11).

We thus find

$$\forall\ x \in R:$$

$$
\begin{aligned}
\mu_{\mathbf{A}(:)\mathbf{B}}(x) &= 0, & x &\le 9/4,\\
&= \frac{8x - 18}{2x + 4} & 9/4 \le x &\le 11/3,\\
&= \frac{-5x + 33}{x + 11}, & 11/3 \le x &\le 33/5,\\
&= 0, & x &\ge 33/5.
\end{aligned}
$$

Remark

Note that $(\mathbf{A}\ (:)\ \mathbf{B})\ (\cdot)\ \mathbf{B} \ne \mathbf{A}$. Indeed,

$$A_\alpha\ (:)\ B_\alpha = \left[\frac{a_1^{(\alpha)}}{b_2^{(\alpha)}},\ \frac{a_2^{(\alpha)}}{b_1^{(\alpha)}},\right]$$

$$
\begin{aligned}
(A_\alpha\ (:)\ B_\alpha)\ (\cdot)\ B_\alpha &= \left[\frac{a_1^{(\alpha)}}{b_2^{(\alpha)}},\ \frac{a_2^{(\alpha)}}{b_1^{(\alpha)}}\right] (\cdot)\ [b_1^{(\alpha)},\ b_2^{(\alpha)}]\\
&= \left[\frac{a_1^{(\alpha)}b_1^{(\alpha)}}{b_2^{(\alpha)}},\ \frac{a_2^{(\alpha)}b_2^{(\alpha)}}{b_1^{(\alpha)}}\right]\\
&\ne\ [a_1^{(\alpha)},\ a_2^{(\alpha)}]; \qquad \text{i.e.,} \ne A_\alpha.
\end{aligned}
$$

This is also true for addition and subtraction.

$$(\mathbf{A}\ (-)\ \mathbf{B})\ (+)\ \mathbf{A} \ne \mathbf{A}.$$

That is,

$$
\begin{aligned}
A_\alpha\ (-)\ B_\alpha &= [a_1^{(\alpha)} - b_2^{(\alpha)},\ a_2^{(\alpha)} - b_1^{(\alpha)}],\\
(A_\alpha\ (-)\ B_\alpha)\ (+)\ B_\alpha &= [a_1^{(\alpha)} - b_2^{(\alpha)},\ a_2^{(\alpha)} - b_1^{(\alpha)}]\ (+)\ [b_1^{(\alpha)},\ b_2^{(\alpha)}]\\
&= [a_1^{(\alpha)} - b_2^{(\alpha)} + b_1^{(\alpha)},\ a_2^{(\alpha)} - b_1^{(\alpha)} + b_2^{(\alpha)}]\\
&\ne [a_1^{(\alpha)},\ a_2^{(\alpha)}].
\end{aligned}
$$

Multiplication of a Fuzzy Number by an Ordinary Number

Let **A** be a fuzzy number in R and k an ordinary number $k \in R_0^+$.

$$\forall\ A \subset R:$$

$$
\begin{aligned}
k \cdot A_\alpha &= [k,\ k]\ (\cdot)\ [a_1^{(\alpha)},\ a_2^{(\alpha)}]\\
&= [ka_1^{(\alpha)},\ ka_2^{(\alpha)}],
\end{aligned}
$$

or

$\forall\ x \in R$:

$$\mu_{k\,\cdot\,\mathbf{A}}(x) = \mu_{\mathbf{A}}(x/k). \tag{1.53}$$

EXAMPLE 1.12

Consider the numerical example shown in Figure 1.14.

$\forall\ x \in R$:

$$\begin{aligned} \mu_{\mathbf{A}}(x) &= 0, & x &\leq -1,\\ &= x/3 + 1/3, & -1 \leq x &\leq 2,\\ &= -x/6 + 8/6, & 2 \leq x &\leq 8,\\ &= 0, & x &\geq 8. \end{aligned}$$

Now let k = 3. Using (1.53) we obtain

$\forall\ x \in R$:

$$\begin{aligned} \mu_{3\,\cdot\,\mathbf{A}}(x) &= 0, & x &\leq -3,\\ &= x/9 + 1/3, & -3 \leq x &\leq 6,\\ &= -x/18 + 4/3, & 6 \leq x &\leq 24,\\ &= 0, & x &\geq 24. \end{aligned}$$

Distributivity in R^+

We now prove that

$\forall\ \mathbf{A}, \mathbf{B}, \mathbf{C} \subset R^+$:

$$(\mathbf{A}\ (+)\ \mathbf{B})\ (\cdot)\ \mathbf{C} = (\mathbf{A}\ (\cdot)\ \mathbf{C})\ (+)\ (\mathbf{B}\ (\cdot)\ \mathbf{C}).$$

In fact,

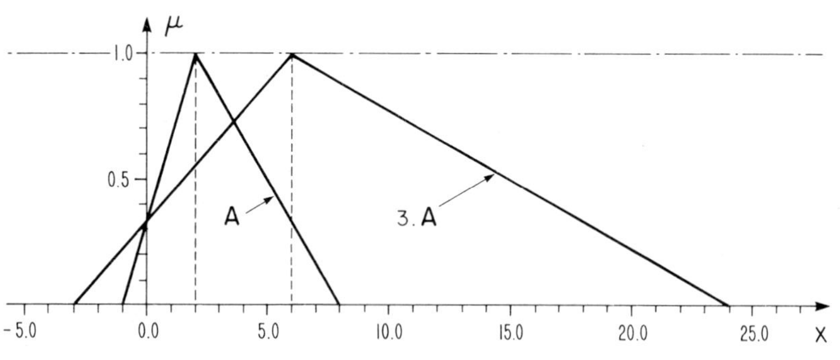

Figure 1.14 Multiplication of fuzzy number by an ordinary number (Example 1.12).

$$\begin{aligned}
&(\mathbf{A}\ (+)\ \mathbf{B})\ (\cdot)\ \mathbf{C}\\
&= ([a_1^{(\alpha)}, a_2^{(\alpha)}]\ (+)\ [b_1^{(\alpha)}, b_2^{(\alpha)}])\ (\cdot)\ [c_1^{(\alpha)}, c_2^{(\alpha)}]\\
&= [a_1^{(\alpha)} + b_1^{(\alpha)}, a_2^{(\alpha)} + b_2^{(\alpha)}]\ (\cdot)\ [c_1^{(\alpha)}, c_2^{(\alpha)}]\\
&= [a_1^{(\alpha)} \cdot c_1^{(\alpha)} + b_1^{(\alpha)} \cdot c_1^{(\alpha)}, a_2^{(\alpha)} \cdot c_2^{(\alpha)} + b_2^{(\alpha)} \cdot c_2^{(\alpha)}]
\end{aligned}$$

and,

$$\begin{aligned}
&(\mathbf{A}\ (\cdot)\ \mathbf{C})\ (+)\ (\mathbf{B}\ (\cdot)\ \mathbf{C})\\
&\quad = ([a_1^{(\alpha)}, a_2^{(\alpha)}]\ (\cdot)\ [c_1^{(\alpha)}, c_2^{(\alpha)}])\\
&\quad\quad + ([b_1^{(\alpha)}, b_2^{(\alpha)}]\ (\cdot)\ [c_1^{(\alpha)}, c_2^{(\alpha)}]\\
&\quad = [a_1^{(\alpha)} \cdot c_1^{(\alpha)} + b_1^{(\alpha)} \cdot c_1^{(\alpha)}, a_2^{(\alpha)} \cdot c_2^{(\alpha)}\\
&\quad\quad + b_2^{(\alpha)} \cdot c_2^{(\alpha)}].
\end{aligned}$$

That is,

$$(\mathbf{A}\ (+)\ \mathbf{B})\ (\cdot)\ \mathbf{C} = (\mathbf{A}\ (\cdot)\ \mathbf{C}) + (\mathbf{B}\ (\cdot)\ \mathbf{C}).$$

On the contrary,

$$(\mathbf{A}\ (\cdot)\ \mathbf{B})\ (+)\ \mathbf{C} \neq (\mathbf{A}\ (+)\ \mathbf{C})\ (\cdot)\ (\mathbf{A}\ (+)\ \mathbf{B}), \qquad \textbf{(1.54)}$$

because

$$(a \cdot b) + c \neq a \cdot c + b \cdot c.$$

1.6 MINIMUM AND MAXIMUM OF FUZZY NUMBERS

Now consider two fuzzy numbers **A** and **B** in R that are not necessarily comparable.

$$A_\alpha = [a_1^{(\alpha)}, a_2^{(\alpha)}]$$

and

$$B_\alpha = [b_1^{(\alpha)}, b_1^{(\alpha)}].$$

If

$\forall\ \alpha \in [0, 1]$:

$$a_1^{(\alpha)} \leq b_1^{(\alpha)} \text{ and } a_2^{(\alpha)} \leq b_2^{(\alpha)}, \qquad \textbf{(1.55)}$$

we can write $\mathbf{A} \leq \mathbf{B}$.

If condition (1.55) or the inverse condition passing from $\leq$ to $\geq$, is not satisfied, than **A** and **B** are not comparable. Fuzzy numbers do not have the structure of total order (linear order): they have only partial order.

We define the *fuzzy minimum* of **A** and **B** as the fuzzy number

$\forall\ \alpha \in [0, 1]$:

$$\begin{aligned} A_\alpha\,(\wedge)\,B_\alpha &= [a_1^{(\alpha)}, a_2^{(\alpha)}]\,(\wedge)\,[b_1^{(\alpha)}, b_2^{(\alpha)}] \\ &= [a_1^{(\alpha)} \wedge b_1^{(\alpha)}, a_2^{(\alpha)} \wedge b_2^{(\alpha)}], \end{aligned} \tag{1.56}$$

and the *fuzzy maximum* of **A** and **B** as the fuzzy number

$\forall\ \alpha \in [0, 1]$:

$$\begin{aligned} A_\alpha\,(\vee)\,B_\alpha &= [a_1^{(\alpha)}, a_2^{(\alpha)}]\,(\vee)\,[b_1^{(\alpha)}, b_1^{(\alpha)}] \\ &= [a_1^{(\alpha)} \vee b_1^{(\alpha)}, a_2^{(\alpha)} \vee b_2^{(\alpha)}]. \end{aligned} \tag{1.57}$$

The same symbols $(\wedge)$ and $(\vee)$ will be used also for representing the minimum and maximum of the fuzzy numbers **A** and **B**.

Figures 1.15 and 1.16 illustrate definitions (1.56) and (1.57). These properties can be presented in another way:

$\forall\ x, y, z \subset R$:

$$\mu_{\mathbf{A}\,(\wedge)\,\mathbf{B}}(z) = \bigvee_{z = x \wedge y} (\mu_{\mathbf{A}}(x) \wedge \mu_{\mathbf{B}}(y))$$

and

$$\mu_{\mathbf{A}\,(\vee)\,\mathbf{B}}(z) = \bigvee_{z = x \vee y} (\mu_{\mathbf{A}}(x) \vee \mu_{\mathbf{B}}(y)).$$

EXAMPLE 1.13

Let us now look at the example illustrated in Figure 1.17(A), where

$\forall\ x \subset R$:

$$\begin{aligned} \mu_{\mathbf{A}}(x) &= 0, & x &\le -2, \\ &= x/2 + 1, & -2 \le x &\le 0, \\ &= -x/6 + 1, & 0 \le x &\le 6, \\ &= 0, & x &\ge 6. \end{aligned}$$

$$\begin{aligned} \mu_{\mathbf{B}}(x) &= 0, & x &\le -4, \\ &= x/7 + 4/7, & -4 \le x &\le 3, \\ &= -x/2 + 5/2, & 3 \le x &\le 5, \\ &= 0, & x &\ge 5. \end{aligned}$$

Then we have

$\forall\ \alpha \in [0, 1]$:

$$A_\alpha = [2\alpha - 2, -6\alpha + 6]$$

$$B_\alpha = [7\alpha - 4, -2\alpha + 5].$$

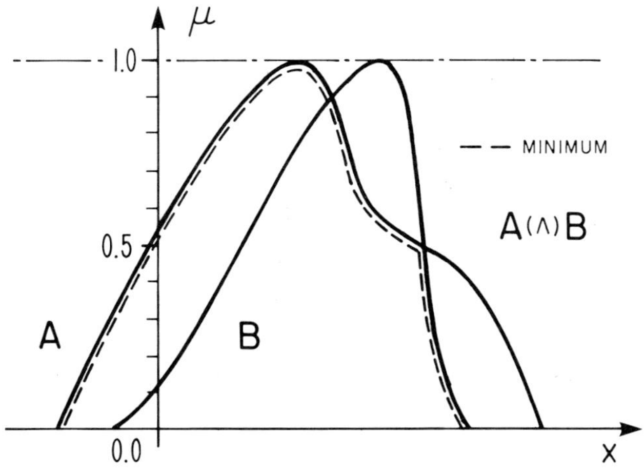

Figure 1.15 Minimum of fuzzy numbers **A** and **B**.

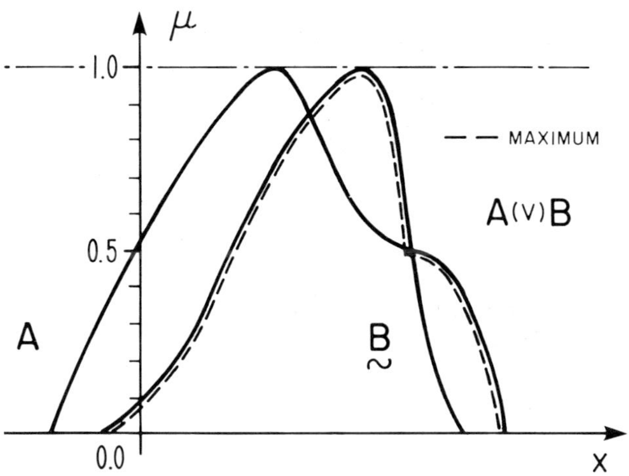

Figure 1.16 Maximum of fuzzy numbers **A** and **B**.

Therefore, for a minimum,

$$A_\alpha\,(\wedge)\,B_\alpha = [(2\alpha - 2) \wedge (7\alpha - 4), (-6\alpha + 6) \wedge (-2\alpha + 5)],$$

we find that

$$\begin{aligned} A_\alpha\,(\wedge)\,B_\alpha &= [7\alpha - 4, -2\alpha + 5], & 0 \le \alpha \le 0.25, \\ &= [7\alpha - 4, -6\alpha + 6], & 0.25 \le \alpha \le 0.40, \\ &= [2\alpha - 2, -6\alpha + 6], & 0.40 \le \alpha \le 1. \end{aligned}$$

From this we obtain the following result:

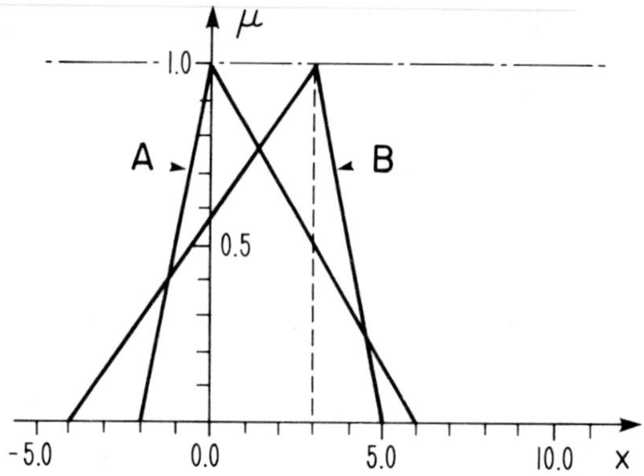

Figure 1.17(A) Two triangular fuzzy numbers **A** and **B** (Example 1.13).

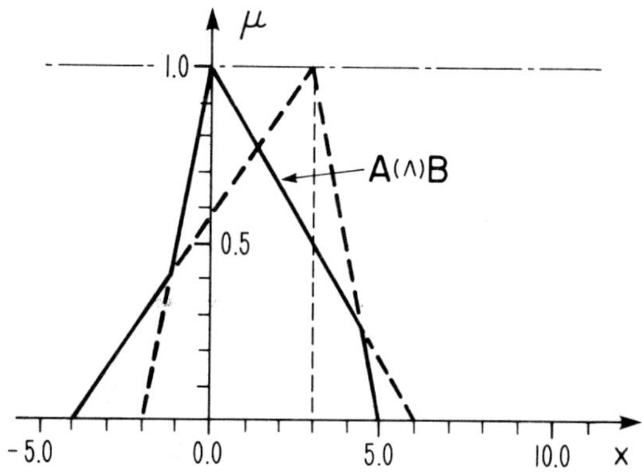

Figure 1.17(B) Minimum of fuzzy numbers **A** and **B** (Example 1.13).

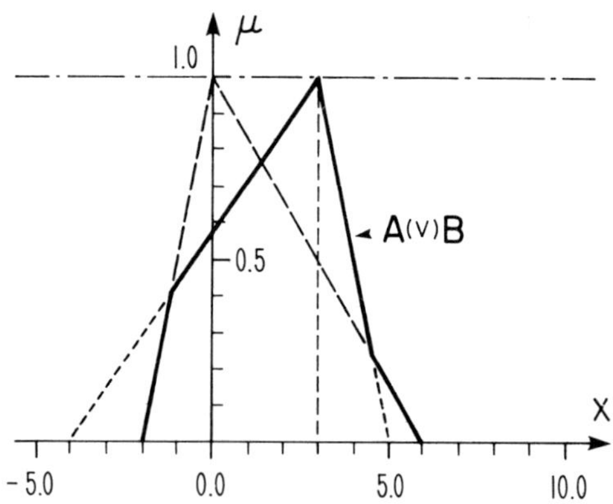

Figure 1.17(C) Maximum of fuzzy numbers **A** and **B** (Example 1.13).

$$
\begin{aligned}
\mu_{\mathbf{A}\,(\vee)\,\mathbf{B}}(x) &= 0, & x &\le -4,\\
&= -x/7 + 4/7, & -4 \le x &\le 1.2,\\
&= x/2 + 1, & -1.2 \le x &\le 0,\\
&= -x/6 + 1, & 0 \le x &\le 4.5,\\
&= 0, & x &\ge 5. \qquad \textbf{(1.58)}
\end{aligned}
$$

Figure 1.17(B) shows a sketch of **A** $(\wedge)$ **B**.

To find the maximum value,

$$
\begin{aligned}
A_\alpha\,(\vee)\,B_\alpha &= [(2\alpha - 2) \vee (7\alpha - 4), \quad (-6\alpha + 6) \vee (-2\alpha + 5)]\\
&= [2\alpha - 2,\ -6\alpha + 6], & 0 \le \alpha &\le 0.25,\\
&= [2\alpha - 2,\ -2\alpha + 5], & 0.25 \le \alpha &\le 0.40,\\
&= [7\alpha - 4,\ -2\alpha + 5], & 0.40 \le \alpha &\le 1.
\end{aligned}
$$

From this we obtain the following result:

$$
\begin{aligned}
\mu_{\mathbf{A}\,(\vee)\,\mathbf{B}}(x) &= 0, & x &\le -2,\\
&= x/2 + 1, & -2 \le x &\le -1.2,\\
&= x/7 + 4/7, & -1.2 \le x &\le 3,\\
&= x/2 + 5/2, & 3 \le x &\le 4.5,\\
&= -x/6 + 1, & 4.5 \le x &\le 6,\\
&= 0, & x &\ge 6. \qquad \textbf{(1.59)}
\end{aligned}
$$

A sketch of this is shown in Figure 1.17(C).

EXAMPLE 1.14

Let us consider another example in z.

A =

−2	−1	0	1	2	3	4
0.2	0.3	0.5	1	0.5	0	0

B =

−2	−1	0	1	2	3	4
0	0.5	1	0.7	0.6	0.4	0.1

Obviously these two fuzzy numbers are not comparable (see Table 1.4).

Alternatively, we will use the minimum set squares and the maximum set squares to obtain the minimum and the maximum values. Thus, we have Table 1.5 (page **42**).

Table 1.4

	−2	−1	0	1	2	3
1				1		
0.9				1		
0.8				1		
0.7				1		
0.6				1		
0.5			1	1	1	
0.4			1	1	1	
0.3		1	1	1	1	
0.2	1	1	1	1	1	
0.1	1	1	1	1	1	
0	1	1	1	1	1	1
	0.2	0.3	0.5	1	0.5	0.0

(∧)

	−1	0	1	2	3	4
1		1				
0.9		1				
0.8		1				
0.7		1	1			
0.6		1	1	1		
0.5	1	1	1	1		
0.4	1	1	1	1	1	
0.3	1	1	1	1	1	
0.2	1	1	1	1	1	
0.1	1	1	1	1	1	1
0	1	1	1	1	1	1
	0.5	1	0.7	0.6	0.4	0.1

=

−2	−1	0	1	2	3	4
		1				
		1				
		1				
		1	1			
		1	1			
	1	1	1	1		
	1	1	1	1		
	1	1	1	1		
1	1	1	1	1		
1	1	1	1	1		
1	1	1	1	1	1	1
0.2	0.5	1	0.7	0.5	0	0

A (∧) **B** =

−2	−1	0	1	2	3	4
0.2	0.5	1	0.7	0.5	0	0

We can very easily validate the following properties in R and Z.

$$\mathbf{A}\ (\wedge)\ \mathbf{B} = \mathbf{B}\ (\wedge)\ \mathbf{A}, \tag{1.60}$$

$$\mathbf{A}\ (\vee)\ \mathbf{B} = \mathbf{B}\ (\vee)\ \mathbf{A}, \tag{1.61}$$

$$(\mathbf{A}\ (\wedge)\ \mathbf{B})\ \wedge\ \mathbf{C} = \mathbf{A}\ (\wedge)\ (\mathbf{B}\ \wedge\ \mathbf{C}), \tag{1.62}$$

$$(\mathbf{A}\ (\vee)\ \mathbf{B})\ \vee\ \mathbf{C} = \mathbf{A}\ (\vee)\ (\mathbf{B}\ (\vee)\ \mathbf{C}), \tag{1.63}$$

$$\mathbf{A}\ (\wedge)\ \mathbf{A} = \mathbf{A}, \tag{1.64}$$

$$\mathbf{A}\ (\vee)\ \mathbf{A} = \mathbf{A}, \tag{1.65}$$

$$\mathbf{A}\ (\vee)\ (\mathbf{A}\ (\wedge)\ \mathbf{B}) = \mathbf{A}, \tag{1.66}$$

Table 1.4 (*continued*)

	−2	−1	0	1	2	3
1				1		
0.9				1		
0.8				1		
0.7				1		
0.6				1		
0.5			1	1	1	
0.4			1	1	1	
0.3		1	1	1	1	
0.2	1	1	1	1	1	
0.1	1	1	1	1	1	
0	1	1	1	1	1	1
	0.2	0.3	0.5	1	0.5	0

(∨)

−1	0	1	2	3	4
	1				
	1				
	1				
	1	1			
	1	1	1		
1	1	1	1		
1	1	1	1	1	
1	1	1	1	1	
1	1	1	1	1	
1	1	1	1	1	1
1	1	1	1	1	1
0.5	1	0.7	0.6	0.4	0.1

=

−2	−1	0	1	2	3	4
			1			
			1			
			1			
			1			
			1	1		
		1	1	1		
		1	1	1	1	
	1	1	1	1	1	
	1	1	1	1	1	
	1	1	1	1	1	1
1	1	1	1	1	1	1
0	0.3	0.5	1	0.6	0.4	0.1

A (∨) **B** =

−2	−1	0	1	2	3	4
0	0.3	0.5	1	0.6	0.4	0.1

$$\mathbf{A}\ (\wedge)\ (\mathbf{A}\ (\vee)\ \mathbf{B}) = \mathbf{A}, \tag{1.67}$$

$$\mathbf{A}\ (\vee)\ (\mathbf{B}\ (\wedge)\ \mathbf{C}) = (\mathbf{A}\ (\vee)\ \mathbf{B})\ (\wedge)\ (\mathbf{A}\ (\vee)\ \mathbf{C}), \tag{1.68}$$

$$\mathbf{A}\ (\wedge)\ (\mathbf{B}\ (\vee)\ \mathbf{C}) = (\mathbf{A}\ (\wedge)\ \mathbf{B})\ (\vee)\ (\mathbf{A}\ (\wedge)\ \mathbf{C}), \tag{1.69}$$

We see from properties (1.60–1.69) that the fuzzy maximum and fuzzy minimum have a distributive lattice structure in R or Z.

Until now we have used the symbols (∧) and (∨) to refer to the minimum and the maximum. We could have used the terms the lower bound and upper bound instead.

Table 1.5

A \ B	−2	−1	0	1	2	3	4
−2	0 / 0.2	0.5 / 0.2	1 / 0.2	0.7 / 0.2	0.6 / 0.2	0.4 / 0.2	0.1 / 0.2
−1	0 / 0.3	0.5 / 0.3	1 / 0.3	0.7 / 0.3	0.6 / 0.3	0.4 / 0.3	0.1 / 0.3
0	0 / 0.5	0.5 / 0.5	1 / 0.5	0.7 / 0.5	0.6 / 0.5	0.4 / 0.5	0.1 / 0.5
1	0 / 1	0.5 / 1	1 / 1	0.7 / 1	0.6 / 1	0.4 / 1	0.1 / 1
2	0 / 0.5	0.5 / 0.5	1 / 0.5	0.7 / 0.5	0.6 / 0.5	0.4 / 0.5	0.1 / 0.5
3	0 / 0	0.5 / 0	1 / 0	0.7 / 0	0.6 / 0	0.4 / 0	0.1 / 0
4	0 / 0	0.5 / 0	1 / 0	0.7 / 0	0.6 / 0	0.4 / 0	0.1 / 0

	−2	−1	0	1	2	3	4
A (∧) **B** =	0.2	0.5	1	0.7	0.5	0	0

A \ B	−2	−1	0	1	2	3	4
−2	0 / 0.2	0.5 / 0.2	1 / 0.2	0.7 / 0.2	0.6 / 0.2	0.4 / 0.2	0.1 / 0.2
−1	0 / 0.3	0.5 / 0.3	1 / 0.3	0.7 / 0.3	0.6 / 0.3	0.4 / 0.3	0.1 / 0.3
0	0 / 0.5	0.5 / 0.5	1 / 0.5	0.7 / 0.5	0.6 / 0.5	0.4 / 0.5	0.1 / 0.5
1	0 / 1	0.5 / 1	1 / 1	0.7 / 1	0.6 / 1	0.4 / 1	0.1 / 1
2	0 / 0.5	0.5 / 0.5	1 / 0.5	0.7 / 0.5	0.6 / 0.5	0.4 / 0.5	0.1 / 0.5
3	0 / 0	0.5 / 0	1 / 0	0.7 / 0	0.6 / 0	0.4 / 0	0.1 / 0
4	0 / 0	0.5 / 0	1 / 0	0.7 / 0	0.6 / 0	0.4 / 0	0.1 / 0

	−2	−1	0	1	2	3	4
A (∨) **B** =	0	0.3	0.5	1	0.6	0.4	0.1

1.7 CONVOLUTION OF FUZZY NUMBERS

The following operations in R (eventually in R^+ only) are called the *max-min convolution.*

$$\forall\ x, y, z \in R:$$

$$\mu_{\mathbf{A}\,(+)\,\mathbf{B}}(z) = \bigvee_{z = x + y} (\mu_{\mathbf{A}}(x) \wedge \mu_{\mathbf{B}}(y)), \tag{1.70}$$

$$\mu_{\mathbf{A}\,(-)\,\mathbf{B}}(z) = \bigvee_{z = x - y} (\mu_{\mathbf{A}}(x) \wedge \mu_{\mathbf{B}}(y)), \tag{1.71}$$

$$\mu_{\mathbf{A}\,(\cdot)\,\mathbf{B}}(z) = \bigvee_{z = x \cdot y} (\mu_{\mathbf{A}}(x) \wedge \mu_{\mathbf{B}}(y)), \tag{1.72}$$

$$\mu_{\mathbf{A}\,(:)\,\mathbf{B}}(z) = \bigvee_{z = x / y} (\mu_{\mathbf{A}}(x) \wedge \mu_{\mathbf{B}}(y)), \tag{1.73}$$

$$\mu_{\mathbf{A}\,(\wedge)\,\mathbf{B}}(z) = \bigvee_{z = x \wedge y} (\mu_{\mathbf{A}}(x) \wedge \mu_{\mathbf{B}}(y)), \tag{1.74}$$

$$\mu_{\mathbf{A}\,(\vee)\,\mathbf{B}}(z) = \bigvee_{z = x \vee y} (\mu_{\mathbf{A}}(x) \wedge \mu_{\mathbf{B}}(y)). \tag{1.75}$$

In Table 1.5 we showed how this convolution is used for the operations $\wedge$ and $\vee$ when **A** and **B** are in Z. We now examine how to use the same process for the operations $(+)$ and $(-)$. For this purpose we again use a numerical example.

EXAMPLE 1.15

A = (1.76)

0	1	2	3	4	5
0.1	0.5	0.9	1	0.6	0.2

B = (1.77)

0	1	2	3	4	5	6
0.2	0.3	1	0.8	0.7	0.1	0.1

For **A** (+) **B**

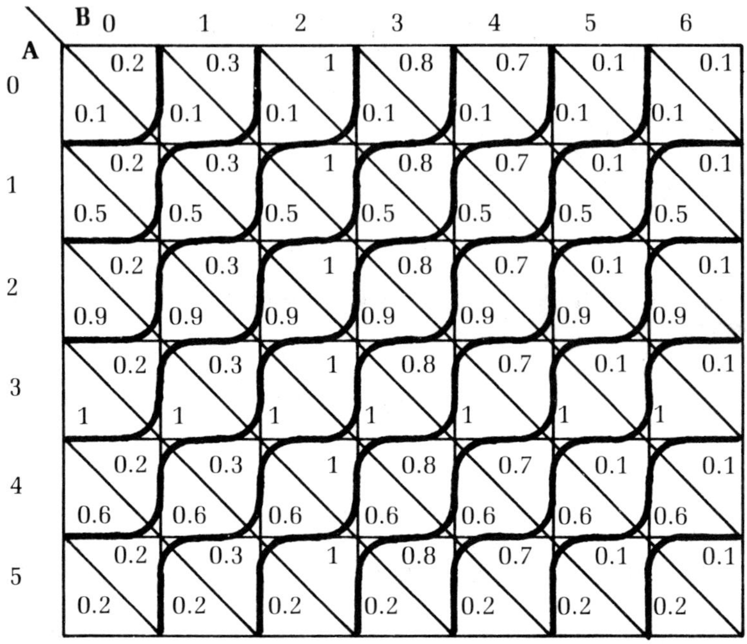

(1.78)

The max-min operations for each diagonal give result shown in Table 1.6. Any type of convolution is possible. For example,

$\forall$ x, y, z $\in$ R: (eventually in R^+ only)

$$\mu_{\mathbf{A}\,(*)\,\mathbf{B}}(z) = \bigvee_{z = x * y} (\mu_{\mathbf{A}}(x) \wedge \mu_{\mathbf{B}}(y)),$$

where $*$ is an operator chosen such that $z = x\,(*)y$. Consequently, $x * y$ can mean $x^2 + y^2$ or any other operation for the couplet (x, y).

We are also free to consider other types of convolutions such as min-max convolution. For example

$\forall$ x, y, z $\in$ R: (eventually in R^+only)

$$\mu_{\mathbf{A}(*)\mathbf{B}}(z) = \bigwedge_{z = x * y} (\mu_{\mathbf{A}}(x) \vee \mu_{\mathbf{B}}(y)).$$

Let us take $* = +$ in the following example.

Table 1.6

	0	1	2	3	4	5	6	7	8	9	10	11
A (+) **B** =	0.1	0.2	0.3	0.5	0.9	1	0.8	0.7	0.6	0.2	0.1	0.1

For **A** (−) **B**

A \ B	0	1	2	3	4	5	6
0	0.2 / 0.1	0.3 / 0.1	1 / 0.1	0.8 / 0.1	0.7 / 0.1	0.1 / 0.1	0.1 / 0.1
1	0.2 / 0.5	0.3 / 0.5	1 / 0.5	0.8 / 0.5	0.7 / 0.5	0.1 / 0.5	0.1 / 0.5
2	0.2 / 0.9	0.3 / 0.9	1 / 0.9	0.8 / 0.9	0.7 / 0.9	0.1 / 0.9	0.1 / 0.9
3	0.2 / 1	0.3 / 1	1 / 1	0.8 / 1	0.7 / 1	0.1 / 1	0.1 / 1
4	0.2 / 0.6	0.3 / 0.6	1 / 0.6	0.8 / 0.6	0.7 / 0.6	0.1 / 0.6	0.1 / 0.6
5	0.2 / 0.2	0.3 / 0.2	1 / 0.2	0.8 / 0.2	0.7 / 0.2	0.1 / 0.2	0.1 / 0.2

	−6	−5	−4	−3	−2	−1	0	1	2	3	4	5
A (−) **B** =	0.1	0.1	0.1	0.5	0.7	0.8	0.9	1	0.6	0.3	0.2	0.2

EXAMPLE 1.16

Using (1.76), (1.77) and (1.78), we obtain, for the min-max convolution,

	0	1	2	3	4	5	6	7	8	9	10	11
A (+) **B** = min-max	0.2	0.3	0.5	0.5	0.6	0.1	0.1	0.8	0.7	0.6	0.2	0.2

This example shows that the arbitrary convolution of a fuzzy number can give a fuzzy subset that is neither convex nor normal. As far as the convolutions given from (1.70–1.75) are concerned, it is easy to prove that such convolutions for fuzzy numbers give fuzzy numbers by using Theorems 1.1 and 1.2.

Recall that the convolution used in the theory of probability is the convolution *sum-product*. That is,

$\forall\ x, y, z \in R$:

$$\underbrace{\mu_{\mathbf{A}(+)\mathbf{B}}(z)}_{\text{Sum-product}} = \sum_{z = x + y} (\mu_{\mathbf{A}}(x) \cdot \mu_{\mathbf{B}}(y)),$$

where

$$\sum_{x} \mu_{\mathbf{A}}(x) = 1 \qquad \text{and} \qquad \sum_{y} \mu_{\mathbf{B}}(y) = 1.$$

Here the μ's are functions that can be probability functions for the random sets A and B. These functions generally are not normal. In the continuous case, the symbol $\int$ takes the place of Σ with corresponding notations.

1.8 DECONVOLUTIONS OF FUZZY NUMBERS

Let us return to (1.22), which is concerned with the max-min convolution for addition.

$\forall\ x, y, z \in R$:

$$\mu_{\mathbf{A}(+)\mathbf{B}}(z) = \bigvee_{z = x + y} (\mu_{\mathbf{A}}(x) \wedge \mu_{\mathbf{B}}(y)). \tag{1.79}$$

Let us define

$$\mathbf{C} = \mathbf{A}\ (+)\ \mathbf{B}. \tag{1.80}$$

In (1.79) **A** and **B** are given and **C** is an unknown. For the inverse problem, we must solve for **B** in (1.80) when **C** and **A** are given. This is called *deconvolution*, which whould not be confused with subtraction. It is defined as follows:

$\forall\ x, y, z \in R$:

$$\mu_{\mathbf{C}(-)\mathbf{B}}(x) = \bigvee_{z = x - y} (\mu_{\mathbf{C}}(x) \wedge \mu_{\mathbf{B}}(y)$$

because (**C** (−) **B**) is not, except in particular cases, a solution of (1.80).

Let us consider now a level $\alpha \in [0, 1]$, where

$$
\begin{aligned}
A_\alpha &= [a_1^{(\alpha)}, a_2^{(\alpha)}],\\
B_\alpha &= [b_1^{(\alpha)}, b_2^{(\alpha)}].\\
C_\alpha &= A_\alpha\,(+)\,B_\alpha = [a_1^{(\alpha)} + b_1^{(\alpha)}, a_2^{(\alpha)} + b_2^{(\alpha)}].\\
C_\alpha\,(-)\,B_\alpha &= [a_1^{(\alpha)} + b_1^{(\alpha)}, a_2^{(\alpha)} + b_2^{(\alpha)}]\,(-)\,[b_1^{(\alpha)}, b_2^{(\alpha)}]\\
&= [a_1^{(\alpha)} + b_1^{(\alpha)} - b_2^{(\alpha)}, a_2^{(\alpha)} + b_2^{(\alpha)} - b_1^{(\alpha)}]\\
&\neq [a_1^{(\alpha)}, a_2^{(\alpha)}] \quad \text{except if } b_1^{(\alpha)} = b_2^{(\alpha)}.
\end{aligned}
$$

In order to establish the deconvolution condition, that is, to determine whether or not (1.80) has a solution, we must consider the following theorem.

THEOREM 1.5

Given any two fuzzy numbers **A** and **C** in R such that for each level α: $(c_2^{(\alpha)} - a_2^{(\alpha)}) \geq (c_1^{(\alpha)} - a_1^{(\alpha)})$, let us consider the following equation:

$$\mathbf{A}\,(+)\,\mathbf{B} = \mathbf{C}. \qquad \textbf{(1.81)}$$

A necessary and sufficient condition for the existence of a fuzzy number **B** satisfying (1.81) is

$\forall\ \alpha, \alpha' \in [0, 1]$:

$$(\alpha' > \alpha) \Rightarrow ((c_1^{(\alpha')} - a_1^{(\alpha')}) \geq (c_1^{(\alpha)} - a_1^{(\alpha)})$$

and

$$(c_2^{(\alpha')} - a_2^{(\alpha')}) \leq (c_2^{(\alpha)} - a_2^{(\alpha)})). \qquad \textbf{(1.82)}$$

If this condition is satisfied, the solution exists and is unique.[7]

$\forall\ \alpha \in [0, 1]$:

$$B_\alpha = [c_1^{(\alpha)} - a_1^{(\alpha)}, c_2^{(\alpha)} - a_2^{(\alpha)}].$$

Proof:

If **A** and **C** are normal, then **B** is also normal. Conversely, if (1.82) is in a nested condition, level by level, and since **A** and **C** are convex, this implies that the convexity of **B** in (1.83) is satisfied. This condition must be checked a priori.

The solution for **B** in (1.83) is normal and convex. It is also unique for **A** and **C**.

EXAMPLE 1.17

Let two fuzzy numbers in N be defined as follows.

	0	1	2	3	4	5	6	7	8	9	10
C =	0.1	0.2	0.6	0.7	0.8	0.9	1	0.5	0.4	0.2	0.1

	0	1	2	3	4	5	6
A =	0.2	0.6	0.8	0.9	1	0.5	0.1

Progressively we can compute

$C_1 = [6, 6]$, $A_1 = [4, 4]$, $B_1 = [2, 2]$,

$C_{0.9} = [5, 6]$, $A_{0.9} = [3, 4]$, $B_{0.9} = [2, 2]$,

$C_{0.8} = [4, 6]$, $A_{0.8} = [2, 4]$, $B_{0.8} = [2, 2]$,

$C_{0.7} = [3, 6]$, $A_{0.7} = [2, 4]$, $B_{0.7} = [1, 2]$,

$C_{0.6} = [2, 6]$, $A_{0.6} = [1, 4]$, $B_{0.6} = [1, 2]$,

$C_{0.5} = [2, 7]$, $A_{0.5} = [1, 5]$, $B_{0.5} = [1, 2]$,

$C_{0.4} = [2, 8]$, $A_{0.4} = [1, 5]$, $B_{0.4} = [1, 3]$,

$C_{0.3} = [2, 8]$, $A_{0.3} = [1, 5]$, $B_{0.3} = [1, 3]$,

$C_{0.2} = [1, 9]$, $A_{0.2} = [0, 5]$, $B_{0.2} = [1, 4]$,

$C_{0.1} = [0, 10]$, $A_{0.1} = [0, 6]$, $B_{0.1} = [0, 4]$,

Condition (1.82) is therefore satisfied (Table 1.7). Let us now check the result **A** (+) **B** = **C**.

Table 1.7

	0	1	2	3	4
1			1		
0.9			1		
0.8			1		
0.7		1	1		
0.6		1	1		
0.5		1	1		
0.4		1	1	1	
0.3		1	1	1	
0.2		1	1	1	1
0.1	1	1	1	1	1
0	1	1	1	1	1
B =	0.1	0.7	1	0.4	0.2

0	1	2	3	4	5	6
0.2	0.6	0.8	0.9	1	0.5	0.1

(+)

0	1	2	3	4
0.1	0.7	1	0.4	0.2

=

0	1	2	3	4	5	6	7	8	9	10
0.1	0.2	0.6	0.7	0.8	0.9	1	0.5	0.4	0.2	0.1

EXAMPLE 1.18

We now consider an example in which the existence of condition (1.82) is not satisfied.

C =

0	1	2	3	4	5	6	7	8	9
0.2	0.3	0.5	0.6	0.8	1	0.9	0.8	0.3	0.1

(1.84)

A =

0	1	2	3	4	5	6
0.4	0.4	0.6	0.6	1	0.8	0.1

(1.85)

$$
\begin{array}{lll}
C_1 = [5, 5], & A_1 = [4, 4], & B_1 = [1, 1], \\
C_{0.9} = [5, 6], & A_{0.9} = [4, 4], & B_{0.9} = [1, 2], \\
C_{0.8} = [4, 7], & A_{0.8} = [4, 5], & B_{0.8} = [0, 2], \\
C_{0.7} = [4, 7], & A_{0.7} = [4, 5], & B_{0.7} = [0, 2], \\
C_{0.6} = [3, 7], & A_{0.6} = [2, 5], & B_{0.6} = [1, 2], \\
C_{0.5} = [2, 7], & A_{0.5} = [2, 5], & B_{0.5} = [0, 2], \\
C_{0.4} = [2, 7], & A_{0.4} = [0, 5], & B_{0.4} = [2, 2], \\
C_{0.3} = [1, 8], & A_{0.3} = [0, 5], & B_{0.3} = [1, 3], \\
C_{0.2} = [0, 8], & A_{0.2} = [0, 5], & B_{0.2} = [0, 3], \\
C_{0.1} = [0, 9], & A_{0.1} = [0, 6], & B_{0.1} = [0, 3].
\end{array}
$$

If we inspect level 0.7 and 0.6 and compare $B_{0.7}$ and $B_{0.6}$: $0 < 1$ and $2 = 2$, we see that it is not compatible with (1.82). Hence the equation

$$\mathbf{A}\ (+)\ \mathbf{B} = \mathbf{C},$$

where **C** is given by (1.84) and **A** by (1.85), does not have a solution for **B**.

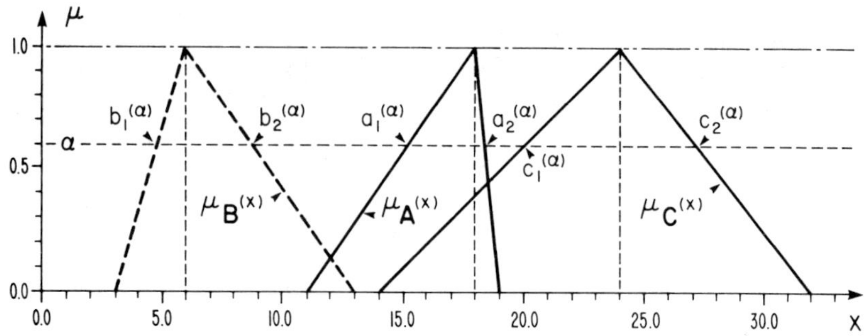

Figure 1.18 Deconvolution for **B** (Example 1.19).

EXAMPLE 1.19

Figure 1.18 shows two fuzzy numbers **C** and **A** with triangular shapes. Let us look for a solution for **B** that concerns the following equation.

$$\mathbf{A}\ (+)\ \mathbf{B} = \mathbf{C}.$$

Fuzzy numbers **A** and **C** are defined by

$\forall\ x \in R$:

$$\begin{aligned}
\mu_{\mathbf{A}}(x) &= 0, & x &\le 11,\\
&= x/7 - 11/7, & 11 \le x &\le 18,\\
&= -x + 19, & 18 \le x &\le 19,\\
&= 0, & x &\ge 19.\\
\mu_{\mathbf{C}}(x) &= 0, & x &\le 14,\\
&= x/10 - 14/10, & 14 \le x &\le 24,\\
&= -x/8 + 4, & 24 \le x &\le 32,\\
&= 0, & x &\ge 32.
\end{aligned}$$

Now we check property (1.82).

$$b_1^{(\alpha)} = c_1^{(\alpha)} - a_1^{(\alpha)} = (10\alpha + 14) - (7\alpha + 11) = (3\alpha + 3),$$
$$b_2^{(\alpha)} = c_2^{(\alpha)} - a_2^{(\alpha)} = (-8\alpha + 32) - (-\alpha + 19) = (-7\alpha + 13).$$

Therefore,

$$[b_1^{(\alpha)}, b_2^{(\alpha)}] = [3\alpha + 3, -7\alpha + 13]. \quad \textbf{(1.86)}$$

It is easy to verify that in (1.86) the lower bound is monotonically increasing with respect to α and the upper bound is monotonically decreasing with respect to α for $\alpha \in [0, 1]$. Thus we see that condition

(1.82) is satisfied. We therefore have the deconvolution for **B** as shown in Figure 1.18.

$\forall\ x \in R$:

$$
\begin{aligned}
\mu_{\mathbf{B}}(x) &= 0, & x &\le 3,\\
&= x/3 - 1, & 3 \le x &\le 6,\\
&= -x/7 + 13/7, & 6 \le x &\le 13,\\
&= 0, & x &\ge 13.
\end{aligned}
$$

EXAMPLE 1.20

In this example fuzzy numbers in Z are used to illustrate another method for obtaining deconvolution.

C =

−7	−6	−5	−4	−3	−2	−1	0	1	2	3	4
0.1	0.4	0.5	0.8	0.8	0.9	1	0.6	0.5	0.2	0.1	0.1

A =

−5	−4	−3	−2	−1	0	1
0.4	0.8	1	0.6	0.2	0.1	0.1

$$
\begin{aligned}
C_1 &= [-1, -1], & A_1 &= [-3, -3], & B_1 &= [2, 2],\\
C_{0.9} &= [-2, -1], & A_{0.9} &= [-3, -3], & B_{0.9} &= [1, 2],\\
C_{0.8} &= [-4, -1], & A_{0.8} &= [-4, -3], & B_{0.8} &= [0, 2],\\
C_{0.7} &= [-4, -1], & A_{0.7} &= [-4, -3], & B_{0.7} &= [0, 2],\\
C_{0.6} &= [-4, 0], & A_{0.6} &= [-4, -2], & B_{0.6} &= [0, 2],\\
C_{0.5} &= [-5, 1], & A_{0.5} &= [-4, -2], & B_{0.5} &= [-1, 3],\\
C_{0.4} &= [-6, 1], & A_{0.4} &= [-5, -2], & B_{0.4} &= [-1, 3],\\
C_{0.3} &= [-6, 1], & A_{0.3} &= [-5, -2], & B_{0.3} &= [-1, 3],\\
C_{0.2} &= [-6, 2], & A_{0.2} &= [-5, -1], & B_{0.2} &= [-1, 3],\\
C_{0.1} &= [-7, 4], & A_{0.1} &= [-5, 1], & B_{0.1} &= [-2, 3].
\end{aligned}
$$

We then have the results given in Table 1.8.

Table 1.8

	−2	−1	0	1	2	3
1.0					1	
0.9				1	1	
0.8			1	1	1	
0.7			1	1	1	
0.6			1	1	1	
0.5		1	1	1	1	1
0.4		1	1	1	1	1
0.3		1	1	1	1	1
0.2		1	1	1	1	1
0.1	1	1	1	1	1	1
B =	0.1	0.5	0.8	0.9	1	0.5

−5	−4	−3	−2	−1	0	1
0.4	0.8	1	0.6	0.2	0.1	0.1

(+)

−2	−1	0	1	2	3
0.1	0.5	0.8	0.9	1	0.5

=

−7	−6	−5	−4	−3	−2	−1	0	1	2	3	4
0.1	0.4	0.5	0.8	0.8	0.9	1	0.6	0.5	0.2	0.1	0.1

Another method for deconvolution uses Example 1.20 and Equation (1.79) directly.

1. $\mu_{\mathbf{C}}(-7) = \mu_{\mathbf{A}}(-5) \wedge \mu_{\mathbf{B}}(-2)$,

or

$$0.1 = 0.4 \wedge \mu_{\mathbf{B}}(-2).$$

Therefore, $\mu_{\mathbf{B}}(-2) = 0.1$.

2. $\mu_{\mathbf{C}}(-6) = (\mu_{\mathbf{A}}(-5) \wedge \mu_{\mathbf{B}}(-1)) \vee (\mu_{\mathbf{A}}(-4) \wedge \mu_{\mathbf{B}}(-2))$,

or

$$0.4 = (0.4 \wedge \mu_{\mathbf{B}}(-1)) \vee (0.8 \wedge 0.1).$$

Therefore, $\mu_{\mathbf{B}}(-1) \geq 0.4$.

3. $\mu_{\mathbf{C}}(-5) = (\mu_{\mathbf{A}}(-5) \wedge \mu_{\mathbf{B}}(0)) \vee (\mu_{\mathbf{A}}(-4) \wedge \mu_{\mathbf{B}}(-1)) \vee (\mu_{\mathbf{A}}(-3) \wedge \mu_{\mathbf{B}}(-2)),$

or

$$0.5 = \overbrace{(0.4 \wedge \mu_{\mathbf{B}}(0))}^{\text{Rejected, smaller than 0.5}} \vee (0.8 \wedge \mu_{\mathbf{B}}(-1)) \vee (1 \wedge 0.1)$$
$$= (\mu_{\mathbf{B}}(0) \leq 0.4) \vee (\mu_{\mathbf{B}}(-1) \geq 0.4) \vee (0.1),$$

which implies that $\mu_{\mathbf{B}}(-1) = 0.5$.

4. $\mu_{\mathbf{C}}(-4) = (\mu_{\mathbf{A}}(-5) \wedge \mu_{\mathbf{B}}(1)) \vee (\mu_{\mathbf{A}}(-4) \wedge \mu_{\mathbf{B}}(0)) \vee (\mu_{\mathbf{A}}(-3) \wedge \mu_{\mathbf{B}}(-1)) \vee (\mu_{\mathbf{A}}(-2) \wedge \mu_{\mathbf{B}}(-2)),$

or

$$0.8 = \underbrace{(0.4 \wedge \mu_{\mathbf{B}}(1))}_{\text{(rejected)}} \vee (0.8 \wedge \mu_{\mathbf{B}}(0)) \vee \underbrace{(1 \wedge 0.5)}_{(0.5)} \vee \underbrace{(0.6 \wedge 0.1)}_{(0.1)},$$

which implies that $\mu_{\mathbf{B}}(0) \geq 0.8$.

5. $\mu_{\mathbf{C}}(-3) = (\mu_{\mathbf{A}}(-5) \wedge \mu_{\mathbf{B}}(2)) \vee (\mu_{\mathbf{A}}(-4) \wedge \mu_{\mathbf{B}}(1)) \vee (\mu_{\mathbf{A}}(-3) \wedge \mu_{\mathbf{B}}(0)) \vee (\mu_{\mathbf{A}}(-2) \wedge \mu_{\mathbf{B}}(-1)) \vee (\mu_{\mathbf{A}}(-1) \wedge \mu_{\mathbf{B}}(-2)),$

or

$$0.8 = \underbrace{(.4 \wedge \mu_{\mathbf{B}}(2))}_{\text{(rejected)}} \vee (0.8 \wedge \mu_{\mathbf{B}}(1)) \vee (1 \wedge \mu_{\mathbf{B}}(0)) \vee$$
$$\underbrace{(0.6 \wedge 0.5)}_{(0.5)} \vee \underbrace{(0.2 \wedge 0.1)}_{(0.1)}$$

Therefore, $\mu_{\mathbf{B}}(0) = 0.8$ and $\mu_{\mathbf{B}}(1) \geq 0.8$

6. $\mu_{\mathbf{C}}(-2) = (\mu_{\mathbf{A}}(-5) \wedge \mu_{\mathbf{B}}(3)) \vee (\mu_{\mathbf{A}}(-4) \wedge \mu_{\mathbf{B}}(2)) \vee (\mu_{\mathbf{A}}(-3) \wedge \mu_{\mathbf{B}}(1)) \vee (\mu_{\mathbf{A}}(-2) \wedge \mu_{\mathbf{B}}(0)) \vee (\mu_{\mathbf{A}}(-1) \wedge \mu_{\mathbf{B}}(-1)) \vee (\mu_{\mathbf{A}}(0) \wedge \mu_{\mathbf{B}}(-2)),$

or

$$0.9 = \underbrace{(0.4 \wedge \mu_{\mathbf{B}}(3))}_{\text{(rejected)}} \vee \underbrace{(0.8 \wedge \mu_{\mathbf{B}}(2))}_{\text{(rejected)}} \vee (1 \wedge \mu_{\mathbf{B}}(1)) \vee \underbrace{(0.6 \wedge 0.8)}_{(0.6)} \vee$$
$$\underbrace{(0.2 \wedge 0.5)}_{(0.2)} \vee \underbrace{(0.1 \wedge (0.1)}_{(0.1)},$$

which implies that $\mu_{\mathbf{B}}(1) = 0.9$.

7. $\mu_{\mathbf{C}}(-1) =$

$$(\mu_{\mathbf{A}}(-5) \wedge \mu_{\mathbf{B}}(4)) \vee (\mu_{\mathbf{A}}(-4) \wedge \mu_{\mathbf{B}}(3)) \vee (\mu_{\mathbf{A}}(-3) \wedge \mu_{\mathbf{B}}(2)) \vee (\mu_{\mathbf{A}}(-2) \wedge \mu_{\mathbf{B}}(1)) \vee (\mu_{\mathbf{A}}(-1) \wedge \mu_{\mathbf{B}}(0)) \vee (\mu_{\mathbf{A}}(0) \wedge \mu_{\mathbf{B}}(-1)) \vee (\mu_{\mathbf{A}}(1) \wedge \mu_{\mathbf{B}}(-2)),$$

or

$$1 = \underbrace{(0.4 \wedge \mu_{\mathbf{B}}(4))}_{\text{(rejected)}} \vee \underbrace{(0.8 \wedge \mu_{\mathbf{B}}(3))}_{\text{(rejected)}} \vee (1 \wedge \mu_{\mathbf{B}}(2)) \vee \underbrace{(0.6 \wedge 0.9)}_{(0.6)}$$

$$\underbrace{(0.2 \wedge 0.8)}_{(0.2)} \vee \underbrace{(0.1 \wedge 0.5)}_{(0.1)} \vee \underbrace{(0.1 \wedge 0.1)}_{(0.1)}.$$

Therefore, $\mu_{\mathbf{B}}(2) = 1$.

8. $\mu_{\mathbf{C}}(0) = (\mu_{\mathbf{A}}(-5) \wedge \mu_{\mathbf{B}}(5)) \vee (\mu_{\mathbf{A}}(-4) \wedge \mu_{\mathbf{B}}(4)) \vee (\mu_{\mathbf{A}}(-3) \wedge \mu_{\mathbf{B}}(3)) \vee (\mu_{\mathbf{A}}(-2) \wedge \mu_{\mathbf{B}}(2)) \vee (\mu_{\mathbf{A}}(-1) \wedge \mu_{\mathbf{B}}(1)) \vee (\mu_{\mathbf{A}}(0) \wedge \mu_{\mathbf{B}}(0)) \vee (\mu_{\mathbf{A}}(1) \wedge \mu_{\mathbf{B}}(-1)) \vee (\mu_{\mathbf{A}}(2) \wedge \mu_{\mathbf{B}}(-2)),$

or

$$0.6 = \underbrace{(0.4 \wedge \mu_{\mathbf{B}}(5))}_{\text{(rejected)}} \vee (0.8 \wedge \mu_{\mathbf{B}}(4)) \vee (1 \wedge \mu_{\mathbf{B}}(3)) \vee \underbrace{(0.6 \wedge 1)}_{(0.6)} \vee$$

$$\underbrace{(0.2 \wedge 0.9)}_{(0.2)} \vee \underbrace{(0.1 \wedge 0.8)}_{(0.1)} \vee \underbrace{(0.1 \wedge 0.5)}_{(0.1)} \vee \underbrace{(0 \wedge 0.1)}_{(0)}.$$

Therefore, $\mu_{\mathbf{B}}(3) \le 0.6$, $\mu_{\mathbf{B}}(4) \le 0.6$.

9. $\mu_{\mathbf{C}}(1) = (\mu_{\mathbf{A}}(-5) \wedge \mu_{\mathbf{B}}(6)) \vee (\mu_{\mathbf{A}}(-4) \wedge \mu_{\mathbf{B}}(5)) \vee (\mu_{\mathbf{A}}(-3) \wedge \mu_{\mathbf{B}}(4)) \vee (\mu_{\mathbf{A}}(-2) \wedge \mu_{\mathbf{B}}(3)) \vee (\mu_{\mathbf{A}}(-1) \wedge \mu_{\mathbf{B}}(2)) \vee (\mu_{\mathbf{A}}(0) \wedge \mu_{\mathbf{B}}(1)) \vee (\mu_{\mathbf{A}}(1) \wedge \mu_{\mathbf{B}}(0)) \vee (\mu_{\mathbf{A}}(2) \wedge \mu_{\mathbf{B}}(-1)),$

or

$$0.5 = \underbrace{(0.4 \wedge \mu_{\mathbf{B}}(6))}_{\text{(rejected)}} \vee (0.8 \wedge \mu_{\mathbf{B}}(5)) \vee (1 \wedge \mu_{\mathbf{B}}(4)) \vee (0.6 \wedge \mu_{\mathbf{B}}(3)) \vee$$

$$\underbrace{(0.2 \wedge 1)}_{(0.2)} \vee \underbrace{(0.1 \wedge 0.9)}_{(0.1)} \vee \underbrace{(0.1 \wedge 0.8)}_{(0.1)} \vee \underbrace{(0 \wedge 0.5)}_{(0)}.$$

Therefore

$$\mu_{\mathbf{B}}(3) = 0.5, \qquad \mu_{\mathbf{B}}(4) \le 0.5, \qquad \mu_{\mathbf{B}}(5) \le 0.5.$$

If this process is continued, one will find

$$\mu_{\mathbf{B}}(x \geq 4) = 0.$$

Thus, the deconvoluted value of **B** is given by

	−2	−1	0	1	2	3
B =	0.1	0.5	0.8	0.9	1	0.5

The use of a table like (1.78) can be very useful for simplifying the computations.

Deconvolution of a Difference or Subtraction

If we let

$\forall\ \alpha \in [0, 1]$: We define the image of **B** as

$$B_{\alpha} = [-b_2^{(\alpha)}, -b_1^{(\alpha)}],$$

then the difference **A** (−) **B** = **C**, becomes **A** (+) $\mathbf{B}^{-}$ = **C**. Hence one can use the deconvolution method, as explained for addition.

Deconvolution of a Product

Let us return to (1.42) for a max-min convolution for an expression involving a product.

$\forall\ x, y, z \in R^{+}$:

$$\mu_{\mathbf{A}\,(\cdot)\,\mathbf{B}}(z) = \bigvee_{z = x \cdot y} (\mu_{\mathbf{A}}(x) \wedge \mu_{\mathbf{B}}(y)), \tag{1.87}$$

with the condition $\mu_{\mathbf{A}}(0) = 0$. Now consider the equation

$$\mathbf{A}\,(\cdot)\,\mathbf{B} = \mathbf{C}, \tag{1.88}$$

where **A** and **C** are given, and **B** is an unknown fuzzy number.

Such an operation is called the *deconvolution of a product*, which should not be confused with division operations and is given by

$\forall\ x, y, z, \in R$:

$$\mu_{\mathbf{C}\,(:)\,\mathbf{B}}(z) = \bigvee_{z = x/y} (\mu_{\mathbf{C}}(x) \wedge \mu_{\mathbf{B}}(y)),$$

because **C** (:) **B** is not, except in a particular case, a solution of equation (1.87). Now, for any level α

$$C_\alpha = A_\alpha\,(\cdot)\,B_\alpha = [a_1^{(\alpha)} \cdot b_1^{(\alpha)},\ a_2^{(\alpha)} \cdot b_2^{(\alpha)}],$$

$$\begin{aligned} C_\alpha\,(:)\,B_\alpha &= [a_1^{(\alpha)} \cdot b_1^{(\alpha)},\ a_2^{(\alpha)} \cdot b_2^{(\alpha)}]\,(:)\,[b_1^{(\alpha)},\ b_2^{(\alpha)}] \\ &= [(a_1^{(\alpha)} \cdot b_1^{(\alpha)})/b_2^{(\alpha)},\ (a_2^{(\alpha)} \cdot b_2^{(\alpha)})/b_1^{(\alpha)} \\ &\neq [a_1^{(\alpha)},\ a_2^{(\alpha)}]. \end{aligned}$$

To solve (1.88) we need an existence and uniqueness condition. We now give a theorem to do this.

THEOREM 1.6

Given any two fuzzy numbers **A** and **C** in R^+ for each level $\alpha \in [0, 1]$:

$$c_2^{(\alpha)}/a_2^{(\alpha)} \geq c_1^{(\alpha)}/a_1^{(\alpha)}, \qquad a_1^{(\alpha)} > 0.$$

Consider now the equation

$$\mathbf{A}\,(\cdot)\,\mathbf{B} = \mathbf{C}. \tag{1.89}$$

A necessary and sufficient condition for the existence of a fuzzy number **B** satisfying (1.89) is

$\forall\ \alpha, \alpha' \in [0, 1]$:

$$(\alpha' > \alpha) \Rightarrow (c_1^{(\alpha')}/a_1^{(\alpha')} \geq (c_1^{(\alpha)}/a_1^{(\alpha)})$$

and

$$c_2^{(\alpha')}/a_2^{(\alpha')} \leq c_2^{(\alpha)}/a_2^{(\alpha)}). \tag{1.90}$$

If this condition is satisfied, the solution exists, and it is unique and is given by

$\forall\ \alpha \in [0, 1]$:

$$B_\alpha = [c_1^{(\alpha)}/a_1^{(\alpha)},\ c_2^{(\alpha)}/a_2^{(\alpha)}].$$

Proof:

If **A** and **C** are normal, then **B** is also normal. On the other hand, (1.90) is a condition of nesting level by level. Since **A** and **C** are convex, this condition implies that the convexity of **B** in (1.90) is satisfied. This condition must be checked a priori. Therefore, the solution for **B** is normal and convex, and is also unique.

EXAMPLE 1.21

Consider two fuzzy numbers **A** and **C** given by the triangular shapes shown in Figure 1.19.

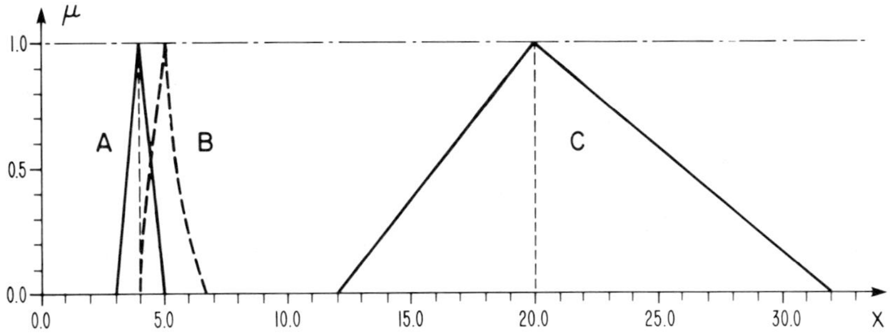

Figure 1.19 Deconvolution for **B** (Example 1.21).

$\forall\ x \in R^+$:

$$
\begin{aligned}
\mu_{\mathbf{A}}(x) &= 0, & x &\le 3,\\
&= x - 3, & 3 \le x &\le 4,\\
&= -x + 5, & 4 \le x &\le 5,\\
&= 0, & x &\ge 5. \qquad \textbf{(1.91)}\\
\mu_{\mathbf{C}}(x) &= 0, & x &\le 12,\\
&= x/8 - 12/8, & 12 \le x &\le 20,\\
&= -x/12 + 32/12, & 20 \le x &\le 32,\\
&= 0, & x &\ge 32. \qquad \textbf{(1.92)}
\end{aligned}
$$

Using (1.91) and (1.92), we have

$$A_\alpha = [\alpha + 3, -\alpha + 5]$$

and

$$C_\alpha = [8\alpha + 12, -12\alpha + 32].$$

In addition,

$$
\begin{aligned}
&[c_1^{(\alpha)}/a_1^{(\alpha)}, c_2^{(\alpha)}/a_2^{(\alpha)}],\\
&\qquad = [(8\alpha + 12)/(\alpha + 3), (-12\alpha + 32)/(-\alpha + 5)],\\
&\qquad = [8 - 12/(\alpha + 3), 12 - 28/(-\alpha + 5)].
\end{aligned}
$$

From this result we see that $[8 - 12/(\alpha + 3)]$ increases with an increase in α while $[12 - 28/(-\alpha + 5)]$ decreases with a decrease in α. As a result, condition (1.90) is satisfied.

Now the solution is given by

$\forall\ x \in R^+$:

$$
\begin{aligned}
\mu_{\mathbf{B}}(x) &= 0, & x &\le 4,\\
&= (12 - 3x)/(x - 8), & 4 \le x &\le 5,\\
&= (32 - 5x)/(12 - x), & 5 \le x &\le 6.4,\\
&= 0, & x &\ge 6.4.
\end{aligned}
$$

Of course, **B** does not have a triangular shape.

As far as the deconvolution of fuzzy max and fuzzy min is concerned, the solution is not unique, and another type of fuzziness is needed to describe it adequately. This results in the use of Φ-fuzzy subsets, as described in section 2.6.

Deconvolution of A (+) A = C and A (·) A = C

Let us consider the following equation:

$$\mathbf{A}\,(+)\,\mathbf{A} = \mathbf{C}, \tag{1.93}$$

where **C** is given. Suppose that **A**, $\mathbf{C} \subset \mathbf{R}$.

At the level α we can write

$\forall\ \alpha \in [0, 1]$:

$$A_\alpha = [a_1^{(\alpha)}, a_2^{(\alpha)}]$$

and

$$C_\alpha = [c_1^{(\alpha)}, c_2^{(\alpha)}].$$

We must then have

$\forall\ \alpha \in [0, 1]$:

$$[a_1^{(\alpha)}, a_2^{(\alpha)}]\,(+)\,[a_1^{(\alpha)}, a_2^{(\alpha)}] = [c_1^{(\alpha)}, c_2^{(\alpha)}],$$

or

$$[2a_1^{(\alpha)}, 2a_2^{(\alpha)}] = [c_1^{(\alpha)}, c_2^{(\alpha)}],$$

or

$$[a_1^{(\alpha)}, a_2^{(\alpha)}] = [c_1^{(\alpha)}/2, c_2^{(\alpha)}/2]. \tag{1.94}$$

The method for solving (1.93) is therefore given by (1.94).

EXAMPLE 1.22

Let us consider the fuzzy number **C** shown in Figure 1.20.

$\forall\ x \in R$:

$$\begin{aligned}\mu_{\mathbf{C}}(x) &= 0, & x &\le -4,\\ &= x/6 + 4/6, & -4 \le x &\le 2,\\ &= -x/8 + 10/8, & 2 \le x &\le 10,\\ &= 0, & x &\ge 10.\end{aligned}$$

We then have

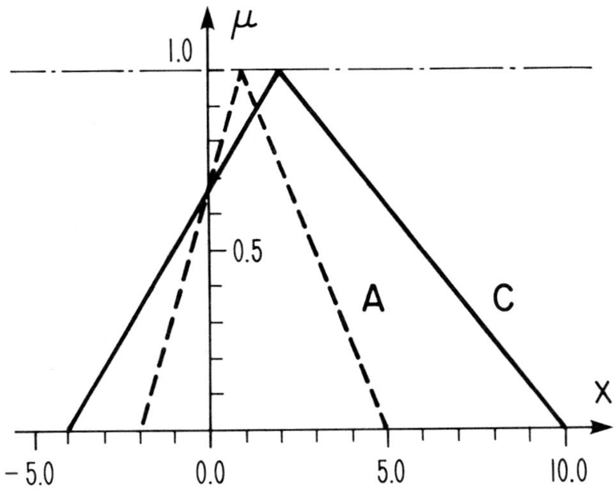

Figure 1.20 Deconvolution for **A** (Example 1.22).

$\forall\ \alpha \in [0, 1]$:

$$C_\alpha = [6\alpha - 4,\ -8\alpha + 10].$$

Using (1.94),

$\forall\ \alpha \in [0, 1]$:

$$A_\alpha = [3\alpha - 2,\ -4\alpha + 5].$$

And

$\forall\ x \in R$:

$$\begin{aligned} \mu_{\mathbf{A}}(x) &= 0, & x &\leq -2, \\ &= x/3 + 2/3, & -2 \leq x &\leq 1, \\ &= -x/4 + 5/4, & 1 \leq x &\leq 5, \\ &= 0, & x &\geq 5. \end{aligned}$$

If we refer to (1.53) we find that

$\forall\ x \in R$:

$$\mu_{\mathbf{A}}(x) = \mu_{2\mathbf{C}}(x) = \mu_{\mathbf{C}}(x/2).$$

More generally we have, for

$\mathbf{A}\ (+)\ \mathbf{A}\ (+) \cdots (+)\ \mathbf{A} = \mathbf{C}$, $\quad k = 1, 2, 3, 4, \ldots, n$,
$\forall\ x \in R$:

$$\mu_{\mathbf{A}}(x) = \mu_{k\mathbf{C}}(x) = \mu_{\mathbf{C}}(x/k),$$

or

$\forall\ \alpha \in [0, 1]$:

$$A_\alpha = [c_1^{(\alpha)}/k,\ c_2^{(\alpha)}/k].$$

The case of **A** (·) **A** = **C** is also trivial, but more interesting.

The number **A** is called square root of **C**, and we limit our study of A to fuzzy numbers in R^+.

At the level α we can write

$\forall\ \alpha \in [0, 1]$:

$$A_\alpha = [a_1^{(\alpha)},\ a_2^{(\alpha)}],$$
$$C_\alpha = [c_1^{(\alpha)},\ c_2^{(\alpha)}].$$

We must have

$\forall\ \alpha \in [0, 1]$:

$$[a_1^{(\alpha)},\ a_2^{(\alpha)}]\ (\cdot)\ [a_1^{(\alpha)},\ a_2^{(\alpha)}] = [c_1^{(\alpha)},\ c_2^{(\alpha)}],$$

or

$$[a_1^{(\alpha)^2},\ a_2^{(\alpha)^2}] = [c_1^{(\alpha)},\ c_2^{(\alpha)}],$$

or

$$[a_1^{(\alpha)},\ a_2^{(\alpha)}] = [\sqrt{c_1^{(\alpha)}},\ \sqrt{c_2^{(\alpha)}}].$$

The method is obvious.

EXAMPLE 1.23

Let, as shown in Figure 1.21,

$\forall\ x \in R^+$:

$$\begin{aligned} \mu_{\mathbf{C}}(x) &= 0, & x &\le 8, \\ &= x/10 - 8/10, & 8 \le x &\le 18, \\ &= -x/14 + 32/14, & 18 \le x &\le 32, \\ &= 0 & x &\ge 32. \end{aligned} \tag{1.95}$$

We can write, then,

$\forall\ \alpha \in [0, 1]$:

$$C_\alpha = [10\alpha + 8,\ -14\alpha + 32].$$

Therefore, using this equation, we obtain

$$A_\alpha = [\sqrt{10\alpha + 8},\ \sqrt{-14\alpha + 32}],$$

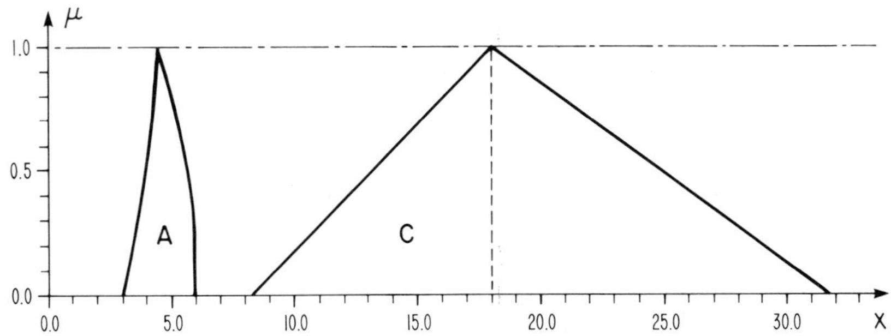

Figure 1.21 Square root of **C** (Example 1.23).

and

$\forall\ x \in R^+$:

$$\begin{aligned} \mu_{\mathbf{A}}(x) &= 0, & x &\le \sqrt{8}, \\ &= x^2/10 - 8/10, & \sqrt{8} \le x &\le \sqrt{18}, \\ &= -x^2/14 + 32/14, & \sqrt{18} \le x &\le \sqrt{32}, \\ &= 0, & x &\ge \sqrt{32}. \end{aligned}$$

This method can be extended without difficulty to roots of order n as follows.

$$\underbrace{\mathbf{A}\,(\cdot)\,\mathbf{A}\,(\cdot)\cdots(\cdot)\,\mathbf{A}}_{n \text{ times}} = \mathbf{C}, \qquad n = 1, 2, 3, 4, \ldots.$$

$\forall\ \alpha \in [0, 1]$:

$$\mathbf{A} = [\sqrt[n]{c_1(\alpha)}, \sqrt[n]{c_2(\alpha)}]$$

We must note now, of course, that fuzzy numbers in N can have their respective roots in R^+ and can therefore pass, if necessary, from N to R^+, by continuous interval functions.

1.9 L-R. THE FUZZY NUMBERS OF DUBOIS AND PRADE

Let us consider a function called the *reference function.*

$\forall\ x' \in R$: $\Phi \in [0, 1]$:

$$\begin{aligned} \Phi(x') &= F_L(x'), & -\infty < x' < 0, \\ &= 1, & x' = 0, \\ &= F_R(x'), & 0 < x' < \infty. \end{aligned} \qquad \textbf{(1.96)}$$

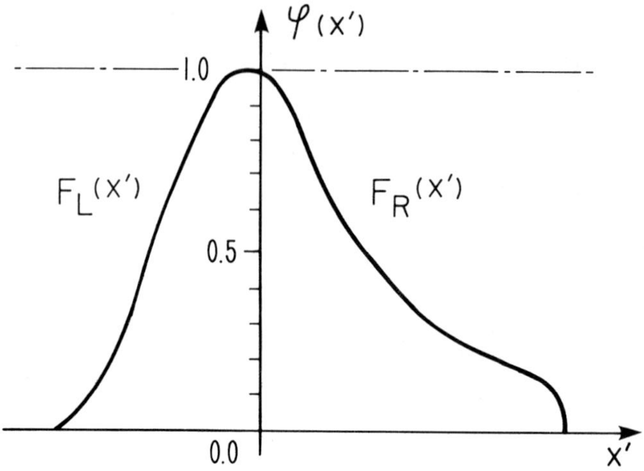

Figure 1.22 A convex and normal L-R fuzzy function $\phi(x')$.

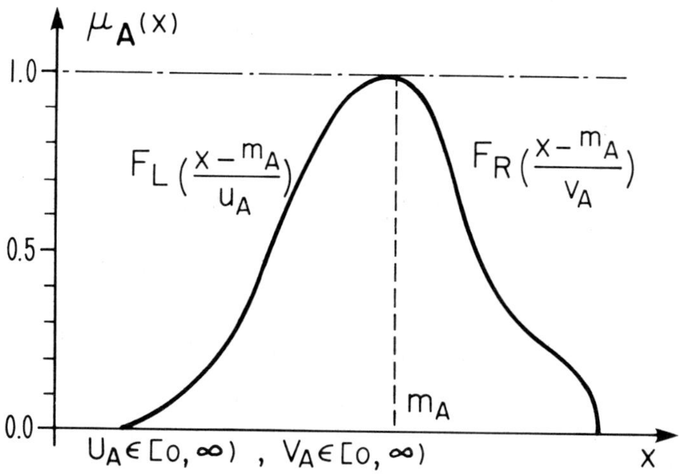

Figure 1.23 Dilatation (F_L) and contraction (F_R) of $\phi(x)$.

We thus impose an increasing monotonicity on $F_L(x')$ and a decreasing monotonicity on $F_R(x')$; that is, $\Phi(x')$ is a normal convex function—a fuzzy number. $F_L(x')$ and $F_R(x')$ are not necessarily symmetric. Figure 1.22 shows such a function $\Phi(x')$.

Using $\Phi(x')$ we may now construct the fuzzy number shown in Figure 1.22:

$\forall\ x \in R$:

$$\begin{aligned}\mu(x) &= F_L((x - m_A)/u_A), & -\infty < x < m_A,\\ &= 1, & x = m_A,\\ &= F_R(x - m_A)/v_A, & m_A < x < \infty.\end{aligned}$$

with $u_A > 0$, $v_A > 0$.

For $m_A < 0$, we have a left translation, and for $m_A > 0$, we have a right translation. If $u_A < 1$, and $v_A < 1$ we have a contraction. For $u_A > 1$, and $v_A > 1$, we have a dilation. Thus, Figure 1.23 represents a dilatation for F_L and a contraction for F_R.

The fuzzy number **B** is constructed in the same way, from (1.96).

$\forall\ x \in R$:

$$\begin{aligned}\mu_{\mathbf{B}}(x) &= F_L((x - m_B)/u_B), & -\infty \le x \le m_B,\\ &= 1, & x = m_B,\\ &= F_R\ ((x - m_B)/v_B), & m_B \le x \le \infty.\end{aligned}$$

We define the fuzzy numbers **A** and **B** using the triplet (m, u, v):

$$\begin{aligned}\mathbf{A} &= (m_A, u_A, v_A).\\ \mathbf{B} &= (m_B, u_B, v_B).\end{aligned}$$

Dubois and Prade[5] proved the max-min convolution for the addition of **A** and **B**; namely,

$\forall\ x \in R$:

$$\mu_{\mathbf{A}\,(+)\,\mathbf{B}}(z) = \bigvee_{z = x + y} (\mu_{\mathbf{A}}(x) \wedge \mu_{\mathbf{B}}(y)),$$

which corresponds to

$$\begin{aligned}\mathbf{A}\,(+)\,\mathbf{B} &= (m_A, u_A, v_A)\,(+)\,(m_B, u_B, v_B)\\ &= (m_A + m_B, u_A + u_B, v_A + v_B).\end{aligned} \qquad \textbf{(1.97)}$$

Let us examine the proof of this. For an arbitrary level α we can write for F_L, as far as **A** is concerned,

$$F_L((x_1 - m_A)/u_A) = \alpha.$$

By taking the inverse of this function we obtain

$$(x_1 - m_A)/u_A = F_L^{-1}(\alpha)$$

or

$$x_1 = m_A + u_A F_L^{-1}(\alpha), \qquad \alpha \in [0, 1].$$

In the same way, considering F_L for **B**, inverse function is given by

$$(y_1 - m_B)/u_B = F_L^{-1}(\alpha)$$

or

$$y_1 = m_B + u_B F_L^{-1}(\alpha),\ \alpha \in [0, 1].$$

Finally,

$$x_1 + y_1 = m_A + m_B + (u_A + u_B) F_L^{-1}(\alpha).$$

In the same manner, considering $F_R^{-1}(\alpha)$,

$$x_2 + y_2 = m_A + m_B + (v_A + v_B) F_R^{-1}(\alpha).$$

Dubois and Prade called this family the *L-R fuzzy numbers.*

EXAMPLE 1.24

Let us consider an *L-R* fuzzy number of reference.

$$\begin{aligned} F_L(x') &= 0, & x' &\le -1, \\ &= \sqrt{1 + x'}, & -1 \le x' &\le 0. \\ F_R(x') &= 1 - x'^2, & 0 \le x' &\le 1, \\ &= 0, & x' &> 1. \end{aligned}$$

This is shown in Figure 1.24.

Now let us take two fuzzy numbers in the same *L-R* family with

$$m_A = 4, \quad u_A = 2, \quad v_A = 3,$$
$$m_B = 8, \quad m_B = 3, \quad v_B = 5.$$

We now have the following fuzzy numbers:

$\forall\ x \in R$:

$$\begin{aligned} \mu_{\mathbf{A}}(x) &= 0, & x &\le 2, \\ &= \sqrt{1 + (x - 4)/2}, & 2 \le x &\le 4, \\ &= 1, & x &= 4, \\ &= 1 - [(x - 4)/3]^2 & 4 \le x &\le 7, \\ &= 0, & x &\ge 7. \\ \mu_{\mathbf{B}}(x) &= 0, & x &\le 5, \\ &= \sqrt{1 + (x - 8)/3}, & 5 \le x &\le 8, \\ &= 1, & x &= 8, \\ &= 1 - [(x - 8)/5]^2, & 8 \le x &\le 13, \\ &= 0, & x &\ge 13. \end{aligned}$$

Finally, using (1.97), we obtain

$$\begin{aligned} \mathbf{A}\,(+)\,\mathbf{B} &= (m_A + m_B,\ u_A + u_B,\ v_A + v_B) \\ &= (4 + 8,\ 2 + 3,\ 3 + 5) \\ &= (12,\ 5,\ 8), \end{aligned}$$

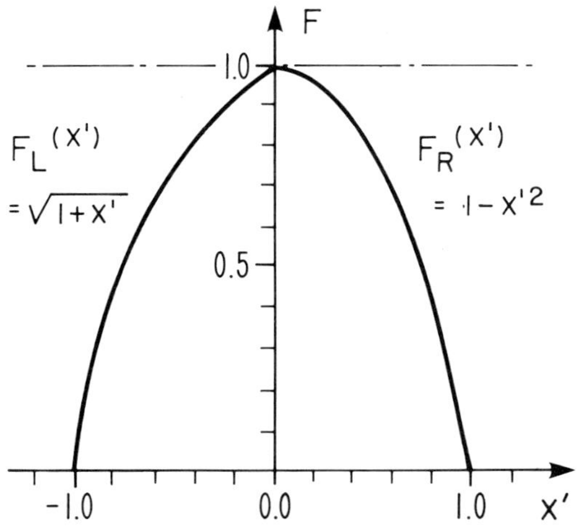

Figure 1.24 A *L-R* fuzzy number (Example 1.24).

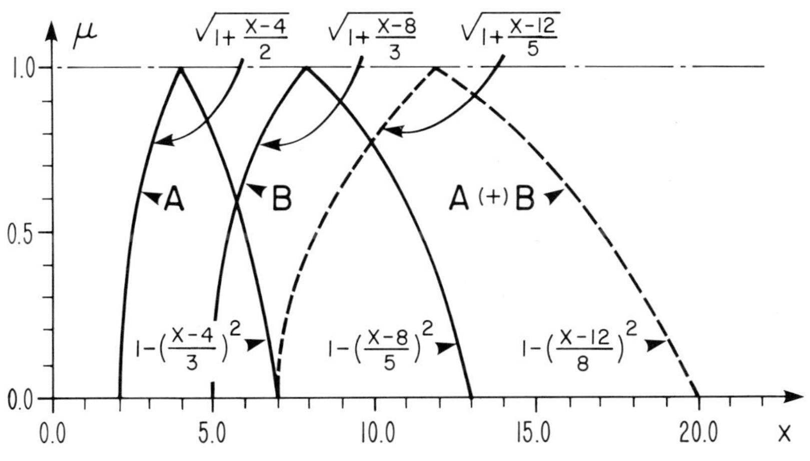

Figure 1.25 Fuzzy addition of two *L-R* fuzzy numbers (Example 1.24).

from which, we have

$\forall\ x \in R$:

$$\mu_{\mathbf{A}\,(+)\,\mathbf{B}}(x) = \begin{cases} 0, & x \le 7, \\ \sqrt{1 + (x-12)/5}, & 7 \le x \le 12, \\ 1, & x = 12, \\ 1 - [(x-12)/8]^2, & 12 \le x \le 20, \\ 0, & x \ge 20. \end{cases}$$

This is shown in Figure 1.25.

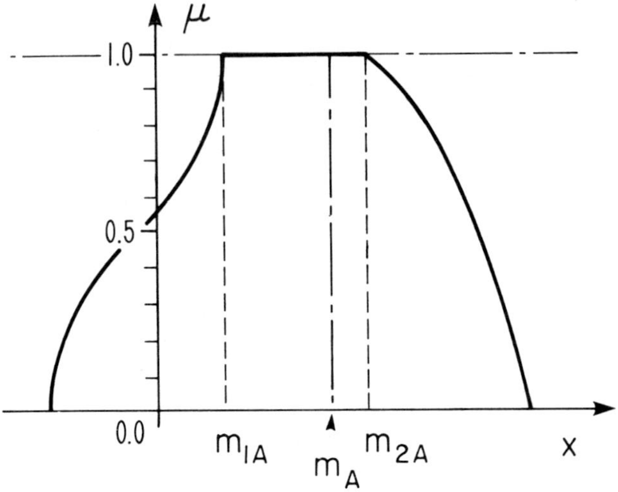

Figure 1.26 *L-R* fuzzy number with a flat.

L-R Fuzzy Numbers with a Flat

If the mode is not single (Fig. 1.26), the *L-R* fuzzy number has a flat spot. This type of fuzzy number can be represented in two different ways.

1. By choosing an arbitary m_A,

$$m_{1A} \le m_A \le m_{2A}.$$

2. By choosing four significant numbers to represent the *L-R* fuzzy number nonunimodally,

$$(m_{1A}, m_{2A}, u_A, v_A).$$

An addition of two numbers with a flat is therefore given by

$$\begin{aligned}\mathbf{A}\,(+)\,\mathbf{B} &= (m_{1A}, m_{2A}, u_A, v_A)\,(+)\,(m_{1B}, m_{2B}, u_B, v_B),\\ &= (m_{1A} + m_{1B}, m_{2A} + m_{2B}, u_A + u_B, v_A + v_B).\end{aligned}$$

Semisymmetric L-R Fuzzy Numbers

Let us consider an *L-R* fuzzy number such that

$\forall\ x' \in R$:

$$\begin{aligned}F_L\,(x') &= F_R(-x'),\\ F_L\,(0) &= F_R(0) = 1.\end{aligned} \qquad \mathbf{(1.98)}$$

In this case, each member of the *L-R* family will be such that

$\forall\ x \in R$:

$$
\begin{aligned}
\mu_{\mathbf{A}}(x) &= F_L((x - m_A)/u_A), && -\infty \le x \le m_A,\\
&= 1, && x = m_A,\\
&= F_L((-x + m_A)/v_A), && m_A \le x \le \infty.
\end{aligned}
$$

the family constructed using (1.98) is called *semisymmetric.*

Figure 1.27 represents a symmetric *L-R* fuzzy number of reference and Figure 1.28 an addition of two semisymmetric *L-R* fuzzy numbers. It is evident that the semisymmetry and *L-R* properties are preserved for addition.

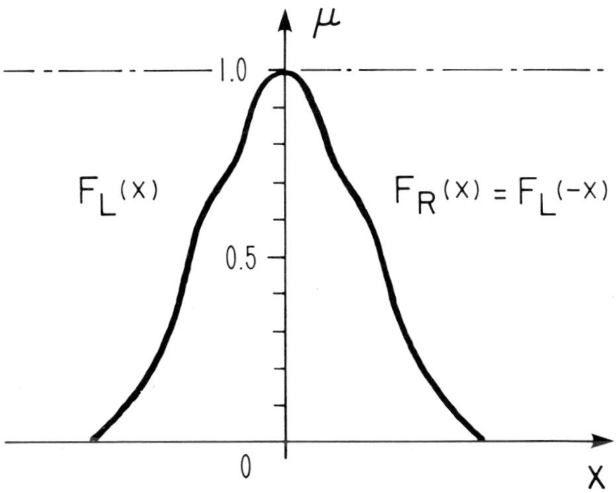

Figure 1.27 A semisymmetric *L-R* fuzzy number.

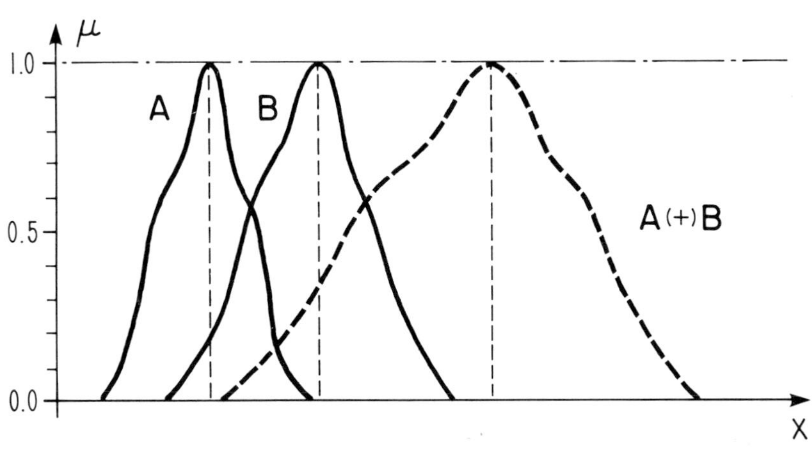

Figure 1.28 Addition of two semisymmetric *L-R* fuzzy numbers.

Image (Opposite) of a Semisymmetric L-R Fuzzy Number

If

$$\mathbf{A} = (m_A, u_A, v_A),$$

then

$$\mathbf{A}^- = (-m_A, v_A, u_A).$$

We see, therefore, that in image, m_A changes its sign and u_A and v_A permute.

Subtraction of Two Semisymmetric L-R Fuzzy Numbers

Subtraction is, of course, the addition of the opposite (image). Using this fact, we obtain

$$\begin{aligned}\mathbf{A}\,(-)\,B &= \mathbf{A}\,(+)\,\mathbf{B}^- = (m_A, u_A, v_A)\,(+)\,(-m_B, v_B, u_B)\\ &= (m_A - m_B, u_A + v_B, v_A + u_B).\end{aligned}$$

Multiplication of L-R Fuzzy Numbers

In this case we consider only the fuzzy numbers in R^+. Except for particular cases, it is easy to show that the multiplication of *L-R* fuzzy numbers does not give an *L-R* fuzzy number. This is true also for division and for fuzzy max and fuzzy min operations.

Multiplication by an Ordinary Number

If

$$\mathbf{A} = (m_A, u_A, v_A),$$

then

$$\begin{aligned}k \cdot \mathbf{A} &= (km_A, ku_A, kv_A), \qquad && k > 0,\\ &= (km_A, -ku_A, -kv_A), && k < 0.\end{aligned}$$

Finally, we conclude that the use of *L-R* fuzzy numbers is mainly useful for addition, subtraction, and multiplication by an ordinary number. For other operations some approximations can be used.

1.10 TRIANGULAR FUZZY NUMBERS

Triangular fuzzy numbers (TFN's) of the type shown in Figure 1.29 are a particular case of semisymmetric *L-R* fuzzy numbers.

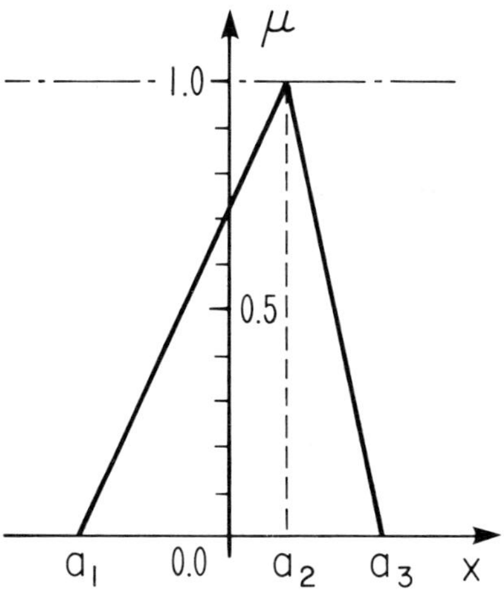

Figure 1.29 A triangular fuzzy number (TFN), $\mathbf{A} = (a_1, a_2, a_3)$.

Let

$\forall\ x, a_1, a_2, a_3 \in R$:

$$\begin{aligned} \mu_{\mathbf{A}}(x) &= 0, & x \le a_1, \\ &= (x - a_1)/(a_2 - a_1), & a_1 \le x \le a_2, \\ &= (a_3 - x)/(a_3 - a_2), & a_2 \le x \le a_3, \\ &= 0, & x \ge a_3. \end{aligned}$$

Assume that a_1, a_2 and a_3 are finite.

A TFN is completely represented by a triplet (not to be confused with the triplet used in the definition of L-R fuzzy numbers) such as

$$\mathbf{A} = (a_1, a_2, a_3).$$

At the level α the interval of confidence is given by

$$\begin{aligned} A_\alpha &= [a_1(\alpha), a_3(\alpha)] \\ &= [a_1 + \alpha(a_2 - a_1), a_3 - \alpha(a_3 - a_2)]. \end{aligned} \tag{1.99}$$

Since TFNs are *L-R* fuzzy numbers that belong to the same family, their addition gives a TFN. This is proved in the following theorem.

THEOREM 1.7

If **A** and **B** are TFN's, then **A** (+) **B** is also a TFN.

Proof:

Let us use the intervals of confidence (1.99).

$$A_\alpha = [a_1 + \alpha(a_2 - a_1), a_3 - \alpha(a_3 - a_2)],$$
$$B_\alpha = [b_1 + \alpha(b_2 - b_1), b_3 - \alpha(b_3 - b_2)].$$

Then

$$A_\alpha\ (+)\ B_\alpha = [a_1 + b_1 + \alpha(a_2 + b_2 - a_1 - b_1),\\ a_3 + b_3 - \alpha_3 + b_3 - a_3 - b_2)].$$

and

$\forall\ x, a_1, a_2, a_3, b_1, b_2, b_3 \in R,\ a_1 \le a_2 \le a_3,\ b_1 \le b_2 \le b_3$:

$$\mu_{\mathbf{A}\,(+)\,\mathbf{B}}(x) = 0 \qquad x \le a_1 + b_1,$$
$$= \frac{x - (a_1 + b_1)}{a_2 + b_2 - a_1 - b_1}, \qquad a_1 + b_1 \le x \le a_2 + b_2,$$
$$= 1, \qquad x = a_2 + b_2,$$
$$= \frac{a_3 + b_3 - x}{a_3 + b_3 - a_2 - b_2}, \qquad a_2 + b_2 \le x \le a_3 + b_3,$$
$$= 0, \qquad x \ge a_3 + b_3.$$

Thus for a summation operation, we are able to write

$$(a_1, a_2, a_3)\ (+)\ (b_1, b_2, b_3) = (a_1 + b_1, a_2 + b_2, a_3 + b_3).$$

Opposite (Image) of a TFN

For $\mathbf{A} = (a_1, a_2, a_3)$, it may be shown that

$$\mathbf{A}^- = (-a_3, -a_2, -a_1),$$

and for the level α

$$A_\alpha^- = [-a_3 + \alpha(a_3 - a_2), -a_1 - \alpha(a_2 - a_1)].$$

Subtraction

Here the opposite (image) is added.

$$\mathbf{A}\ (-)\ \mathbf{B}^- = \mathbf{A}\ (+)\ \mathbf{B}^- = (a_1, a_2, a_3)\ (+)\ (-b_3, -b_2, -b_1)$$
$$= (a_1 - b_3, a_2 - b_2, a_3 - b_1).$$

And, for the level α, we have

$$A_\alpha\ (-)\ B_\alpha = A_\alpha\ (+)\ B_\alpha^-$$
$$= [a_1 - b_3 + \alpha(a_2 - b_2 - a_1 + b_3), a_3 - b_1 - \alpha(a_3 - b_1 - a_2 + b_2)].$$

Also,

$\forall\ x, a_1, a_2, a_3, b_1, b_2, b_3 \in R, \qquad a_1 \le a_2 \le a_3,\ b_1 \le b_2 \le b_3$:

$$
\begin{aligned}
\mu_{\mathbf{A}\,(-)\,\mathbf{B}}(x) &= 0, && x \le (a_1 - b_3),\\
&= \frac{x-(a_1-b_3)}{a_2-b_2-(a_1-b_3)}, && a_1 - b_3 \le x \le (a_2 - b_2),\\
&= 1, && x = a_2 - b_2,\\
&= \frac{a_3-b_1-x}{a_3-b_1-(a_2-b_2)}, && a_2 - b_2 \le x \le a_3 - b_1,\\
&= 0, && x \ge (a_3 - b_1).
\end{aligned}
$$

Hence, if **A** and **B** are TFN's, $\mathbf{A}^-$, $\mathbf{B}^-$, $\mathbf{A}\,(-)\,\mathbf{B}$ are also TFN's. However, this is not true for $\mathbf{A}\,(\cdot)\,\mathbf{B}$, $\mathbf{A}^{-1}$,$\mathbf{B}^{-1}$, $\mathbf{A}\,(:)\,\mathbf{B}$, $\mathbf{A}\,(\wedge)\,\mathbf{B}$ and $\mathbf{A}\,(\vee)\,\mathbf{B}$.

We now examine a theorem that is concerned with multiplication by an ordinary number.

THEOREM 1.8

If **A** is a TFN, then $k \cdot \mathbf{A}$ $(k \in R)$ is also a TFN, where

$\forall\, x \in R$:

$$\mu_{k\,\cdot\,\mathbf{A}}(x) = \mu_{\mathbf{A}}(x/k), \qquad k \neq 0.$$

Proof:

Supposing that $k > 0$, we may write from (1.99)

$\forall\, \alpha \in [0, 1]$:

$$
\begin{aligned}
k \cdot A_\alpha = [&ka_1 + \alpha(ka_2 - ka_1),\\
&ka_3 - \alpha(ka_3 - ka_2)]. \qquad \textbf{(1.100)}
\end{aligned}
$$

And from (1.100) we have

$\forall\, x \in R$:

$$
\begin{aligned}
\mu_{k\,\cdot\,\mathbf{A}}(x) &= 0, && x \le ka_1,\\
&= \frac{x-ka_1}{ka_2-ka_1}, && ka_1 \le x \le ka_2,\\
&= 1, && x = ka_2,\\
&= \frac{ka_3-x}{ka_3-ka_2}, && ka_2 \le x \le ka_3,\\
&= 0, && x \ge ka_3.
\end{aligned}
$$

Alternatively,

$\forall\ x \in R$:

$$\begin{aligned}
\mu_{k \cdot \mathbf{A}}(x) &= 0, && (x/k) \leq a_1, \\
&= \frac{x/k - a_1}{a_2 - a_1}, && a_1 \leq (x/k) \leq a_2, \\
&= 1, && (x/k) = a_2, \\
&= \frac{a_3 - x/k}{a_3 - a_2}, && a_2 \leq (x/k) \leq a_3, \\
&= 0, && (x/k) \geq a_3.
\end{aligned}$$

If $k < 0$, the proof is similar. If $k = 0$, then $k \cdot \mathbf{A} = (0, 0, 0)$, which is the ordinary number 0, which is a particular case of a TFN.

Thus we summarize the results as follows:

$$\begin{aligned}
k \cdot \mathbf{A} &= (ka_1, ka_2, ka_3), && k > 0, \\
&= (ka_3, ka_2, ka_1), && k < 0, \\
&= (0, 0, 0), && k = 0.
\end{aligned}$$

Shifting to the Right and to the Left

A TFN is shifted to the right ($r > 0$), or to the left ($r < 0$), using the following operations.

$\forall\ x \in R$:

$$\mu_{\mathbf{A}\,(+)\,r}(x) = \mu_{\mathbf{A}}\,(x - r)$$

or

$$\begin{aligned}
\mathbf{A}\,(+)\,r &= (a_1, a_2, a_3)\,(+)\,(r, r, r) \\
&= (a_1 + r, a_2 + r, a_3 + r).
\end{aligned}$$

Also

$$\begin{aligned}
A_\alpha\,(+)\,r &= [a_1 + \alpha(a_2 - a_1), a_3 - \alpha(a_3 - a_2)] + [r, r] \\
&= [a_1 + r + \alpha(a_2 - a_1), a_3 + r - \alpha(a_3 - a_2)].
\end{aligned}$$

Deconvolution of TFN

In the following equation let

$$\mathbf{A}\ (+)\ \mathbf{B} = \mathbf{C},$$

where **A** and **C** are given TFN's. By Theorem 1.5 we know that there exists one and only one fuzzy number **B** that is the solution if and only if the following conditions are satisfied:

1. $\forall\ \alpha \in [0, 1]$:

$$c_3^{(\alpha)} - a_3^{(\alpha)} \geq c_1^{(\alpha)} - a_1^{(\alpha)}.$$ [8]

2. $\forall\ \alpha, \alpha' \in [0, 1]$:

$$(\alpha' > \alpha) \Rightarrow (c_1^{(\alpha')} - a_1^{(\alpha')}) \geq (c_1^{(\alpha)} - a_1^{(\alpha)})$$

and

$$(c_3^{(\alpha')} - a_3^{(\alpha')}) \leq (c_3^{(\alpha)} - a_3^{(\alpha)}). \qquad \textbf{(1.101)}$$

In this case the solution is

$\forall\ \alpha \in [0, 1]$:

$$B_\alpha = [c_1^{(\alpha)} - a_1^{(\alpha)}, c_3^{(\alpha)} - a_3^{(\alpha)}].$$

We will now examine the following theorem. If **A** and **C** are TFN's, then **B** is also a TFN, under a certain condition.

THEOREM 1.9

If **A** and **C** are TFN's, and

$$\mathbf{A}\ (+)\ \mathbf{B} = \mathbf{C}, \qquad \textbf{(1.102)}$$

then if **B** exists, it is a TFN subject to the following condition:

$$(c_2 - c_1) \geq (a_2 - a_1) \text{ and } (c_3 - c_2) \geq (a_3 - a_2), \qquad \textbf{(1.103)}$$

or, equivalently, if

$$(c_3 - a_3) \geq (c_2 - a_2) \geq (c_1 - a_1).$$

Proof:

Since **A** and **C** are assumed to be TFN's, we have

$$A_\alpha = [a_1 + \alpha(a_2 - a_1), a_3 - \alpha(a_3 - a_2)].$$

$$C_\alpha = [c_1 + \alpha(c_2 - c_1), c_3 - \alpha(c_3 - c_2)].$$

$$c_1^{(\alpha)} = c_1 + \alpha(c_2 - c_1).$$

$$c_3^{(\alpha)} = c_3 - \alpha(c_3 - c_2). \qquad \textbf{(1.104)}$$

$$a_1^{(\alpha)} = a_1 + \alpha(a_2 - a_1).$$

$$a_3^{(\alpha)} = a_3 - \alpha(a_3 - a_2). \qquad \textbf{(1.105)}$$

$$c_1^{(\alpha')} = c_1 + \alpha'(c_2 - c_1).$$

$$c_3^{(\alpha')} = c_3 - \alpha'(c_3 - c_2). \quad \textbf{(1.106)}$$

$$a_1^{(\alpha')} = a_1 + \alpha'(a_2 - a_1).$$

$$a_3^{(\alpha')} = a_3 - \alpha'(a_3 - a_2). \quad \textbf{(1.107)}$$

$$c_1^{(\alpha')} - a_1^{(\alpha')} = c_1 - a_1 + \alpha'(c_2 - a_2 - (c_1 - a_1)). \quad \textbf{(1.108)}$$

$$c_1^{(\alpha)} - a_1^{(\alpha)} = c_1 - a_1 + \alpha(c_2 - a_2 - (c_1 - a_1)). \quad \textbf{(1.109)}$$

Now from (1.108) and (1.109), we have

$$(c_1^{(\alpha')} - a_1^{(\alpha')}) - (c_1^{(\alpha)} - a_1^{(\alpha)}) = (\alpha' - \alpha)(c_2 - c_1 - (a_2 - a_1)).$$

To satisfy condition (1.101), we must have

$$(c_2 - c_1) \geq (a_2 - a_1).$$

Taking again (1.104), (1.105), (1.106), and (1.107), we have the following:

$$c_3^{(\alpha')} - a_3^{(\alpha')} = c_3 - a_3 - \alpha'(c_3 - a_3 - (c_2 - a_2)),$$

$$c_3^{(\alpha)} - a_3^{(\alpha)} = c_3 - a_3 - \alpha(c_3 - a_3 - (c_2 - a_2)).$$

Hence

$$(c_3^{(\alpha')} - a_3^{(\alpha')}) - (c_3^{(\alpha)} - a_3^{(\alpha)}) = -(\alpha' - \alpha)(c_3 - c_2 - (a_3 - a_2).$$

Using (1.101) again, we have

$$(c_3 - c_2) \geq (a_3 - a_2).$$

We thus have obtained (1.103) again.

The solution of (1.102) can be written using the special notation of TFN.

$$\mathbf{B} = (c_1 - a_1, c_2 - a_2, c_3 - a_3)$$

We now give an example.

EXAMPLE 1.25

We use Example 1.19 again, but with the special TFN notation. Let **A** and **C** be given by

$$\mathbf{A} = (a_1, a_2, a_3) = (11, 18, 19),$$

$$\mathbf{C} = (c_1, c_2, c_3) = (14, 24, 32).$$

Checking, we find that

$$(32 - 19) > (24 - 18) > (14 - 11).$$

We see that deconvolution of **A** and **C** exists (see Example 1.19) and that it is a TFN **B**. We also find that

$$\mathbf{B} = (c_1 - a_1, c_2 - a_2, c_3 - a_3),$$
$$= (3, 6, 13).$$

This can be seen from Figure 1.30.

Trapezoidal Fuzzy Numbers (T_rFN)

Trapezoidal fuzzy numbers, represented by T_rFN's as shown in Figure 1.31, are a special case of the *L-R* fuzzy numbers and the TFN's

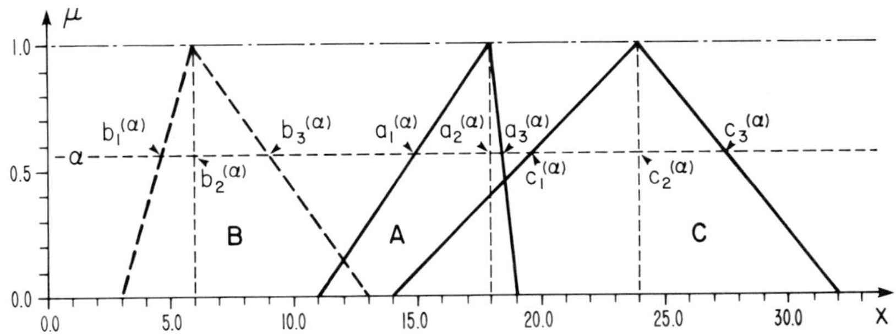

Figure 1.30 Deconvolution of TFN's for **B** (Example 1.25).

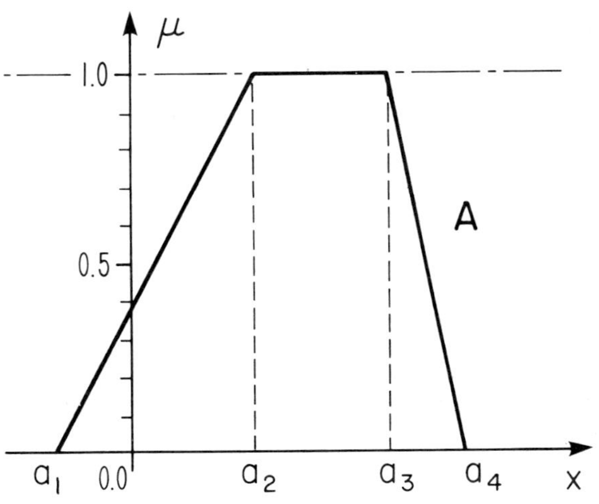

Figure 1.31 A trapezoidal fuzzy number (T_rFN).

are a special case of the T_rFN's. For this reason, we use four points instead of three for T_rFN. Thus

$$\mathbf{A} = (a_1, a_2, a_3, a_4).$$

The process of addition is very simple as can be seen from the following.

$$\mathbf{A}\,(+)\,\mathbf{B} = (a_1, a_2, a_3, a_4)\,(+)\,(b_1, b_2, b_3, b_4),$$
$$= (a_1 + b_1, a_2 + b_2, a_3 + b_3, a_4 + b_4).$$

The opposite (image) is given by

$$\mathbf{A}^- = (-a_4, -a_3, -a_2, -a_1).$$

The subtraction process is given by

$$\mathbf{A}\,(-)\,\mathbf{B} = (a_1 - b_4, a_2 - b_3, a_3 - b_2, a_4 - b_1).$$

Certain operations such as $(\cdot)$, $(:)$, $(\wedge)$, $(\vee)$ do not yield T_rFN's except for some special cases.

NOTES

1. The concept of confidence interval in statistics is well known. In order to avoid any confusion here we use the phrase *interval of confidence*. The phrase *segment of confidence* may also be used.
2. From Latin *praesumptio*. A presumption always supposes some parts of the proof are used like arguments.
3. The referential set is an ordinary set and here fuzzy subsets of this set are considered. We can use both fuzzy sets or fuzzy subsets.
4. Here in (1.23), (1.24) and later in (1.27) we use a symbolic writing.
5. In Appendix A, we present multiplication in R for intervals of confidence, and in Appendix B the extension is made to fuzzy numbers.
6. Fuzzy integers can be nonunimodal. This means these numbers are normal for several joined values. The rules for these numbers are applicable in the same way.
7. If we consider a fuzzy number with a flat the solution can be nonunique, but this depends on the way a fuzzy number is defined. Like a generalization of an interval of confidence using the concept of level of presumption, the solution is unique; like a normal and convex fuzzy subset the solution can be considered like nonunique (a flat gives a discontinuity).
8. For a TFN index 3 takes the place of index 2.

REFERENCES

[1] L. A. Zadeh, 1965, Fuzzy sets, *Inf. Control* **8**:338-353.
[2] S. Nahmias, 1977, Fuzzy variables, *Fuzzy Sets and System* **1**(2):97-110.
[3] S. Nahmias, 1979, Fuzzy variables in a random environment, in *Advances in Fuzzy Set Theory and Applications,* M. M. Gupta, R. Ragade, R. Yager, eds., North-Holland, Amsterdam, pp. 165-180.
[4] D. Dubois and H. Prade, 1982, A class of fuzzy measures based on triangular norms—A general framework for the combination of uncertain information. *Int. J. Gen. Syst.* **8**(1):43-61.
[5] D. Dubois and H. Prade, 1980, *Fuzzy Sets and Systems: Theory and Applications,* Academic Press, New York.

CHAPTER 2

A LARGE SAMPLE OF SOME NOVEL CONCEPTS AND TOOLS

2.1 INTRODUCTION

The uncertainty that arises from human thought processes and the randomness associated with experiments are often confused by social scientists and even mathematicians, scientists, and engineers. The fact that this confusion exists is unfortunate, but understandable. For example, some observations obtained from a system are precise, while some are measurable only in statistical sense, and others cannot be measured at all. Some of the data obtained in this manner are hybrid; that is, their components are not homogeneous but a blend of precise and fuzzy information.

Hybrid data lead to hybrid operations and hybrid numbers, which are very recent developments and appear to be suitable for dealing with large-scale systems. An introduction to these concepts is given in this chapter. We will show that these concepts can be associated with Monte Carlo method.

2.2 UNCERTAINTY AND RANDOMNESS: HYBRID NUMBERS

A fuzzy number is not a measurement. It is a function $\mu_{\mathbf{A}}(x)$ and corresponds to both convex and normal fuzzy subsets. In other words, a fuzzy number is a subjective valuation assigned by one or more human operators.

In this chapter we examine the addition of fuzzy and random data. In order to do this, we denote a fuzzy number by a boldface letter, for example, **A**. An ordinary capital letter, for example, A, is used to denote a random variable, and a lowercase letter, say a, is used to denote a special value of a random variable.

We begin our analysis by reviewing the addition of fuzzy and random variables. In the referential set R, we have a fuzzy number **A** with membership function $\mu_{\mathbf{A}}(x)$ and a random variable L whose probability is given by the density function $f_L(x)$, where x is a value of L in R. We want to add **A** and L.

Suppose that $f_L(x)$ has a convex shape. Let the maximum value of $f_L(x)$ be

$$\max_{x} f_L(x).$$

Divide the function $f_L(x)$ by its maximum value and define the new function $\mu_{\mathbf{L}}(x)$, which is both convex and normal. By this process we substitute the fuzzy number **L** for the random variable L, which allows **L** to be added to any other fuzzy number by the operation of max-min convolution. We therefore have

$$\mu_{\mathbf{L}}(x) = \frac{f_L(x)}{\max_{x} f_L(x)}, \tag{2.1}$$

and addition with a fuzzy number **A**,

$\forall\ x, y, z \in R$:

$$\mu_{\mathbf{A}(+)\mathbf{L}}(z) = \bigvee_{z=x+y} (\mu_{\mathbf{A}}(x) \wedge \mu_{\mathbf{L}}(y)). \tag{2.2}$$

EXAMPLE 2.1

A random variable L with the probability density [Fig. 2.1(A)]

$\forall\ x \in R$:

$$\begin{aligned} f_L(x) &= 0, & x &\le 4, \\ &= x/12 - 4/12, & 4 \le x &\le 7, \\ &= -x/20 + 12/20, & 7 \le x &\le 12, \\ &= 0, & x &\ge 12, \end{aligned}$$

may be transformed into **L** by normalization, as shown in Figure 2.1(B). This transformation gives the membership function

$\forall\ x \in R$:

$$\begin{aligned} \mu_{\mathbf{L}}(x) &= 0, & x &\le 4, \\ &= x/3 - 4/3, & 4 \le x &\le 7, \\ &= -x/5 + 12/5, & 7 \le x &\le 12, \\ &= 0, & x &\ge 12. \end{aligned}$$

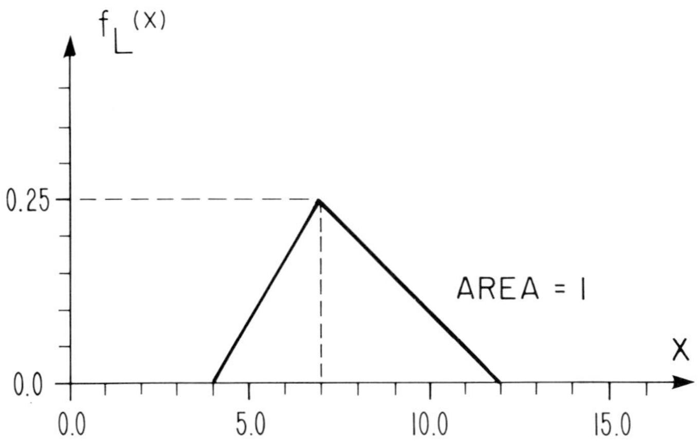

Figure 2.1(A) A random variable $f_L(x)$ (Example 2.1).

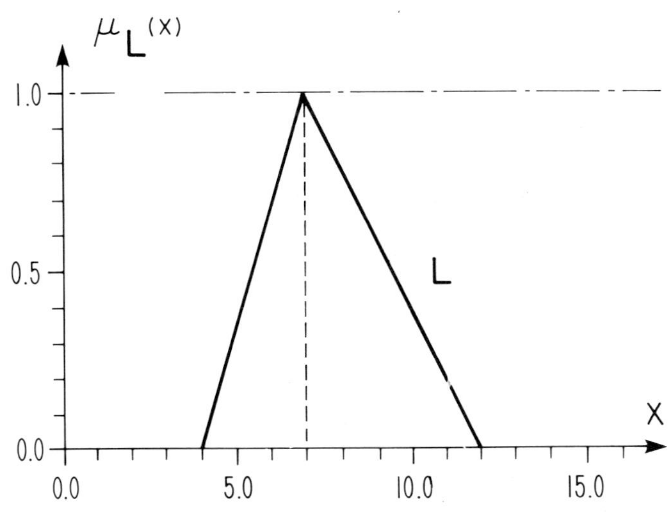

Figure 2.1(B) The normalized random variable (Example 2.1).

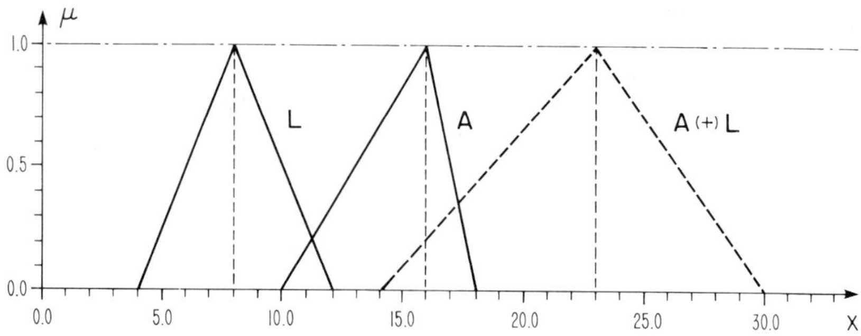

Figure 2.2 Addition of fuzzy and random variables, **A** + **L** (Example 2.1).

Now we consider the fuzzy number (see Fig. 2.2),

$$\begin{aligned} \mu_{\mathbf{A}}(x) &= 0, & x &\le 10,\\ &= x/6 - 10/6, & 10 \le x &\le 16,\\ &= -x/2 + 18/2, & 16 \le x &\le 18,\\ &= 0, & x &\ge 18, \end{aligned}$$

and compute **A** (+) **L** as shown in Figure 2.2.

$\forall\ x \in R$:

$$\begin{aligned} \mu_{\mathbf{A}\,(+)\,\mathbf{L}}(x) &= 0, & x &\le 14,\\ &= x/9 - 11/9, & 14 \le x &\le 23,\\ &= -x/7 + 30/7, & 23 \le x &\le 30,\\ &= 0, & x &\ge 30. \end{aligned}$$

There is, however, a serious disadvantage to this procedure. Basically, we have transformed a measurement of an objective value to a valuation of a subjective value, which results in the loss of information. Although this procedure is mathematically correct, it decreases the amount of information that is available in the original data, and we should avoid it. A procedure that achieves the same objective but does not lose information, is given by introducing the concept of a hybrid number.

The Concept of a Hybrid Number

Let us consider a fuzzy number $\mathbf{A} \subset R$. For a move to the right, $\ell > 0$, or a move to the left, $\ell < 0$, we can write

$\forall\ \alpha \in [0, 1]$:

$$\begin{aligned} \mathbf{A}\,(+)\,\ell &= [a_1^{(\alpha)}, a_2^{(\alpha)}]\,(+)\,[\ell, \ell\,]\\ &= [a_1^{(\alpha)} + \ell\,, a_2^{(\alpha)} + \ell\,], \end{aligned}$$

or

$\forall$ x, y, z $\in$ R:

$$\mu_{\mathbf{A}(+)\ell}(z) = \bigvee_{z=x+y} (\mu_{\mathbf{A}}(x) \wedge \mu_{\ell}(y)),$$

where

$$\mu_{\ell}(y) = 0, \quad y \neq \ell,$$
$$= 1, \quad y = \ell.$$

Let ℓ not be a specific ordinary number but a random variable L with a probability density $f(\ell)$. Then the fuzzy number **A** will move randomly according to the law $f(\ell)$. The couplet (**A**, L), called a *hybrid number,* represents the addition of a fuzzy number to a random variable without altering the characteristics of either and without decreasing the amount of information available.

A hybrid number can be represented as

$$\mathbf{A}_f = A(\mu, f), \qquad \text{or} \qquad \mathbf{A}_f = \mathbf{A}\,[+]\,L,$$

where L is the random variable with probability density $f(\ell)$.

The probability density of $(A_\alpha\ (+)\ \ell)$ is the density of $L = \ell$.

$$g(A_\alpha\ (+)\ \ell) = g([a_1^{(\alpha)} + \ell,\ a_2^{(\alpha)} + \ell]) = f(\ell).$$

If $[\ell_1, \ell_2]$ is the segment within which L lies, the extreme positions of $\mathbf{A}_f$ are those shown in Figure 2.3. In Figure 2.3, $\ell_1 < 0$ and $\ell_2 > 0$, of course ℓ_1 and ℓ_2 can be positive, negative, or zero, and the function $f(\ell)$ may or may not be convex.

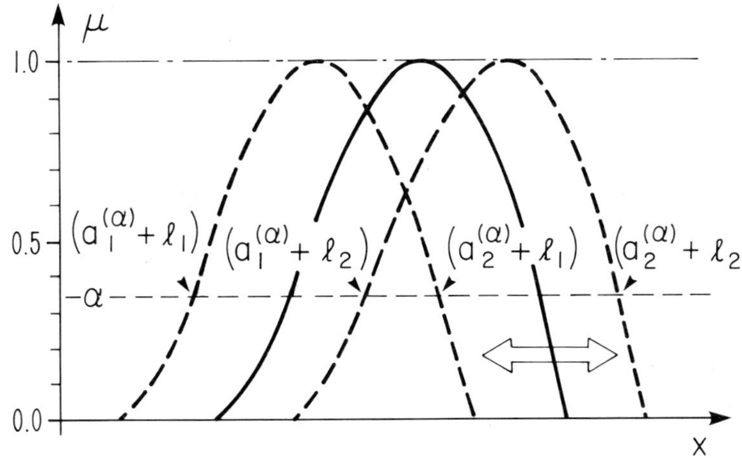

Figure 2.3 The hybrid numbers.

Addition of Hybrid Numbers

Consider two hybrid numbers in R: $(\mathbf{A}_1, L_1)$ and $(\mathbf{A}_2, L_2)$, where L_1 and L_2 have the densities $f_1(\ell_1)$ and $f_2(\ell_2)$, respectively. We now define addition by the *hybrid convolution*:

$$(\mathbf{A}_1, L_1)\,[+]\,(\mathbf{A}_2, L_2) = (\mathbf{A}_1\,(+)\,\mathbf{A}_2,\ L_1\,(+)'\,L_2)$$
$$= (\mathbf{A}, L),$$

where $(+)$ represents the max-min convolution for addition and $(+)'$ represents the sum-product convolution for addition.

We may also write

$\forall\ x, y, z \in R$:

$$\mu_{\mathbf{A}_1\,(+)\,\mathbf{A}_2}(z) = \bigvee_{z = x + y} (\mu_{\mathbf{A}_1}(x) \wedge \mu_{\mathbf{A}_2}(y))$$

and

$$f(\ell) = \int_R f_1(\ell - \ell_2) \cdot f_2(\ell_2)\, d\ell_2$$

$$= \int_R f_1(\ell_1) \cdot f_2(\ell - \ell_1)\, d\ell_1.$$

Because a fuzzy number is a special case of a hybrid number, $\mathbf{A} = (\mathbf{A}, 0)$, where 0 is the trivial random variable with probabilities

$$\begin{aligned} \mathrm{pr}\,(\ell) &= 1, & \ell &= 0, \\ &= 0, & \ell &\neq 0. \end{aligned}$$

A random variable is also a special case of a hybrid number: $L = (0, L)$, where 0 is the trivial fuzzy number with certainty

$$\begin{aligned} \mu_0(x) &= 1, & x &= 0, \\ &= 0, & x &\neq 0. \end{aligned}$$

Because the operations $(+)$ and $(+)'$ are commutative and associative, the operations $[+]$ also have these properties. The neutral for $[+]$ is $(0, 0)$, which shows that hybrid numbers have a commutative monoidal structure. The following examples show that hybrid addition can be carried out in the same way as ordinary numbers.

EXAMPLE 2.2

In manufacturing problems, one tends to add several kinds of costs such as a fuzzy cost $\mathbf{C}_1$, a random cost C_2, and a fixed cost c_3. The total cost can be expressed as

$$\mathbf{C}_1 \, [+] \, C_2 \, [+] \, c_3 = (\mathbf{C}_1, 0) \, [+] \, (0, C_2) \, [+] \, (0, c_3)$$
$$= (\mathbf{C}_1, C_2 \, (+)' \, c_3)$$
$$= (\mathbf{C}_1 \, (+) \, c_3, C_2),$$

because an ordinary number is a trivial case of a fuzzy number and also a trivial case of a random variable. The ordinary number c_3 provides a move for $\mathbf{C}_1$ or for C_2. In both cases, the results are the same.

A variant of this is $c_3 = c_3^{(1)} + c_3^{(2)}$, which can be written as

$$(\mathbf{C}_1 \, (+) \, c_3, C_2) = (\mathbf{C}_1, C_2 \, (+)' \, c_3)$$
$$= (\mathbf{C}_1 \, (+) \, c_3^{(1)}, C_2 \, (+)' \, c_3^{(2)}).$$

We now study an example in N.

EXAMPLE 2.3

$\mathbf{C}_1 =$

0	1	2	3	4	5	6	7	≥8
0	0.1	0.3	0.4	1	0.8	0.7	0.3	0

$\text{pr}(c_2) =$

0	1	2	3	4	5	6	≥7
0.1	0.1	0.1	0.3	0.2	0.1	0.1	0

and

$c_3 = 3$

$\text{pr}(c_2 + c_3) =$ **(2.3)**

2≤	3	4	5	6	7	8	9	≥10
0	0.1	0.1	0.1	0.3	0.2	0.1	0.1	0

Now we have the results shown in Table 2.1 (on page **86**).

Is the family of fuzzy numbers probabilistic? The fuzzy number **M** of Table 2.1 is M at the maximum value of the membership function. Hence the fuzzy numbers of row 1 are represented by **4**, those of row 2 by **5**, those of row 8 by **11**, etc.

In (2.3) we considered the referential set

$$E = \{0, 1, 2, 3, 4, 5, 6\} \subset N,$$

where the integers with the null probability are deleted. The family corresponding to this set is

$$\Delta = \{\emptyset, \{0\}, \{1\}, \ldots, \{6\}, \{0, 1\}, \{0, 2\}, \ldots, \{5, 6\}, \ldots, \{0, 1, 2\}, \{0, 1, 3\}, \ldots, E\}.$$

Table 2.1

1. pr

0	1	2	3	4	5	6	7	≥8
0	0.1	0.3	0.4	1	0.8	0.7	0.3	0

= 0,

2. pr

1	2	3	4	5	6	7	8	≥9
0	0.1	0.3	0.4	1	0.8	0.7	0.3	0

= 0,

3. pr

2	3	4	5	6	7	8	9	≥10
0	0.1	0.3	0.4	1	0.8	0.7	0.3	0

= 0,

4. pr

3	4	5	6	7	8	9	10	≥11
0	0.1	0.3	0.4	1	0.8	0.7	0.3	0

= 0.1,

5. pr

4	5	6	7	8	9	10	11	≥12
0	0.1	0.3	0.4	1	0.8	0.7	0.3	0

= 0.1,

6. pr

5	6	7	8	9	10	11	12	≥13
0	0.1	0.3	0.4	1	0.8	0.7	0.3	0

= 0.1,

7. pr

6	7	8	9	10	11	12	13	≥14
0	0.1	0.3	0.4	1	0.8	0.7	0.3	0

= 0.3,

8. pr

7	8	9	10	11	12	13	14	≥15
0	0.1	0.3	0.4	1	0.8	0.7	0.3	0

= 0.2,

9. pr

8	9	10	11	12	13	14	15	≥16
0	0.1	0.3	0.4	1	0.8	0.7	0.3	0

= 0.1,

10. pr

9	10	11	12	13	14	15	16	≥17
0	0.1	0.3	0.4	1	0.8	0.7	0.3	0

= 0.1,

11. pr

10	11	12	13	14	15	16	17	≥18
0	0.1	0.3	0.4	1	0.8	0.7	0.3	0

= 0.

By construction, this family satisfies the Borel-Kolmogorov axioms in probability theory.

Consider the referential set created by the fuzzy numbers in Table 2.1 from row 4 to row 10, that is, the fuzzy numbers with probability nonzero.

$$E' = \{\mathbf{7, 8, 9, 10, 11, 12, 13}\}.$$

The corresponding family Δ' is

$$\Delta' = \{\emptyset, \{\mathbf{7}\}, \{\mathbf{8}\}, \ldots, \{\mathbf{13}\}, \{\mathbf{7, 8}\}, \{\mathbf{7, 9}\}, \ldots, \{\mathbf{12, 13}\}, \ldots, \{\mathbf{7, 8, 9}\}, \{\mathbf{7, 8, 10}\}, \ldots, \mathbf{E'}\}.$$

Families Δ and Δ' are isomorphic and Δ' satisfies the axioms of probability theory as well.

E is a referential set that is totally ordered (a linear order) and, by construction, E' and its fuzzy number are ordered by the previously defined operations ($\wedge$) or ($\vee$).

EXAMPLE 2.4

We have the triangular fuzzy number in R^+ shown in Figure 2.4.

$$\begin{aligned} \mu_{\mathbf{A}}(x) &= 0, & x \le 2, \\ &= x/2 - 1, & 2 \le x \le 4, \\ &= -x + 5, & 4 \le x \le 5, \\ &= 0, & x \ge 5. \end{aligned}$$

A random variable L whose probability density is $f(\ell)$ (Fig. 2.5) is given by

$$\begin{aligned} f(\ell) &= 0, & \ell \le 4, \\ &= (1/22)/(\ell - 4), & 4 \le \ell \le 8, \\ &= (2/11)([-\ell + 15]/7), & 8 \le \ell \le 15, \\ &= 0, & \ell \ge 15. \end{aligned}$$

The domain of L is $[4, 15]$.

The TFN can be represented by a triplet, (2, 4, 5), and for every ℓ for L: $(\ell - 2, \ell, \ell + 1)$. Hence the density of probability for $\mathbf{A}\,[+]\,L$ is

$$\begin{aligned} g(\mathbf{A}\,[+]\,L) &= 0, & (\ell - 2, \ell, \ell + 1) \le (2, 4, 5) \\ &= 1/22)\,(\ell - 4), & (2, 4, 5) \le (\ell - 2, \ell, \ell + 1) \le (6, 8, 9) \\ &= (2/11)([-\ell + 15]/7), & (6, 8, 9) \le (\ell - 2, \ell, \ell + 1) \\ & & \le (13, 15, 16) \\ &= 0, & (\ell - 2, \ell, \ell + 1) \ge (13, 15, 16). \end{aligned}$$

This is represented in Figure 2.4.

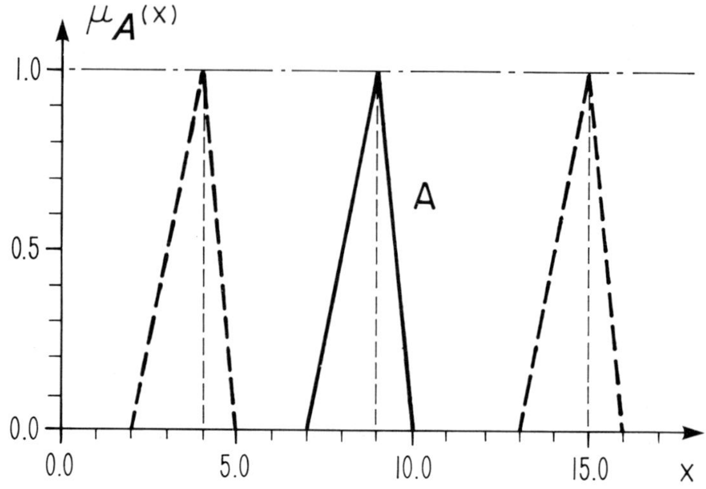

Figure 2.4 A triangular fuzzy number **A** (Example 2.4).

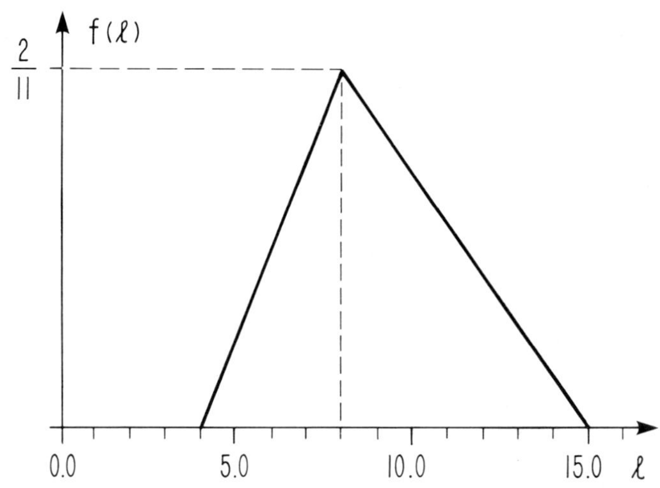

Figure 2.5 A random variable $f(\ell)$ (Example 2.4).

Mathematical Expectation of a Hybrid Number

A function $\phi(x)$ in R that is nonnegative and monotonically increasing is

$\forall, x_1, x_2 \in R$:

$$(x_2 \geq x_1) \Rightarrow (\phi(x_2) \geq \phi(x_1)).$$

For a closed interval R, $[a_1^{(\alpha)}, a_2^{(\alpha)}]$, we have

$$[\phi(a_1^{(\alpha)}), \phi(a_2^{(\alpha)})] \subset R,$$

and for $\ell \in R$,

$$[\phi(a_1^{(\alpha)} + \ell), \phi(a_2^{(\alpha)} + \ell)] \subset R. \tag{2.5}$$

If ℓ is a value of the random variable L the lower and upper bounds of (2.5) depend only on ℓ for a given level α. We can now compute the mathematical expectations for each bound:

$$[\phi(a_1^{(\alpha)} + \ell), \phi(a_2^{(\alpha)} + \ell]$$

$$= [\int_{\ell_1}^{\ell_2} \phi(a_1^{(\alpha)} + \ell)\, f(\ell)\, d\ell, \int_{\ell_1}^{\ell_2} \phi(a_2^{(\alpha)} + \ell)\, f(\ell)\, d\ell].$$

THEOREM 2.1

The membership function of the mathematical expectation of a hybrid number $(\mathbf{A}, L)$ is the membership function of $\mathbf{A}$ shifted by the mathematical expectation of L.

Proof:

Let us use the intervals of confidence for level α.

$$\xi_\alpha(\mathbf{A}\,[+]\,L) = [\int_{\ell_1}^{\ell_2} (a_1^{(\alpha)} + \ell)\, f(\ell)\, d\ell, \int_{\ell_1}^{\ell_2} (a_2^{(\alpha)} + \ell)\, f(\ell)\, d\ell]$$

$$= [a_1^{(\alpha)} \int_{\ell_1}^{\ell_2} f(\ell)\, d\ell + \int_{\ell_1}^{\ell_2} \ell\, f(\ell)\, d\ell,\ a_2^{(\alpha)} \int_{\ell_1}^{\ell_2} f(\ell)\, d\ell + \int_{\ell_1}^{\ell_2} f(\ell)\, d\ell]$$

$$= [a_1^{(\alpha)} + \xi(L), a_2^{(\alpha)} + \xi(L)].$$

Hence, in a hybrid sum, if the random variables satisfy their mathematical expectation, they will have the same effect as ordinary numbers, shifting the sum of fuzzy numbers. It is important to mention this result even though it is trivial.

Remark

Using the mathematical expectation of an hybrid number instead of the hybrid number itself generally decreases the quality of information, and should be avoided, if possible, in any sequential process of optimization.

Other Operations with Hybrid Numbers

Multiplication by an Ordinary Number

We now study multiplication by an ordinary number $k \neq 0$, $k \in R$:

$\forall\ x \in R$:

$$\mu_{k\mathbf{A}}(x) = \mu_{\mathbf{A}}(x/k), \qquad k \neq 0,$$

or

$\forall\ \alpha \in [0, 1]$:

$$\begin{aligned} k \cdot A_\alpha &= [k \cdot a_1^{(\alpha)}, k \cdot a_2^{(\alpha)}], \qquad k > 0, \\ &= [k \cdot a_2^{(\alpha)}, k \cdot a_1^{(\alpha)}], \qquad k < 0. \end{aligned}$$

If $f(\ell)$ is the density of L, then the density of $k \cdot L$ is given by

$$g(k \cdot \ell) = (1/k)f(\ell), \qquad \text{if } k > 0.$$

For $k \neq 0$,

$$k \cdot (\mathbf{A}, 0) = (k \cdot \mathbf{A}, 0),$$

$$k \cdot (0, L) = (0, k \cdot L)$$

with $g(k \cdot \ell) = (1/k)f(\ell), \qquad k > 0$

and

$$k \cdot (\mathbf{A}, L) = (k \cdot \mathbf{A}, k \cdot L)$$

with $g(k \cdot \ell) = (1/k) \cdot f(\ell), \qquad k > 0.$

If $k \in N_0$, we must be careful in using these results. If $2 \cdot \mathbf{A}$ is equivalent to $\mathbf{A}\ (+)\ \mathbf{A}$, we can see the effect on the corresponding intervals of confidence of level α:

$$\begin{aligned} [a_1^{(\alpha)}, a_2^{(\alpha)}]\ (+)\ [a_1^{(\alpha)}, a_2^{(\alpha)}] &= [a_1^{(\alpha)} + a_1^{(\alpha)}, a_2^{(\alpha)} + a_2^{(\alpha)}] \\ &= [2 \cdot a_1^{(\alpha)}, 2 \cdot a_2^{(\alpha)}]. \end{aligned}$$

The case is very different for probability densities; for example, for $2 \cdot L$ the density function is given by

$$g(2 \cdot \ell) = (1/2)\ f(\ell).$$

If $L = L_1 + L_2$, and L_1 and L_2 have the same density $f(\ell)$, then

$$g(\ell) = \int_R f(\ell - \ell_1) \cdot f(\ell_1)\, d\ell_1$$

$$= \int_R f(\ell - \ell_2) \cdot f(\ell_2)\, d\ell_2.$$

This problem is well known to anyone who is familiar with probability theory. Another problem of concern is that of discrete values. For this we must write

$$\mathrm{pr}_k\,(k \cdot \ell) = \mathrm{pr}\,(\ell), \qquad \ell \in N, \qquad k \in N_0,$$

where pr is the discrete probability law relative to L and pr_k is relative to $k \cdot L$.

Image of Hybrid Numbers

The image (opposite) of addition is easily described for a fuzzy number and for a random variable.

$\forall\ x, y \in R$:

$$\mu_{(-\mathbf{A})}(y) = \mu_{\mathbf{A}}(-x)$$

and

$\forall \ell, m \in R$:

$$g(m) = f(-\ell), \qquad \text{where } M = -L,$$

which shows that subtraction is the opposite of addition.

At this point we must note the difference between subtraction and deconvolution. Deconvolution consists of finding $\mathbf{A}_2$ in $\mathbf{A} = \mathbf{A}_1\ (+)\ \mathbf{A}_2$ when $\mathbf{A}_1$ and $\mathbf{A}$ are given.

Similarly, in the case of random variables, we must be careful not to confuse the random variable $(L - L_1)$ with the sum-product deconvolution, which consists of finding L_2 if $L = L_1\ (+)'\ L_2$, where L_1 and L_2 are given.

The multiplication of fuzzy numbers by the max-min convolution is similar to obtaining the probability density for the product of two random variables. These are known, but very complicated, operations. The same applies to the divisions of fuzzy numbers or random variables. In this case, division must not be confused with deconvolution for multiplication.

Hybrid numbers have many more interesting features, but we do not discuss them in this book.

2.3 SHEAF AND EXPECTATION OF FUZZY NUMBERS

We have n observations (n finite) of the same phenomenon, each resulting in a fuzzy number $\mathbf{A}_i$: $i = 1, 2, \ldots, n$,. The set of $\mathbf{A}_i$ in the same referential set E constitutes a *sheaf of fuzzy numbers.*

THEOREM 2.2

Let a sheaf of n fuzzy numbers be $\mathbf{A}_i \in R$, $i = 1, 2, 3, \ldots, n$, then

$$A_{i,\,\alpha} = [a_{1,\,i}^{(\alpha)}, a_{2,\,i}^{(\alpha)}], \qquad \alpha \in [0, 1],$$

for the interval of confidence at the level α of $\mathbf{A}_i$. Then, defining the mean as

$$a_1^{m(\alpha)} = (1/n) \sum_{i=1}^{n} a_{1,\,i}^{(\alpha)}$$

$$a_2^{m(\alpha)} = (1/n) \sum_{i=1}^{n} a_{2,\,i}^{(\alpha)},$$

the mean interval of confidence at level α of the mean fuzzy number $\mathbf{A}^m$ is described as

$$A_\alpha^m = [a_1^{m(\alpha')}, a_2^{m(\alpha)}].$$

Proof:

It is sufficient to demonstrate that the mean fuzzy number $\mathbf{A}^m$ obtained is convex and normal. For each $A_{i,\,\alpha}$ we can write

$$(\alpha' > \alpha) \Rightarrow [a_{1,\,i}^{(\alpha')}, a_{2,\,i}^{(\alpha)}] \subset [a_{1,\,i}^{(\alpha)}, a_{2,\,i}^{(\alpha)}], \qquad \alpha \in [0, 1], \quad i = 1, 2, \ldots, n,$$

or

$$a_{1,\,i}^{(\alpha)} \geq a_{1,\,i}^{(\alpha')}, \qquad a_{2,\,i}^{(\alpha)} \leq a_{2,\,i}^{(\alpha')}.$$

Taking into account that the intervals of level α are added by the convolution max-min for addition, it follows that

$$\sum_{i=1}^{n} a_{1,\,i}^{(\alpha)} \geq \sum_{i=1}^{n} a_{1,\,i}^{(\alpha')} \tag{2.6}$$

and

$$\sum_{i=1}^{n} a_{2,i}^{(\alpha)} \le \sum_{i=1}^{n} a_{2,i}^{(\alpha')}. \tag{2.7}$$

If we divide (2.6) and (2.7) by the number of observations n, we have

$$a_1^{m(\alpha)} \ge a_1^{m(\alpha')} \qquad \text{and} \qquad a_2^{m(\alpha)} \le a_2^{m(\alpha')},$$

which proves convexity.

With reference to normality, we have

$$A_{i,\,\alpha=1} = [a_{1,i}^{(\alpha=1)}, a_{2,1}^{(\alpha=1)}] \ne \emptyset, \qquad i = 1, 2, \ldots, n,$$

and

$$\left[\sum_{i=1}^{n} a_{1,i}^{(\alpha=1)}, \qquad \sum_{i=1}^{n} a_{2,i}^{(\alpha=1)}\right] \ne \emptyset.$$

Therefore

$$[a_1^{m(\alpha=1)}, a_2^{m(\alpha=1)}] \ne \emptyset,$$

which proves the normality.

EXAMPLE 2.5

Consider a sheaf of three TFN's in $R^{(+)}$, as shown in Figure 2.6(A).

1. $\forall\, x \in R^+$:

$$\begin{aligned}
\mu_{\mathbf{A}_1}(x) &= 0, & x &\le 4,\\
&= x/5 - 4/5, & 4 \le x &\le 9,\\
&= -x + 10, & 9 \le x &\le 10,\\
&= 0, & x &\ge 10
\end{aligned}$$

or

$$A_{1,\,\alpha} = [4 + 5\alpha, 10 - \alpha].$$

2. $\forall\, x \in R^+$:

$$\begin{aligned}
\mu_{\mathbf{A}_2}(x) &= 0, & x &\le 5,\\
&= x/3 - 5/3, & 5 \le x &\le 8,\\
&= -x/5 + 13/5, & 8 \le x &\le 13,\\
&= 0, & x &\ge 13
\end{aligned}$$

or

$$A_{2,\,\alpha} = [5 + 3\alpha, 13 - 5\alpha].$$

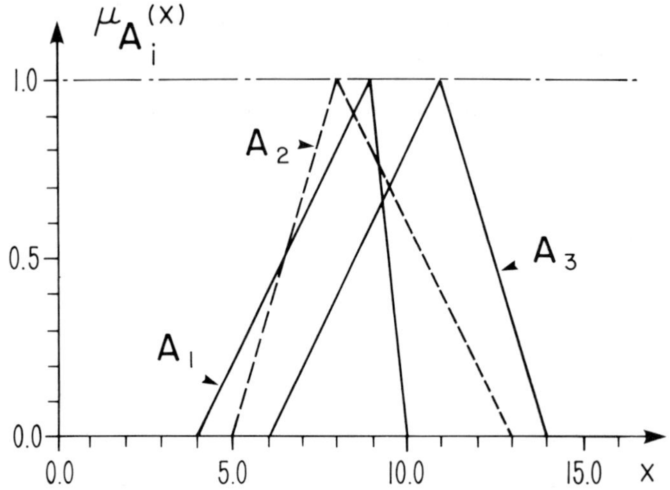

Figure 2.6(A) A sheaf of three TFN's in R^+ (Example 2.5).

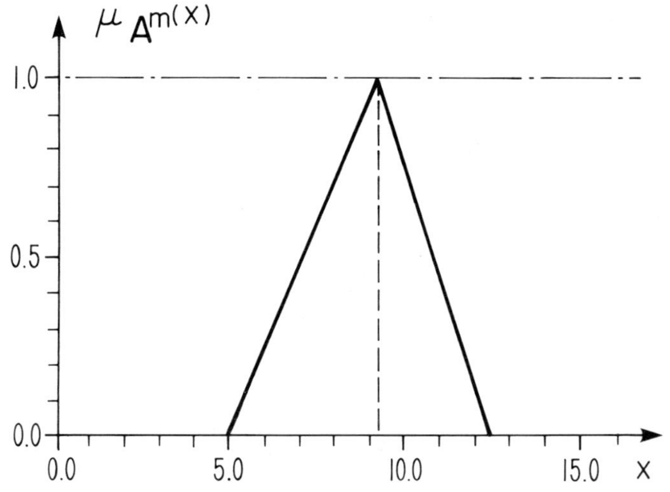

Figure 2.6(B) The mean of three TFN's (Example 2.5).

3. $\forall\ x \in R^+$:

$$\mu_{\mathbf{A}_3}(x) = \begin{cases} 0, & x \leq 6, \\ x/5 - 6/5, & 6 \leq x \leq 11, \\ -x/3 + 14/3, & 11 \leq x \leq 14, \\ 0, & x \geq 14 \end{cases}$$

or

$$A_{3,\alpha} = [6 + 5\alpha, 14 - 3\alpha].$$

By the addition and computation of the mean, we find that

$$\begin{aligned}\mathbf{A}_\alpha^m &= (1/3)\,([4 + 5\alpha, 10 - \alpha]\,(+)\,[5 + 3\alpha, 13 - 5\alpha] \\ &\quad (+)\,[6 + 5\alpha, 14 - 3\alpha]) \\ &= [5 + 13\alpha/3, 37/3 - 3\alpha].\end{aligned}$$

as shown in Figure 2.6(*B*),

$\forall\ x \in R^+$:

$$\begin{aligned}\mu_{\mathbf{A}}^m(x) &= 0, & x &< 5, \\ &= 3x/13 - 15/13, & 5 \le x &\le 28/3, \\ &= 37/9 - 3x/9, & 28/3 \le x &\le 37/3, \\ &= 0, & x &\ge 37/3.\end{aligned}$$

These results can also be obtained by using the triplets for TFN's.

$$\begin{aligned}\mathbf{A}_1 &= (4, 9, 10), \\ \mathbf{A}_2 &= (5, 8, 13), \\ \mathbf{A}_3 &= (6, 11, 14), \\ \mathbf{A}^m &= (1/3)\,(\mathbf{A}_1\,(+)\,\mathbf{A}_2\,(+)\,\mathbf{A}_3) \\ &= (1/3)\,(4 + 5 + 6, 9 + 8 + 11, 10 + 13 + 14) \\ &= (5, 28/3, 37/3),\end{aligned}$$

The results are shown in Figure 2.6(*B*).

Mathematical Expectation of a Sheaf

To begin, we present a theorem for the case of a sheaf of fuzzy data with a finite numbers of fuzzy numbers.

THEOREM 2.3

Consider a sheaf of n fuzzy numbers $\mathbf{A}_i$, $i = 1, 2, \ldots, n$, and the corresponding probability law pr (i), for $i = 1, 2, \ldots, n$. Then, if $\overline{\mathbf{A}}$ represents the fuzzy subset defined at the level α by

$$\begin{aligned}\overline{A}_\alpha &= \left[\sum_{i=1}^{n} a_{1,i}^{(\alpha)}\,\mathrm{pr}\,(i),\ \sum_{i=1}^{n} a_{2,i}^{(\alpha)}\,\mathrm{pr}\,(i)\right] \\ &= [\overline{a}_1^{(\alpha)}, \overline{a}_2^{(\alpha)}],\end{aligned}$$

$\overline{\mathbf{A}}$ is a fuzzy number called the *expected fuzzy number* of the sheaf, which is both convex and normal.

EXAMPLE 2.6

Taking the three fuzzy numbers in the sheaf given in Example 2.1 (Fig. 2.1A) and considering the probability laws

$$\text{pr}(1) = 1/5, \qquad \text{pr}(2) = 3/5, \qquad \text{pr}(3) = 1/5,$$

we find that

$$\begin{aligned}\bar{\mathbf{A}} &= \left(\frac{1}{5}\right)(4, 9, 10)\ (+)\left(\frac{3}{5}\right)(5, 8, 13)\ (+)\left(\frac{1}{5}\right)(6, 11, 4)\\ &= \left(\frac{4}{5}, \frac{9}{5}, \frac{10}{5}\right)(+)\left(\frac{15}{5}, \frac{24}{5}, \frac{39}{5}\right)(+)\left(\frac{6}{5}, \frac{11}{5}, \frac{14}{5}\right)\\ &= \left(5, \frac{44}{5}, \frac{63}{5}\right).\end{aligned}$$

Let us now examine the general case of a sheaf that has a nonfinite number of fuzzy numbers such that this family $\mathcal{F}(\mathbf{A})$ is probabilistic with a density $f(\mathbf{A})$, $\mathbf{A} \subset \mathbf{R}$, where each $\mathbf{A}$ depends on the parameter λ, and the probabiity law is given on λ. Hence

$$\begin{aligned}\bar{\mathbf{A}}_\alpha &= \left[\int a_1^{(\alpha)}(\lambda)\, f(\lambda)\, d\lambda, \int a_2^{(\alpha)}(\lambda)\, f(\lambda)\, d\lambda\right]\\ &= [\bar{a}_1^{(\alpha)}, \bar{a}_2^{(\alpha)}].\end{aligned}$$

A probability law that has a convex weighting preserves the convexity and normality of the fuzzy subsets. To illustrate this property we again use the data of Example 2.4 and determine the mathematical expectation of the hybrid number.

EXAMPLE 2.7

From (2.4) we find

$$A_\alpha = [2 + 2\alpha, 5 - \alpha]$$

and

$$A_\alpha\ [+]\ \ell = [2 + 2\alpha + \ell, 5 - \alpha + \ell].$$

Therefore, the expectation is given by

$$\xi(A_\alpha\ [+]\ \ell) = \left[2 + 2\alpha + \int_R \ell f(\ell)\, d, 5 - \alpha + \int_R \ell f(\ell)\, d\ell\right].$$

Computing the integral we obtain

$$\bar{\ell} = \int \ell f(\ell)\, d\ell,$$

$$= \int_{\ell=4}^{8} \left(\frac{1}{22}\right)(\ell - 4)\,\ell\, d\ell + \int_{\ell=8}^{15} \left(\frac{2}{11}\right)\left(\frac{-\ell + 15}{7}\right)\ell\, d\ell,$$

$$= 9.$$

Therefore,

$$\begin{aligned} \xi(A_\alpha\,[+]\,(\ell)) &= [2 + 2\alpha + 9,\ 5 - \alpha + 9] \\ &= [11 + 2\alpha,\ 14 - \alpha]. \end{aligned}$$

Or, as shown in Figure 2.7,

$\forall\ x \in R$:

$$\begin{aligned} \mu_{\xi(\mathbf{A}\,[+]\,L)}(x) &= 0, & x \le 11, \\ &= x/2 - 11/2, & 11 \le x \le 13, \\ &= -x + 14, & 13 \le x \le 14, \\ &= 0, & x \ge 14. \end{aligned}$$

When the sheaf is formed by fuzzy numbers in Z, the mathematical expectation is in R instead of in Z.

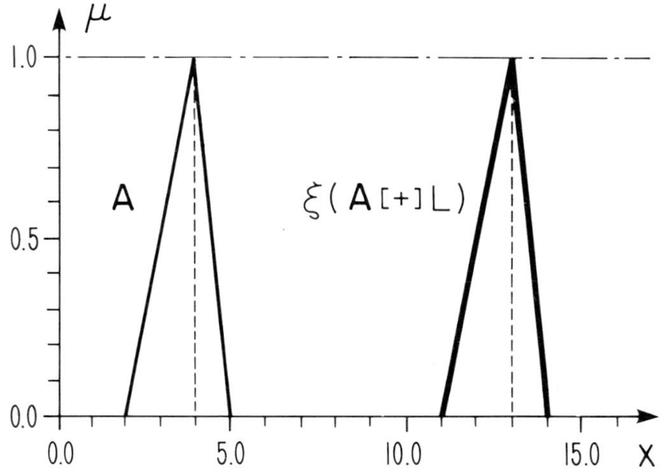

Figure 2.7 Mathematical expectation of hybrid number (Example 2.7).

Comparison Between Addition by Normalization and Addition by Hybridization

In Section 2.2 [see Eqns. (2.1) and (2.2)] we added a fuzzy number and a random variable by normalization, in which the random variable was transformed into a fuzzy number. Another method uses the concept of hybrid numbers and mathematical expectation. In the first method we used modes and in the second mathematical, expectation. The problem remains of how to choose the mode or the expectation for a given law in a way that represents this law. In practice, expectation is preferred to mode in many situations. For purposes of comparison, we again consider Example 2.3.

EXAMPLE 2.8

Figure 2.8 shows the normalized sum **A** (+) $f(\ell)$, and Figure 2.9 shows

$$\mathbf{A}\ (+)\ \xi(L). \tag{2.8}$$

A comparison of these two figures shows an obvious interest in (2.8).

We recall that the relative position between mode and expectation depends on the nature of $f(\ell)$, which is seen from Figures 2.10(*A*) and 2.10(*B*).

Still another well-known property is the following. If $L_1, L_2, \ldots, L_k$ are random variables in the same referential set, then

$$\xi(L_1 + L_2 + \cdots + L_k) = \xi(L_1) + \xi(L_2) + \cdots + \xi(L_k).$$

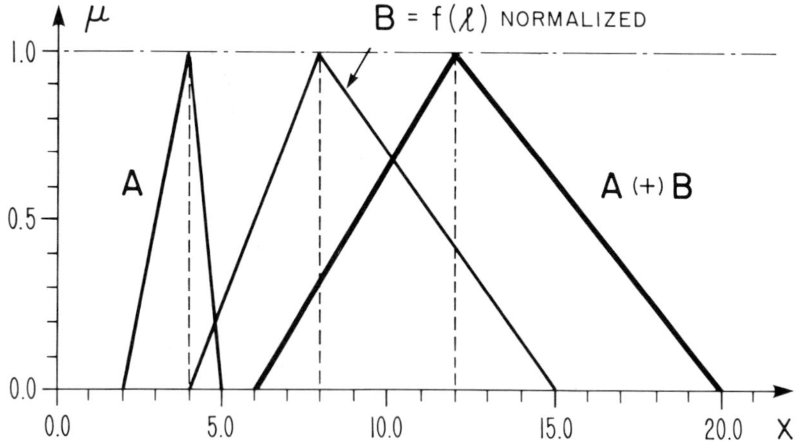

Figure 2.8 Normalized sum **A** (+) $f(\ell)$ of hybrid numbers (Example 2.8).

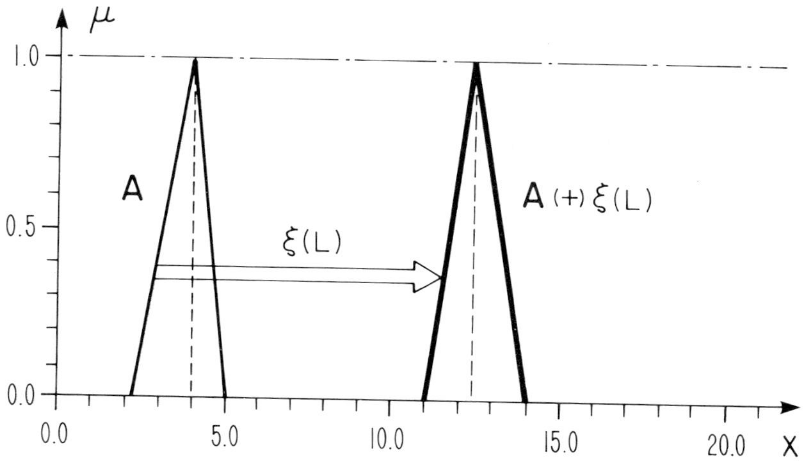

Figure 2.9 Sum of expectation in hybrid numbers (Example 2.8).

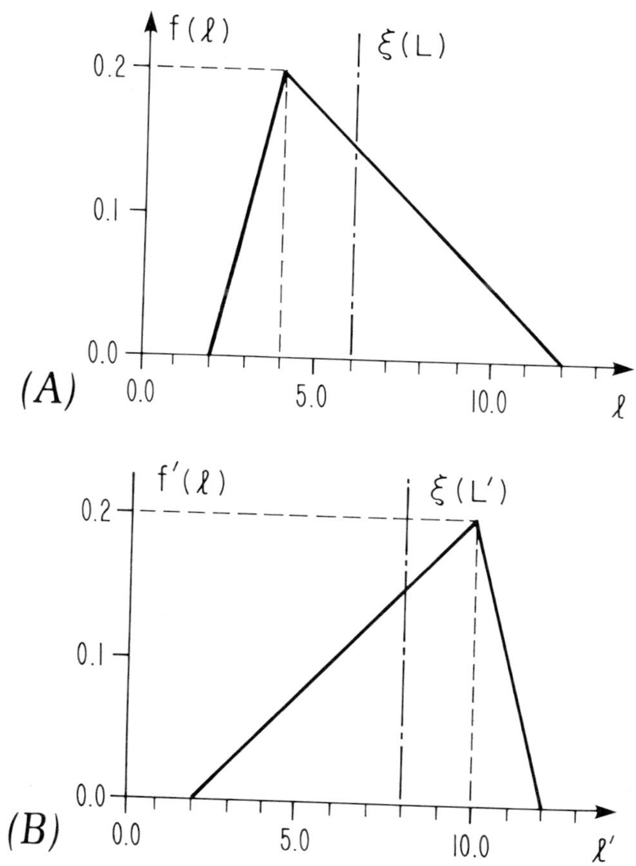

Figure 2.10 Two $f(\ell)$ having different expectations (Example 2.8).

Although independence of hybrid numbers is interesting, we do not discuss it here.

2.4 DISTANCE BETWEEN TWO FUZZY NUMBERS: THE DISSEMBLANCE INDEX

Let us consider three intervals of confidence in R, namely,

$$A = [a_1, a_2], \quad \textbf{(2.9)}$$
$$B = [b_1, b_2], \quad \textbf{(2.10)}$$
$$C = [c_1, c_2], \quad \textbf{(2.11)}$$

Any concept of distance must satisfy the following properties: A numerical function $d(X, Y) \in R$, $(X, Y) \in E \times E$ is a *distance* if and only if

$\forall\ X, Y, Z, \in E$:

$$d(X, Y) \geq 0, \quad \textbf{(2.12)}$$
$$(X = Y) \Rightarrow (d(X, Y) = 0), \quad \textbf{(2.13)}$$
$$d(X, Y) = d(Y, X), \quad \textbf{(2.14)}$$
$$d(X, Z) \leq d(X, Y) * d(Y, Z), \quad \textbf{(2.15)}$$

where $*$ is an operator associated with the notion of distance. This concept of distance is different, of course, from our usual concept of "metric" where we assume that $(X = Y) \Leftrightarrow (d(X, Y) = 0)$. If this were not so, in metric we would not be able to find $d(X, Y) = 0$ when $X \neq Y$.

To consider the concept of *distance to the left*, let

$$\Delta_\ell\ (A, B) = | a_1, b_1 |.$$

Similarly, for the *distance to the right*, let

$$\Delta_r(A, B) = | a_2, b_2 |.$$

We must now check the conditions from (2.12–2.15) for this concept of distance. Let

$\forall\ A, B, C \subset R$:

1. $\Delta_\ell\ (A, B) \geq 0,\qquad$ because $| a_1 - b_1 | \geq 0$.

2. $(A = B) \Rightarrow (\Delta_\ell\ (A, B) = 0),\qquad$ because

$$(a_1 = b_1) \Rightarrow (| a_1 - b_1 | = 0).$$

3. $\Delta_\ell\,(A, B) = \Delta_\ell\,(B, A)$, because $|a_1 - b_1| = |b_1 - a_1|$.

4. $\Delta_\ell\,(A, C) \le \Delta_\ell\,(A, B) + \Delta_\ell\,(B, C)$, because

$$|a_1 - c_1| \le |a_1 - b_1| + |b_1 - c_1|.$$

The proof for Δ_r may be carried out in the same way.

Now we consider the concept of distance $\Delta(A, B)$ as follows:

$$\Delta(A, B) = \Delta_\ell(A, B) + \Delta_r(A, B).$$

It is easy to prove that conditions (2.12–2.15) are also satisfied for this condition.

Suppose now that any interval like (2.9), (2.10), and (2.11) is a subset of $[\beta_1, \beta_2] \subset R$. We define

$$\delta(A, B) = \{1/[2(\beta_2 - \beta_1)]\}\, d(A, B).$$

as the *normalized distance*. The number 2 in the denominator permits us to write

$$0 \le \delta(A, B) \le 1. \tag{2.16}$$

We have two fuzzy numbers **A** and **B** in R as shown in Figure 2.11. For each level α, we can write

$$\delta(A_\alpha, B_\alpha) = \{1/[2(\beta_2 - \beta_1)]\}\, d(A_\alpha, B_\alpha),$$

where β_1 and β_2 are given any convenient values in order to surround all $\mathbf{A}_{\alpha=0}$ and $\mathbf{B}_{\alpha=0}$ (Fig. 2.11).

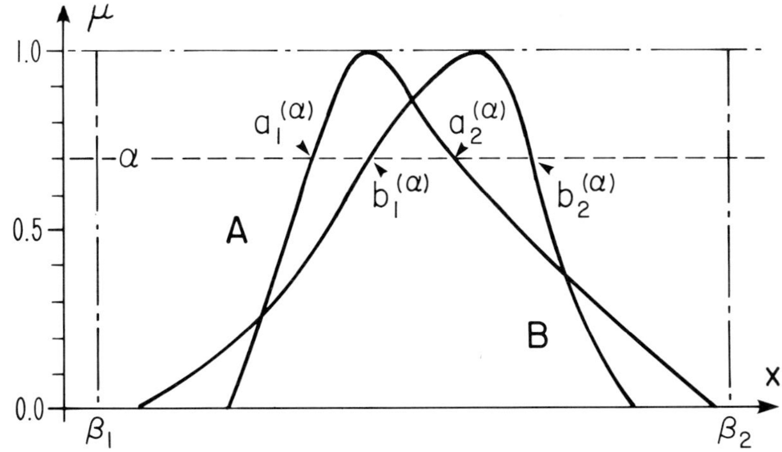

Figure 2.11 Concepts of distance and dissemblance index between two fuzzy numbers **A** and **B**.

If we proceed to integrate from $\alpha = 0$ to $\alpha = 1$, we obtain a distance by the summation of distances that satisfies equation (2.16),

$$\delta(\mathbf{A}, \mathbf{B}) = \int_{\alpha=0}^{1} \delta(A_\alpha, B_\alpha)\, d\alpha$$

$$= 1/2\,(\beta_2 - \beta_1) \int_{\alpha=0}^{1} \Delta(A_\alpha, B_\alpha)\, d\alpha$$

$$= 1/2\,(\beta_2 - \beta_1) \int_{\alpha=0}^{1} (|a_1^{(\alpha)} - b_1^{(\alpha)}| + |a_2^{(\alpha)} - b_2^{(\alpha)}|)\, d\alpha. \quad \textbf{(2.17)}$$

Equation (2.17) gives the distance between two fuzzy numbers; it may also be called the *dissemblance index* of **A** and **B**.

EXAMPLE 2.9

Let two triangular fuzzy numbers in R, as shown in Figure 2.12(A), on page **104**, be

$\forall\ x \in R$:

$$\mu_{\mathbf{A}}(x) = \begin{cases} 0, & x \le 2, \\ x/8 - 2/8, & 2 \le x \le 10, \\ -x/3 + 13/3, & 10 \le x \le 13, \\ 0, & x \ge 13, \end{cases}$$

or

$\forall\ \alpha \in [0, 1]$:

$$A_\alpha = [2 + 8\alpha,\ 13 - 3\alpha]$$

and

$$\mu_{\mathbf{B}}(x) = \begin{cases} 0, & x \le 5, \\ x/2 - 5/2, & 5 \le x \le 7, \\ -x/8 + 15/8, & 7 \le x \le 15, \\ 0, & x \ge 15. \end{cases}$$

or

$$B_\alpha = [5 + 2\alpha,\ 15 - 8\alpha].$$

Note that **A** and **B** intersect at the following two points:

$2 + 8\alpha = 5 + 2\alpha$ gives: $\alpha = 0.5$ and $x = 6$,

and

$$13 - 3\alpha = 15 - 8\alpha \qquad \text{gives:} \qquad \alpha = 0.4 \quad \text{and} \quad x = 11.8.$$

We then have

$$a_1^{(\alpha)} - b_1^{(\alpha)} = 2 + 8\alpha - 5 - 2\alpha = -3 + 6\alpha$$

and

$$a_2^{(\alpha)} - b_2^{(\alpha)} = 13 - 3\alpha - 15 + 8\alpha = -2 + 5\alpha.$$

And

$$\int_{\alpha=0}^{0.5} (-3 + 6\alpha)\, d\alpha + \int_{\alpha=0.5}^{1} (6\alpha - 3)\, d\alpha + \int_{\alpha=0}^{0.4} (-2 + 5\alpha)\, d\alpha + \int_{\alpha=0.4}^{1} (5\alpha - 2)\, d\alpha = 2.8.$$

If we set $\beta_1 = 2$ and $\beta_2 = 16$, we find

$$\delta(\mathbf{A}, \mathbf{B}) = [1/2(16 - 2)]\,[2.8] = 0.1.$$

It is interesting that with TFN's, distance computations can be carried out more rapidly by adding areas of triangles (Fig. 2.12(*B*)]:

area *ABC* + area *CDE* + area *DEF* + area *FGH*

$$\begin{aligned} &= (3) \cdot (0.5)/2 + (3) \cdot (0.5)/2 + \\ &\quad (3) \cdot (0.6)/2 + (2) \cdot (0.4)/2, \\ &= 2.8. \end{aligned}$$

From Figure 2.12(*B*) we see that the triangle *DEI* is taken into account twice, once in Δ_l and a second time in Δ_r. This duplication disappears only if

$$[a_1^{(1)}, a_2^{(1)}] = [b_1^{(1)}, b_2^{(1)}].$$

In fact, we can ask when does $\delta(\mathbf{A}, \mathbf{B}) = 1$. It takes place when

$\forall\ \alpha \in [0, 1]$:

$$A_\alpha = [\beta_1, \beta_1],$$
$$B_\alpha = [\beta_2, \beta_2].$$

This is an extreme case. One point to note is that if we wish to compare two fuzzy numbers it is often useful to normalize the result with respect to δ.

We know that a normalized distance can be considered as a dissemblance index.[1] In Example 2.10 we study the resemblance between fuzzy numbers in a sheaf.

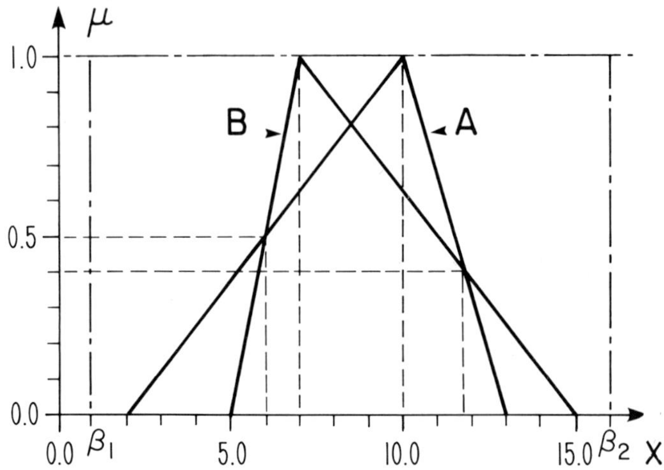

Figure 2.12(A) Two TFN's (Example 2.9).

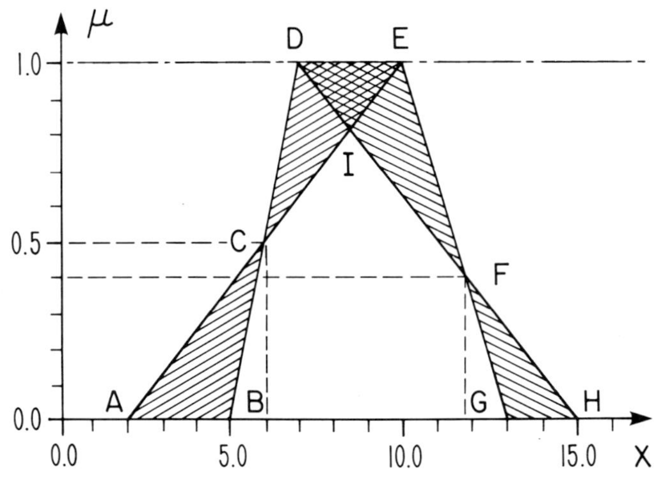

Figure 2.12(B) Distance δ (**A, B**) (using area method) (Example 2.9).

EXAMPLE 2.10

Let us compute the distances $\delta(X, Y)$, where **X**, **Y** $\in$ {**A, B, C, D, E, F**}, as shown in Figure 2.13. We obtain the dissemblance relation in Table 2.2.

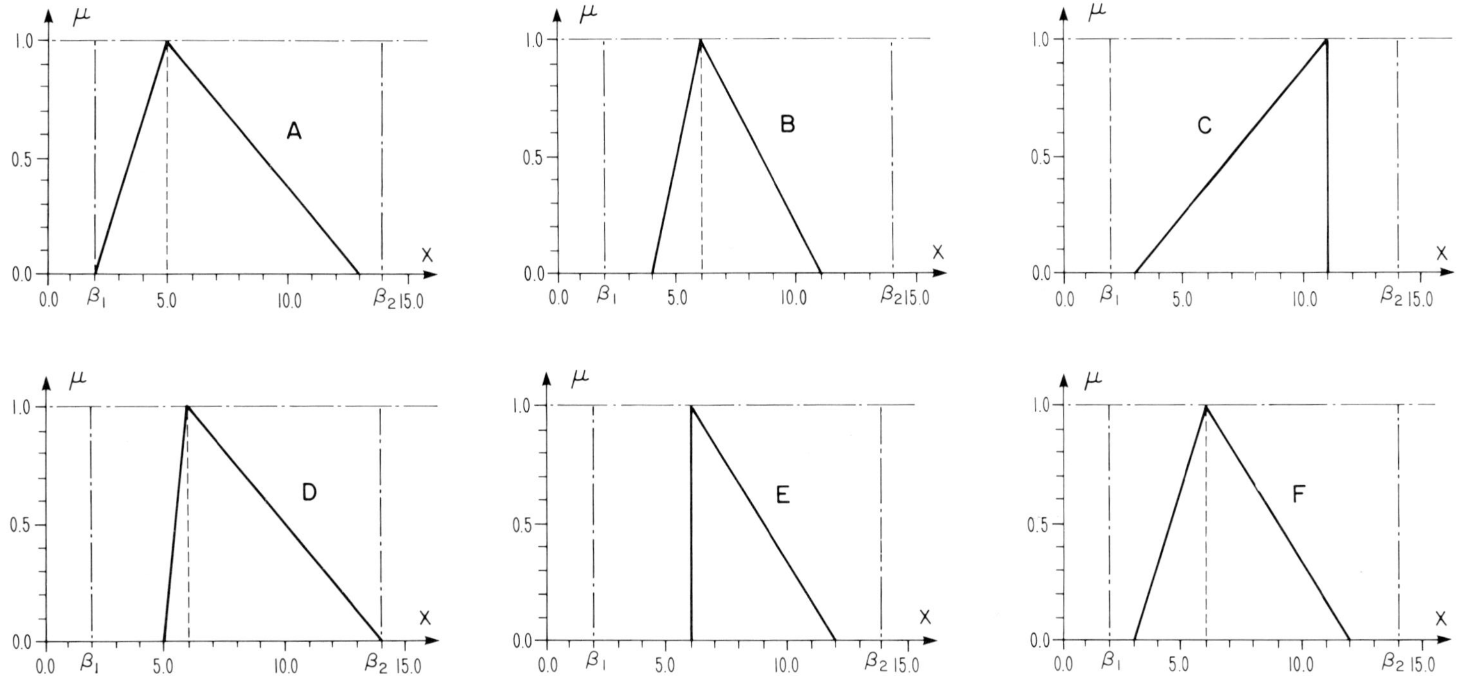

Figure 2.13 TFN's (**A, B, C, D, E, F**) (Example 2.10).

Table 2.2

$\delta(\mathbf{x}, \mathbf{y})$	**A**	**B**	**C**	**D**	**E**	**F**
A	0	0.09	0.24	0.08	0.13	0.06
B	0.09	0	0.19	0.08	0.08	0.08
C	0.24	0.19	0	0.17	0.17	0.19
D	0.08	0.08	0.17	0	0.06	0.08
E	0.13	0.08	0.17	0.06	0	0.06
F	0.06	0.08	0.19	0.08	0.06	0

To study this situation level by level, we decompose each level into the maximum subrelations of similitude* as shown in Tables 2.3–2.9.

By inspecting Tables 2.3–2.9 we conclude that for $\delta \leq 0.06$, **A** and **F** are very similar, as are **D** and **E**, and **E** and **F**. This does not

Table 2.3

	A	B	C	D	E	F
A	1					1
B		1				
C			1			
D				1	1	
E				1	1	1
F	1				1	1

	A	F
A	1	1
F	1	1

	B
B	1

	C
C	1

	D	E
D	1	1
E	1	1

	E	F
E	1	1
F	1	1

$\delta \leq 0.06$

Table 2.4

	A	B	C	D	E	F
A	1			1		1
B		1		1	1	1
C			1			
D	1	1		1	1	1
E		1		1	1	1
F	1	1		1	1	1

	A	D	F
A	1	1	1
D	1	1	1
F	1	1	1

	B	D	E	F
B	1	1	1	1
D	1	1	1	1
E	1	1	1	1
F	1	1	1	1

	C
C	1

$\delta \leq 0.08$

Table 2.5

	A	B	C	D	E	F
A	1	1		1		1
B	1	1		1	1	1
C			1			
D	1	1		1	1	1
E		1		1	1	1
F	1	1		1	1	1

	A	B	D	E
A	1	1	1	1
B	1	1	1	1
D	1	1	1	1
E	1	1	1	1

	B	D	E	F
B	1	1	1	1
D	1	1	1	1
E	1	1	1	1
F	1	1	1	1

	C
C	1

$\delta \leq 0.09$

Table 2.6

	A	B	C	D	E	F
A	1	1		1	1	1
B	1	1		1	1	1
C			1			
D	1	1		1	1	1
E	1	1		1	1	1
F	1	1		1	1	1

	A	B	D	E	F
A	1	1	1	1	1
B	1	1	1	1	1
D	1	1	1	1	1
E	1	1	1	1	1
F	1	1	1	1	1

	C
C	1

$\delta \leq 0.13$

Table 2.7

	A	B	C	D	E	F
A	1	1		1	1	1
B	1	1		1	1	1
C			1	1	1	
D	1	1	1	1	1	1
E	1	1	1	1	1	1
F	1	1		1	1	1

	A	B	D	E	F
A	1	1	1	1	1
B	1	1	1	1	1
D	1	1	1	1	1
E	1	1	1	1	1
F	1	1	1	1	1

	C	D	E
C	1	1	1
D	1	1	1
E	1	1	1

$\delta \leq 0.18$

Table 2.8

	A	B	C	D	E	F
A	1	1		1	1	1
B	1	1	1	1	1	1
C		1	1	1	1	1
D	1	1	1	1	1	1
E	1	1	1	1	1	1
F	1	1	1	1	1	1

	A	B	D	E	F
A	1	1	1	1	1
B	1	1	1	1	1
D	1	1	1	1	1
E	1	1	1	1	1
F	1	1	1	1	1

	B	C	D	E	F
B	1	1	1	1	1
C	1	1	1	1	1
D	1	1	1	1	1
E	1	1	1	1	1
F	1	1	1	1	1

$\delta \leq 0.19$

Table 2.9

	A	B	C	D	E	F
A	1	1	1	1	1	1
B	1	1	1	1	1	1
C	1	1	1	1	1	1
D	1	1	1	1	1	1
E	1	1	1	1	1	1
F	1	1	1	1	1	1

$\delta \leq 0.24$

mean, however, that **D** and **F**, and **A** and **E** are also similar; the resemblance is not necessarily transitive.

In fuzzy numbers the concept of distance plays the same role as does divergence for the case of ordinary numbers. However, we must remember that fuzzy numbers are not linearly ordered.

2.5 COMPARISONS BETWEEN HYBRID NUMBERS: THEIR FUZZY OPTIMUM

As we have shown in Equations (1.60–1.69) for fuzzy numbers, the operators ($\wedge$) and ($\vee$) have a lattice configuration. Random variables, however, do not have such a configuration for these operators. For hybrid numbers we can come up with several processes for comparison.

In section 2.2, Theorem 2.1, we saw that the mathematical expectation of a hybrid number is a fuzzy number $\{\mathbf{A}\ (+)\ \xi(L)\}$; that is, the fuzzy number $\mathbf{A}$ is shifted by $\xi(L)$.

$$\mathbf{A}'_1 = \mathbf{A}_1 + \xi(L_1),$$
$$\mathbf{A}'_2 = \mathbf{A}_2\ (+)\ \xi(L_2).$$

Thus the hybrid numbers $\mathbf{A}_1\,[+]\,L_1$ and $\mathbf{A}_2\,[+]\,L_2$ become fuzzy numbers and we are able to compute their fuzzy optimum.

Fuzzy Minimum

$\forall$ x, y, z $\in$ R:

$$\mu_{\mathbf{A}'_1\,(\wedge)\,\mathbf{A}'_2}(x) = \bigvee_{z = x \wedge y} (\mu_{\mathbf{A}'_1}(x) \wedge \mu_{\mathbf{A}'_2}(y)),$$

or

$\forall\ \alpha \in [0, 1]$:

$$\begin{aligned} A'_{1\alpha}(\wedge)\,A'_{2\alpha} &= [a_{1,1}{}^{(\alpha)} + \xi(L_1),\ a_{1,2}{}^{(\alpha)} + \xi(L_1)]\,(\wedge) \\ &\quad [a_{2,1}{}^{(\alpha)} + \xi(L_2),\ a_{2,2}{}^{(\alpha)} + \xi(L_2)] \\ &= [(a_{1,1}{}^{(\alpha)} + \xi(L_1)) \wedge (a_{2,1}{}^{(\alpha)} + \xi(L_2)), \\ &\quad (a_{1,2}{}^{(\alpha)} + \xi(L_1)) \wedge (a_{2,2}{}^{(\alpha)} + \xi(L_2))]. \end{aligned}$$

Fuzzy Maximum

$\forall$ x, y, x, z $\in$ R:

$$\mu_{\mathbf{A}'_1\,(\vee)\,\mathbf{A}'_2}(z) = \bigvee_{z = x \vee y} (\mu_{\mathbf{A}'_1}(x) \wedge \mu_{\mathbf{A}'_2}(y)),$$

or

$\forall\ \alpha \in [0, 1]$:

$$\begin{aligned} A'_{1\alpha}(\vee)\,A'_{2\alpha} &= [a_{1,1}{}^{(\alpha)} + \xi(L_1),\ a_{1,2}{}^{(\alpha)} + \xi(L_1)]\,(\vee) \\ &\quad [a_{2,1}{}^{(\alpha)} + \xi(L_2),\ a_{2,2}{}^{(\alpha)} + \xi(L_2)] \\ &= [a_{1,1}{}^{(\alpha)} + \xi(L_1)) \vee (a_{2,1}{}^{(\alpha)} + \xi(L_2)), \\ &\quad (a_{1,2}{}^{(\alpha)} + \xi(L_1)) \vee (a_{2,2}{}^{(\alpha)} + \xi(L_2))]. \end{aligned}$$

EXAMPLE 2.11

Consider the following example. (See Figures 2.14–2.16.) For

$$A_{1\alpha} = [2 + 6\alpha,\ 9 - \alpha] \text{ and } \xi(L_1) = 8,$$

$$A'_{1\alpha} = A_{1\alpha}\ (+)\ \xi(L_1) = [10 + 6\alpha,\ 17 - \alpha] \qquad \text{(Fig. 2.14)}.$$

For

$$A_{2\alpha} = [1 + 2\alpha,\ 7 - 4\alpha] \text{ and } \xi(L_2) = 11,$$

$$A'_{2\alpha} = A_{2\alpha}\ (+)\ \xi(L_2) = [12 + 2\alpha,\ 18 - 4\alpha] \qquad \text{(Fig. 2.15)}.$$

Now the minimum of the hybrid numbers is given by

$$\begin{aligned} A'_{1\alpha}\ (\wedge)\ A'_{2\alpha} &= (A_{1\alpha}\ (+)\ \xi(L_1))\ (\wedge)\ (A_{2\alpha}\ (+)\ \xi(L_2)) \\ &= [(10 + 6\alpha) \wedge (12 + 2\alpha),\ (17 - \alpha) \wedge (18 - 4\alpha)]. \end{aligned}$$

Looking at Figure 2.16, we see that intersections are located at $\alpha = 1/2$ at the left and $\alpha = 1/3$ at the right. Thus we obtain the following results for fuzzy minimum:

$$\begin{aligned} A'_{1\alpha}\ (\wedge) A'_{2\alpha} &= [10 + 6\alpha,\ 17 - \alpha], & 0 \le \alpha \le 1/3, \\ &= [10 + 6\alpha,\ 18 - 4\alpha], & 1/3 \le \alpha \le 1/2, \\ &= [12 + 2\alpha,\ 18 - 4\alpha], & 1/2 \le \alpha \le 1. \end{aligned}$$

Similarly, for the fuzzy maximum we obtain:

$$\begin{aligned} A'_{1\alpha}\ (\vee)\ A'_{2\alpha} &= [12 + 2\alpha,\ 18 - 4\alpha], & 0 \le \alpha \le 1/3, \\ &= [12 + 2\alpha,\ 17 - \alpha], & 1/3 \le \alpha \le 1/2, \\ &= [10 + 6\alpha,\ 17 - \alpha], & 1/2 \le \alpha \le 1. \end{aligned}$$

For the purpose of finding order between hybrid numbers, we propose the following procedure.

Arbitrary Order Relation Between Two Hybrid Numbers

Consider two hybrid numbers ($\mathbf{A}_1$ [+] L_1) and ($\mathbf{A}_2$ [+] L_2), where fuzzy numbers $\mathbf{A}_1$ and $\mathbf{A}_2$ are assumed to be unimodal (the case where fuzzy numbers are not unimodal is examined later). Let us represent by a_{M_1} the ordinary number that corresponds to the mode $\mathbf{A}_1$ and by a_{M_2} that number that corresponds to the mode $\mathbf{A}_2$, $\xi(L_1)$ the expectation of L_1, and $\xi(L_2)$ that of L_2.

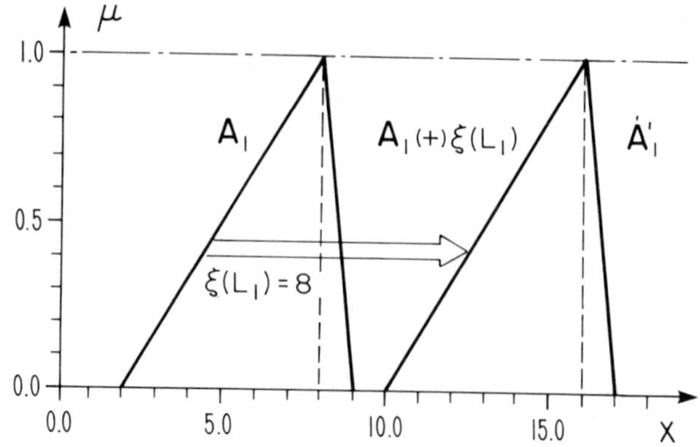

Figure 2.14 Addition of fuzzy numbers $\mathbf{A}_1$ and $\xi(L_1)$ (Example 2.11).

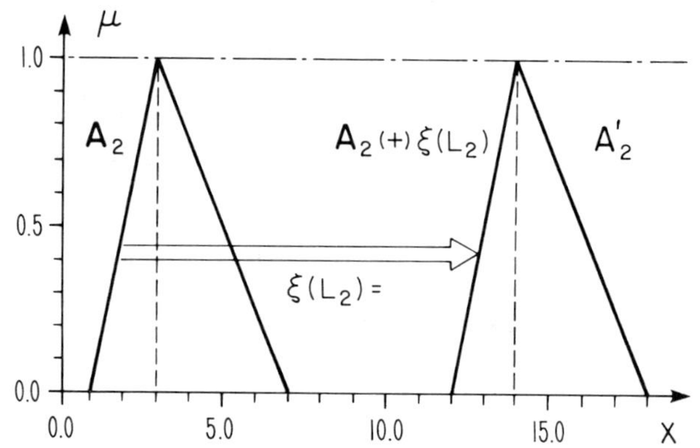

Figure 2.15 Addition of fuzzy numbers $\mathbf{A}_2$ and $\xi(L_2)$ (Example 2.11).

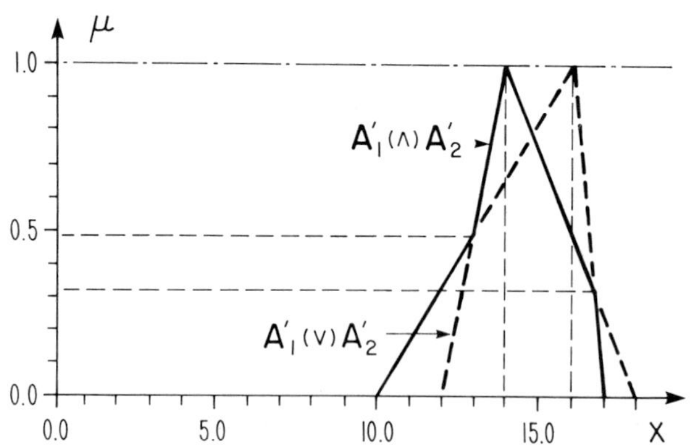

Figure 2.16 Fuzzy maximum and minimum of $\mathbf{A'}_1$ and $\mathbf{A'}_2$ (Example 2.11).

Let us then write arbitrarily

$$((a_{M_1} + \xi(L_1) < (a_{M_2} + \xi(L_2)) \Rightarrow (\mathbf{A}_1 \, [+] \, L_1) < (\mathbf{A}_2 \, [+] \, L_2)). \quad \mathbf{(2.18)}$$

This induced order relation corresponds to the concept of giving the same weight to the mode of **A** and to the expectation of L. In uncertainty the maximum of presumption corresponds to the mode, and in random theory the mathematical expectation is that value that is the most representative, if only one value is given.

If a fuzzy number is not unimodal, that is, it has a flat spot, an arbitrary point along the plate is chosen (for instance, at the center).

For a sequential optimization problem we propose the following methods:

1. Use the fuzzy minimum (or maximum). The description of the strategies, however, generally will be difficult but perfectly correct from the mathematical point of view.
2. Use the arbitrary order relation (2.18), which is not indisputable from the mathematical point of view. The advantage of this process is that the strategies are simple to describe.

A variant to 1 and 2 takes into account only the a_M's, or only the $\delta(L)$'s, or both with a convex weighting. Many other similar rules can be postulated.

EXAMPLE 2.12

Let us consider a graph (Fig. 2.17) in which directed links carry fuzzy numbers, random variables, or hybrid numbers. In order to simplify computations, we assume that all fuzzy numbers are TFN's (Fig. 2.18) and that all the random numbers have Gaussian distributions (Fig. 2.19). Each TFN is given by a triplet and the random variables by their respective mean $\bar{\ell}$ and variance $\sigma_L{}^2$. The resulting computations will be very easy and explicit. For hybrid numbers, on their directed links, the sum is $a_2 + \bar{\ell}$ (see Table 2.10 and Fig. 2.17).

We now optimize, taking into account only the values $a_2 + \bar{\ell}$. In the graph of Figure 2.17 we look for the path or paths that give the maximal value from the origin point A to the terminal point I. Using sequential optimization by means of dynamic programming (see Fig. 2.20), we obtain the optimal path shown in Figure 2.21, which gives the maximal value 37. (Figs. 2.20 and 2.21 are on page **116.**)

Now let us consider the hybrid addition that corresponds to this optimal path. TFN's add themselves by their respective triplet, and so do random variables by their respective means and variances,

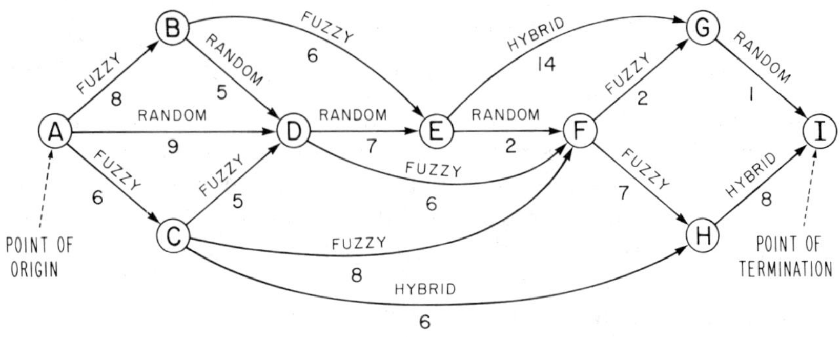

Figure 2.17 An optimization problem involving fuzzy, random, and hybrid numbers (Example 2.12).

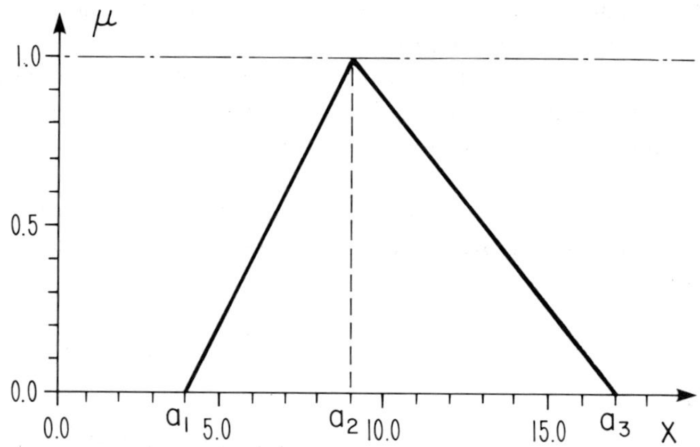

Figure 2.18 TFN distribution of fuzzy numbers (Example 2.12).

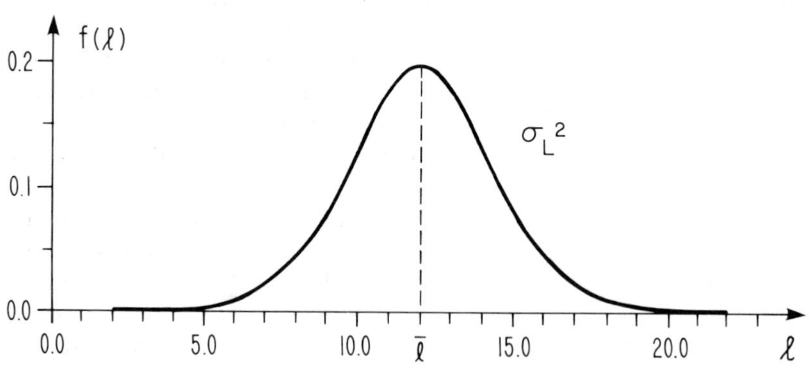

Figure 2.19 Gaussian distribution of random numbers (Example 2.12).

Table 2.10

Link	Fuzzy Number	Random Variable		$a_2 + \bar{\ell}$
	(a_1, a_2, a_3)	$\bar{\ell}$	$\sigma_L{}^2$	
AB	(5, 8, 11)			8
AC	(3, 6, 10)			6
AD		9	5	9
BD		5	1	5
BE	(6, 6, 6)			6
CD	(3, 5, 9)			5
CF	(8, 8, 11)			8
CH	(4, 2, 6)	4	2	6
DE		7	3	7
DF	(5, 6, 7)			6
EF		2	1	2
EG	(5, 8, 9)	6	3	14
FG	(2, 2, 4)			2
FH	(4, 7, 8)			7
GI		1	1	1
HI	(3, 5, 6)	3	1	8

since the distribution is assumed to be Gaussian. We can therefore write the cost of each path:

Path *AB*: (5, 8, 11).

Path *ABD*: (5, 8, 11) [+] (5, 1).

Path *ABDE*: ((5, 8, 11) [+] (5, 1)) [+] (7, 3)
= (5, 8, 11) [+] (12, 4).

Path *ABDEF*: ((5, 8, 11) [+] (12, 4)) [+] (2, 1)
= (5, 8, 11) [+] (14, 5).

Path *ABDEFH*: ((5, 8, 11) [+] (14, 5)) [+] (4, 7, 8)
= (9, 15, 19) [+] (14, 5).

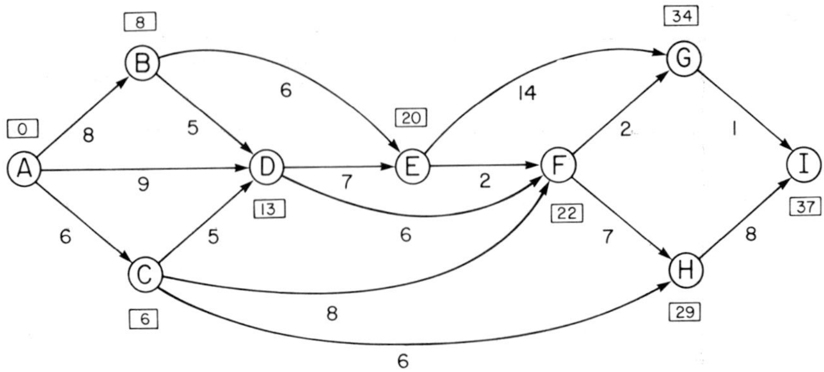

Figure 2.20 Sequential optimization for optimum (maximum) path (Example 2.12).

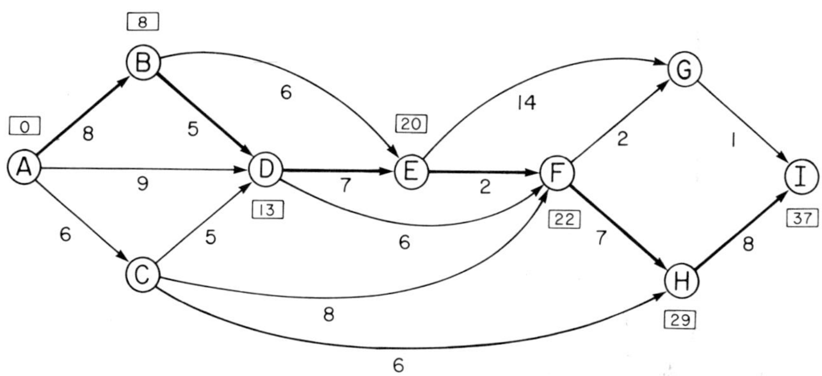

Figure 2.21 Optimum path (Example 2.12).

Path *ABDEFHI*: $((9, 15, 19)\ [+]\ (14, 5))\ [+]\ ((3, 5, 6)\ [+]\ (3, 1))$
$= (12, 20, 25)\ [+]\ (17, 6)$
$= (29, 37, 42)\ [+]\ (0, 6).$

The outcome can be written in the standard deviation of

$$\sqrt{6} = 2.449.$$

Therefore the cost is a fuzzy number with random variables with distribution.

$$(29 - \sqrt{6}, 37 - \sqrt{6}, 42 - \sqrt{6}) \leq (a_1, a_2, a_3) \leq (29 + \sqrt{6}, 37 + \sqrt{6}, 42 + \sqrt{6}),$$

or

$$(26.551, 34.551, 39.551) \leq (a_1, a_2, a_3) \leq (31.449, 39.449, 44.449).$$

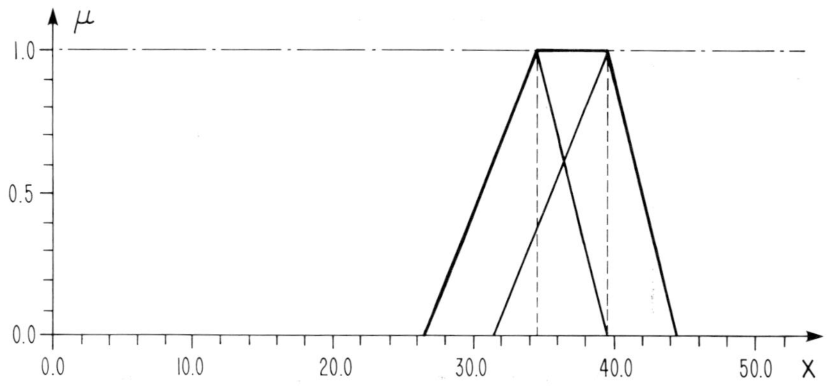

Figure 2.22 Fuzzy distribution of the cost of optimum path (Example 2.12).

Figure 2.22 shows the result, which can be represented by a trapezoidal fuzzy number that is the envelope of the sheaf, where the randomness is given by mean and standard deviation as follows:

$$(26.551, 34.551, 39.449, 44.449)$$

As we have seen, at the level $\alpha = 1$ of presumption, the interval of confidence when randomness is given by the mean and standard deviation is [34.551, 39.449], and at the level 0 it is [26.551, 44.449].

The process we have proposed seems to be convenient for approximate optimization when fuzzy and random data are involved. Of course, many other variations can be proposed. For instance, to look for the maximum (or minimum), induce a total order by the distance of each $\mathbf{A}_i$ [+] $\xi(L_i)$ to its maximum (or minimum). The maximum is, in fact, a supremum in this case and minimum is an infermum.

2.6 SIMULATION OF HYBRID DATA

Before giving detailed descriptions of how to simulate hybrid data, we give a very simple example that shows what types of mistakes must be avoided. First, let us consider two random numbers with equiprobability law; that is, the same value of probability density as represented in Figures 2.23(*A*) and 2.23(*B*). Their convolution sum-product for $X = X_1 + X_2$ will give a random variable where the shape of the density will be triangular instead of a constant [Fig. 2.23(C)]. If we carry on the computation with more than two random variables $X_1 + X_2 + \cdots + X_n$, the sum will achieve a Gaussian shape that obeys the Laplace-Gauss central limit theorem.

On the other hand, if we have two fuzzy numbers whose member-

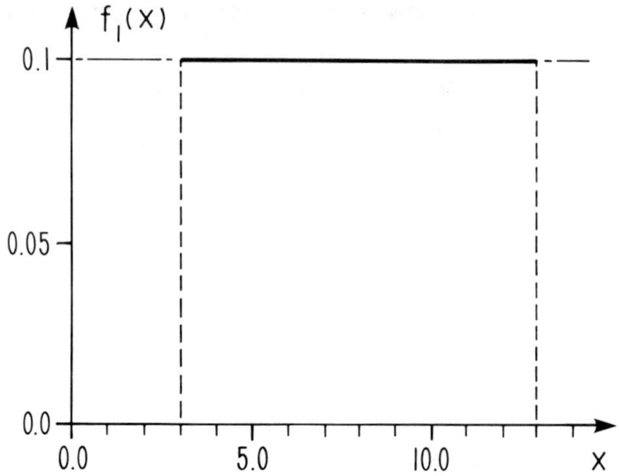

Figure 2.23(A) Random variable X_1 with constant probability distribution $f_1(x)$.

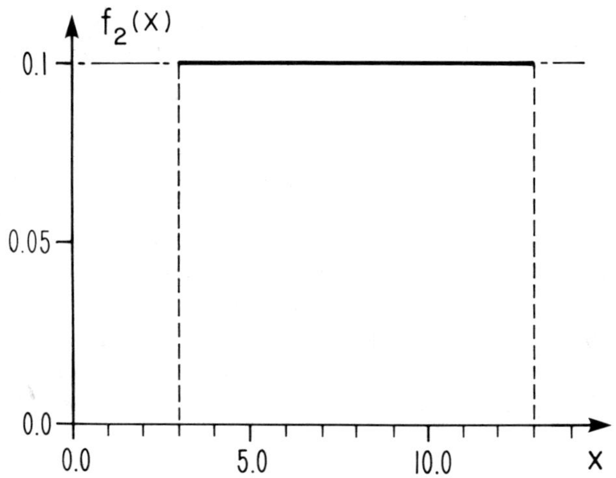

Figure 2.23(B) Random variable X_2 with constant probability distribution $f_2(x)$.

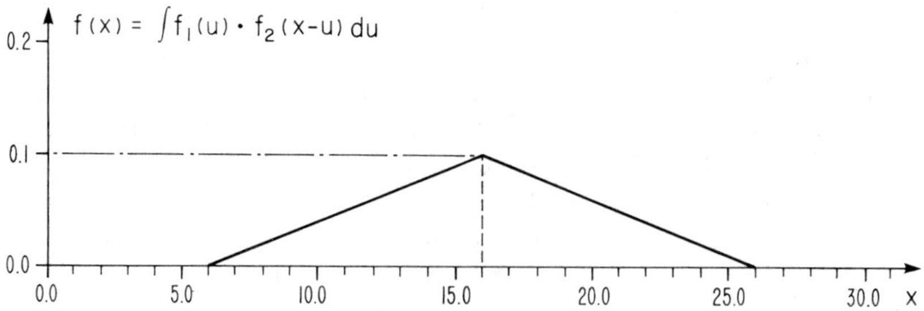

Figure 2.23(C) Probability distribution of $X = X_1 + X_2$.

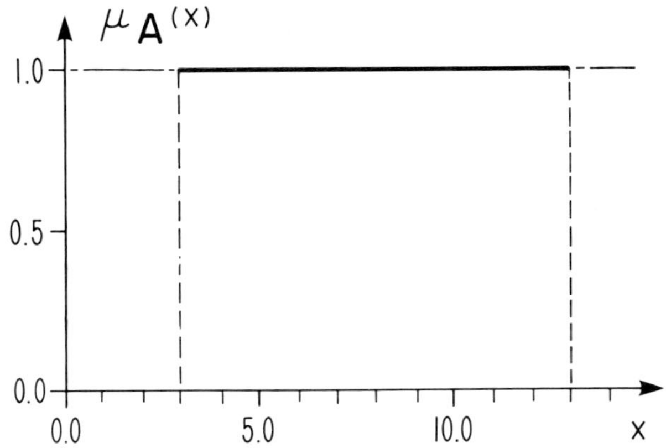

Figure 2.24(A) Fuzzy number **A** with a constant distribution.

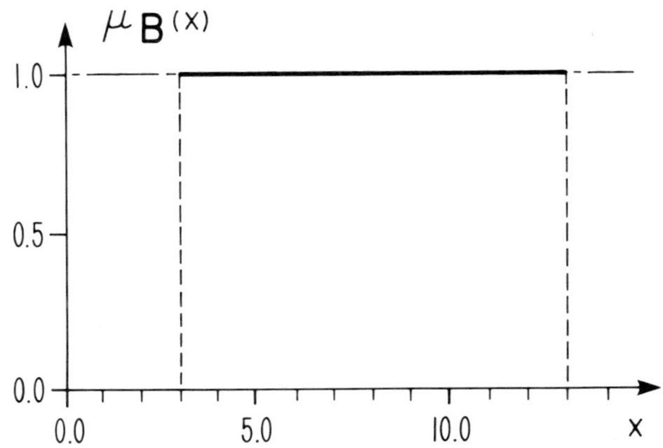

Figure 2.24(B) Fuzzy number **B** with a constant distribution.

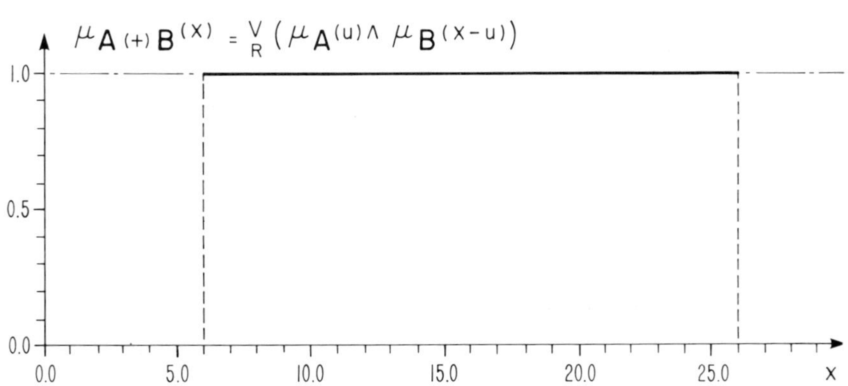

Figure 2.24(C) Max-min convolution of **A** and **B**.

ship functions are constant, as shown in Figures 2.24(*A*) and 2.24(*B*) and lies in the range of our gusses. Then their max-min convolution for addition [see Fig. 2.24(C)], will give a membership function that is also a constant. Similarly, with *n* fuzzy numbers of the same kind, $\mathbf{A}_1 \ (+) \ \mathbf{A}_2 \ (+) \cdots (+) \ \mathbf{A}_n$, we obtain a membership function that stays constant.

We have discussed this example so that we may recall the importance of being careful when we consider uncertainty. Laplace said "in the lack of knowledge for a probability law, I consider each possible event with the same probability." These words certainly apply here as well, and we treat the uncertainty like equiprobable randomness. This hypothesis is, of course, limited to a few cases. In fact, uncertainty is generally a concept associated with the interval of confidence or fuzziness. Uncertainty in the sense of Laplace and uncertainty in the sense of interval of confidence are different concepts. Mathematical operations for these two different concepts are not the same.

We see then, that fuzzy numbers cannot be simulated like random variables. Nevertheless, fuzzy subsets can be simulated with the use of random numbers for union, intersection, complementation, and any Boolean operation extended to membership functions, but not through max-min convolution. But, as we will see later, the simulation of fuzzy numbers and hybrid numbers is possible and useful.

To begin, let us see how to simulate an hybrid number using a Monte Carlo method. Figure 2.25 shows how to find a sample from a table of equiprobable random numbers that belong to [0, 1]. We assume that the law to be simulated is also equiprobable. From

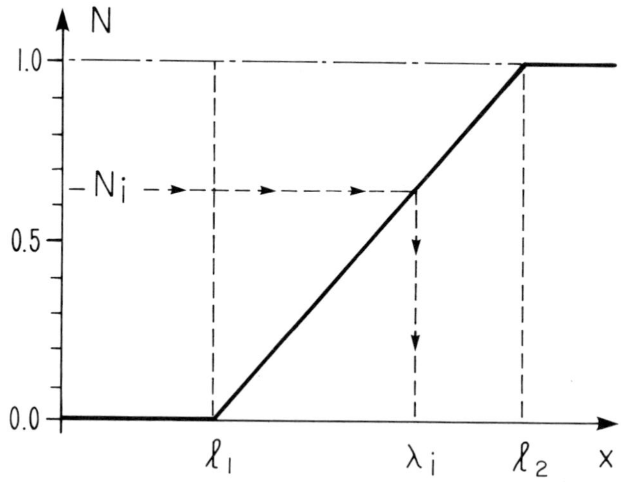

Figure 2.25 Generation of random $N_i \in [\ell_1, \ell_2]$.

any number N_i from the table we obtain a number $\lambda_i \in [\ell_1, \ell_2] \in R$ in such a way that any λ_i will have the same probability as any other. From this result, with the aid of an example, we can show how to add a fuzzy number to a random variable with equiprobable density. We note in Figure 2.25 that we can obtain λ_i using the equation

$$\lambda_i = \ell_1 + N_i(\ell_2 - \ell_1).$$

EXAMPLE 2.13

Let the TFN **A**

$\forall\, x \in R$:

$$\begin{aligned} \mu_{\mathbf{A}}(x) &= 0, & x &\le 2,\\ &= x/3 - 2/3, & 2 \le x &\le 5,\\ &= -x/2 + 7/2, & 5 \le x &\le 7,\\ &= 0, & x &\ge 7. \end{aligned} \tag{2.19}$$

On the other hand, consider a random variable with a density function

$$\begin{aligned} f(\ell) &= 0, & \ell &\le 4,\\ &= 1/5, & 4 \le \ell &\le 9,\\ &= 0, & \ell &> 9. \end{aligned} \tag{2.20}$$

A random sample of (2.20) may be obtained with the equation $\lambda_i = 4 + 5N_i$.

Table 2.11 is a table of equiprobable five-digit random numbers, similar to those published by the Rand Corporation. A small sample of random numbers is given in column 1, with corresponding λ_i, in column 2.

If we add a random number taken from column 2 in Table 2.11 to the fuzzy number **A** given by 2.19, we obtain a simulation sample of a random sheaf given by **A** [+] L.

Let the first random number from Table 2.11 be 5.17290. We then have

$\forall\, x \in R$:

$$\begin{aligned} \mu_{\mathbf{A}\,(+)\,\lambda_1}(x) &= 0, & x &\le 7.17290,\\ &= (x - 5.17290)/3 - 2/3, & 7.17290 \le x &\le 10.17290,\\ &= (-x + 5.17290)/2 + 7/2, & 10.17290 \le x &\le 12.17290,\\ &= 0, & x &\ge 12.17290, \text{ etc.} \end{aligned}$$

Figure 2.26 shows one sample of the sheaf for some random numbers from the population.

Next we give an example that illustrates another kind of fuzzy number and a Laplace-Gauss probability law.

Table 2.11 Equiprobable Five-Digit Random Numbers

(1)	(2)	(1)	(2)	(1)	(2)
$10^5 \cdot N_i$	λ_i	$10^5 \cdot N_i$	λ_i	$10^5 \cdot N_i$	λ_i
23458	5.17290	46191	6.30955	25486	5.27430
57782	6.88910	12805	4.64025	09306	4.46530
67537	7.37685	50502	6.52510	24721	5.23605
38813	5.94065	08810	4.44050	04187	4.20935
00377	4.01885	19572	4.97860	62361	7.11805
93873	8.69365	48024	6.80120	59105	6.95525
97813	8.89065	58206	6.91020	39338	5.96690
10039	4.50195	52251	6.61255	59358	6.96790
25457	5.27285	06804	4.34020	—	—
28716	5.43580	83959	8.19795	—	—
03954	4.19770	20974	5.04870		
14799	4.73995	73104	7.65520		
63187	7.15935	15009	4.75045		

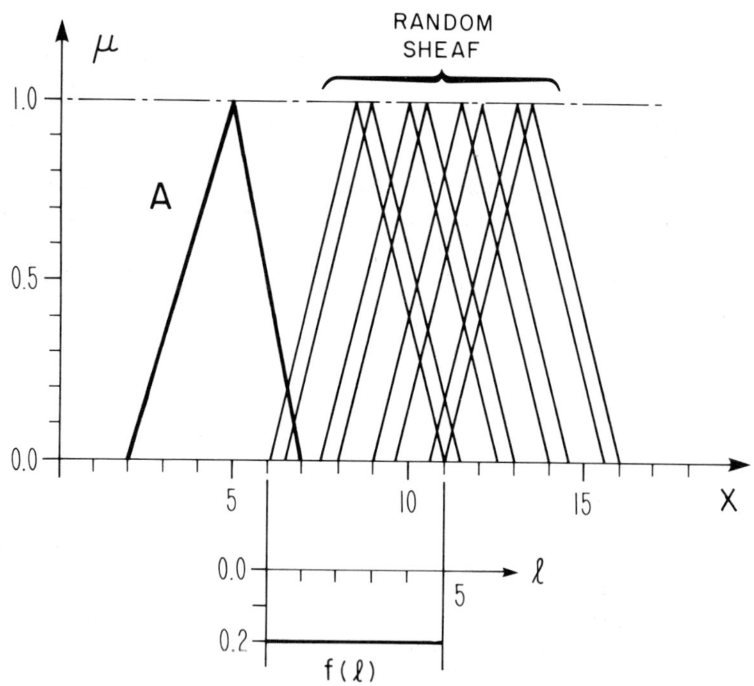

Figure 2.26 A sample of sheaf from random numbers (Example 2.13).

EXAMPLE 2.14

Consider the fuzzy number shown in Figure 2.27.
$\forall\ x \in R$:

$$\begin{aligned} \mu_{\mathbf{A}}(x) &= 0, & x \le 2, \\ &= 1 - [(x-5)/3)]^2, & 2 \le x \le 8, \\ &= 0, & x \ge 8. \end{aligned}$$

At the level α we have

$$x = 5 \pm 3\sqrt{1-\alpha}.$$

Hence

$\forall\ \alpha \in [0, 1]$:

$$A_\alpha = [5 - 3\sqrt{1-\alpha}, 5 + 3\sqrt{1-\alpha}].$$

Figure 2.28 shows the Laplace-Gaussian distribution where $\ell = 0$ and $\sigma_L = 1$; that is

$$F(x) = \left(\frac{1}{\sqrt{2\pi}}\right) \int_{-\infty}^{\infty} \exp\left(\frac{-u^2}{2}\right) du, \tag{2.21}$$

which will be used to provide a sample of this law.

Consider the following sequence:

$$\begin{gathered} 0.308, 2.537, 1.220, -1.250, \ldots, \\ 1.464, -1.210, -0.957, 0.265, 0.724, 0.555, \ldots, \end{gathered} \tag{2.22}$$

which is completed using Equation (2.21).

The sample from (2.22) coresponds to the sample from A_α [+] L:

$$\begin{gathered} [5.308 - 3\sqrt{1-\alpha}, 5.308 + 3\sqrt{1-\alpha}] \\ [7.537 - 3\sqrt{1-\alpha}, 7.537 + 3\sqrt{1-\alpha}] \\ [6.220 - 3\sqrt{1-\alpha}, 6.220 + 3\sqrt{1-\alpha}] \\ [3.750 - 3\sqrt{1-\alpha}, 3.750 + 3\sqrt{1-\alpha}] \\ [6.464 - 3\sqrt{1-\alpha}, 6.464 + 3\sqrt{1-\alpha}] \ldots \text{etc.} \end{gathered}$$

This example has been chosen to show how simulations may be obtained from intervals of confidence at each level α.

Of course, hybrid simulations, like random simulations, must be reserved for cases where analytical methods cannot be found or are too difficult to use.

Combining the theory of fuzzy numbers with simulation methods using hybrid numbers is a good method for solving many problems in system theory and operations research.

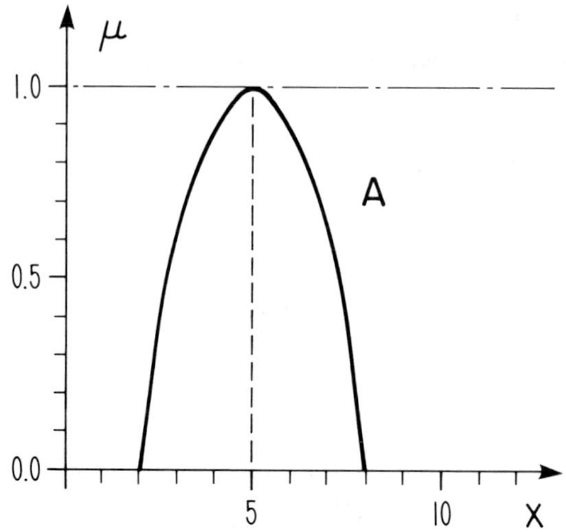

Figure 2.27 Fuzzy number (Example 2.14).

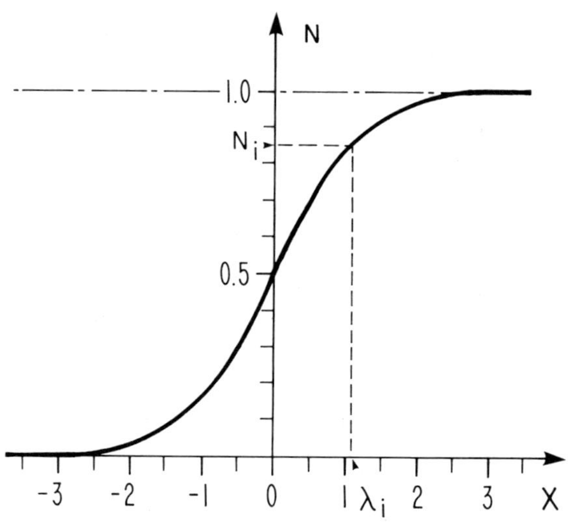

Figure 2.28 Laplace-Gaussian distribution for random number N_1 (Example 2.14).

2.7 INTERVAL OF CONFIDENCE WITH FUZZY NUMBERS OF TYPE 2

In section 1.2 we used the concept of the interval of confidence as the starting point for the theory of fuzzy numbers. We now begin this section by introducing the concept of the interval of confidence of type 2.

Let us assume that the lower and upper bounds of an interval of confidence, instead of being ordinary numbers, are fuzzy numbers that themselves have intervals of confidence. That is,

$$\mathcal{A} = [[a_1, a'_1], [a_2, a'_2]].$$

This concept is even more uncertain than that specified in section 1.2 because the bounds are uncertain. In discussing some properties that are concerned with such an extension of the interval of confidence, we must note that if an interval of confidence in R is a subset of R it is not a subset of $\mathcal{A}$. Instead, it will be a subset with an uncertain border.

We assert, for each $\mathcal{A}$,

$\forall\ a_1, a'_1, a'_2, a'_2 \in R$:

$$a_1 \leq a'_1 \leq a_2 \leq a'_2.$$

When $a_1 = a'_1$ and $a_2 = a'_2$, the interval of confidence of *type* 2 becomes an interval of *type* 1, as discussed in section 1.2. Now, if $a_1 = a'_1 = a_2 = a'_2$, we obtain an interval of confidence of *type* 0; that is, an ordinary number.

We now discuss some properties of intervals of confidence of *type* 2.

$$\mathcal{A} = [[a_1, a'_1], [a_2, a'_2]],$$

$$\mathcal{B} = [[b_1, b'_1], [b_2, b'_2]],$$

Addition is defined as

$$\begin{aligned}\mathcal{A}\ (+)\ \mathcal{B} &= [[a_1, a'_1]\ (+)\ [b_1, b'_1]], [a_2, a'_2]\ (+)\ [b_2, b'_2]] \\ &= [[a_1 + b_1, a'_1 + b'_1], [a_2 + b_2, a'_2 + b'_2]].\end{aligned}$$

It is obvious that this operation is associative and commutative. As a result, a neutral exists.

$$[[0, 0], [0, 0]] = [0, 0] = 0,$$

hence the set of intervals of confidence of *type* 2 have a commutative monoidal structure for addition.

We now examine subtraction.

$$\begin{aligned}\mathcal{A}\ (-)\ \mathcal{B} &= [[a_1, a'_1], [a_2, a'_2]]\ (-)\ [[b_1, b'_1], [b_2, b'_2]] \\ &= [[a_1, a'_1]\ (-)\ [b_2, b'_2]], [a_2, a'_2]\ (-)\ [b_1, b'_1] \\ &= [[a_1 - b'_2, a'_1 - b_2], [a_2 - b'_1, a'_2 - b_1]].\end{aligned}$$

Note that

$$a_1 - b'_2 \leq a'_1 - b_2 \leq a_2 - b'_1 \leq a'_2 - b_1. \tag{2.23}$$

Since, by hypothesis,

$$a_1 \leq a'_1 \leq a_2 \leq a'_2,$$

and

$$b_1 \leq b'_1 \leq b_2 \leq b'_2,$$

and from this

$$(a_1 \leq a'_1 \text{ and } b_2 \leq b'_2) \Rightarrow (a_1 - b'_2 \leq a'_1 - b_2).$$

We can now define the *opposite (image)* for *type 2* fuzzy numbers.

$$\begin{aligned} \mathcal{A}^- &= [[0, 0], [0, 0] (-) [[a_1, a'_1], [a_2, a'_2] \\ &= [[-a'_2, -a_2], [-a'_1, -a_1]]. \end{aligned}$$

The addition of $\mathcal{A}$ with the *image* $\mathcal{A}^-$ becomes

$$\mathcal{A}\ (+)\ \mathcal{A}^- = [[a_1 - a'_2, a'_1 - a_2], [a_2 - a'_1, a'_2 - a_1]] \neq 0.$$

From the above we see that this monoid is not a group.

Now let us investigate *multiplication* (in R^+ for simplicity).

$$\begin{aligned} \mathcal{A}\ (\cdot)\ \mathcal{B} &= [[a_1, a'_1], [a_2, a'_2]] (\cdot) [[b_1, b'_1], [[b_2, b'_2,]] \\ &= [[a_1, a'_1] (\cdot) [b_1, b'_1], [a_2, a'_2] (\cdot) [b_2, b'_2]] \\ &= [[a_1 b_1, a'_1 b'_1], [a_2 b_2, a'_2 b'_2]]. \end{aligned}$$

It is easy to check associativity and commutativity in R^+. The neutral is as follows:

$$[[1, 1], [1, 1]] = [1, 1] = 1.$$

Hence in R^+ we find a commutative monoid for multiplication.

Division is defined by

$$\begin{aligned} \mathcal{A}\ (:)\ \mathcal{B} &= [[a_1, a'_1], [a_2, a'_2]] (:) [[b_1, b'_1], [b_2, b'_2]] \\ &= [[a_1, a'_1] (:) [b_2, b'_2], [a_2, a'_2] (:) [b_1, b'_1] \\ &= [[a_1/b'_2, a'_1/b_2], [a_2/b'_1, a'_2/b_1]], \end{aligned}$$

with

$$b_1, b'_1, b_2, b'_2 > 0.$$

It is easy to prove that

$$a_1/b'_2 \leq a'_1/b_2 \leq a_2/b'_1 \leq a'_2/b_1.$$

The *inverse* of $\mathcal{A}$ is defined as

$$\begin{aligned} \mathcal{A}^{-1} &= [[1, 1], [1, 1]] (:) [[a_1, a'_1], [a_2, a'_2] \\ &= [[1/a'_2, 1/a_2], [1/a'_1, 1/a_1], \end{aligned}$$

with

$$a_1, a'_1, b_1, b'_1 > 0.$$

Note that

$$\mathcal{A}\ (\cdot)\ \mathcal{A}^{-1} \neq 1.$$

Multiplication by an interval K of *type* 1 gives

$$\begin{aligned} K\ (\cdot)\ \mathcal{A} &= [k_1, k_2]\ (\cdot)\ [[a_1, a'_1], [a_2, a'_2] \\ &= [[k_1 \cdot a_1, k_1 \cdot a'_1], [k_2 \cdot a_2, k_2 \cdot a'_2]. \end{aligned}$$

Multiplication by an interval of K *type* 0 gives

$$\begin{aligned} K \cdot \mathcal{A} &= [k, k]\ (\cdot)\ [[a_1, a'_1], [a_2, a'_2]] \\ &= [[ka_1, ka'_1], [ka_2, ka'_2]]. \end{aligned}$$

We now consider the minimum and maximum in R.

$$\mathcal{A}\ (\wedge)\ \mathcal{B} = [[a_1 \wedge b_1, a'_1 \wedge b'_1], [a_2 \wedge b_2, a'_2 \wedge b'_2]],$$

$$\mathcal{A}\ (\vee)\ \mathcal{B} = [[a_1 \vee b_1, a'_1 \vee b'_1], [a_2 \vee b_2, a'_2 \vee b'_2]].$$

Note that a finite neutral does not exist. Associativity and commutativity properties are applicable for both $(\wedge)$ and $(\vee)$ operations.

Fuzzy Numbers of Type 2

Let us consider a sequence of intervals of confidence of *type* 2 that depends on α (see Fig. 2.29).

$\forall\ \alpha \in [0, 1],\ \forall\ a_1^{(\alpha)}, a'^{(\alpha)}_1, a_2^{(\alpha)}, a'^{(\alpha)}_2 \in R$:

$$\mathcal{A}_\alpha = [[a_1^{(\alpha)}, a'^{(\alpha)}_1], [a_2^{(\alpha)}, a'^{(\alpha)}_2]], \tag{2.24}$$

with the following constraints:

1. $a_1^{(\alpha)} \leq a'^{(\alpha)}_1 \leq a_2^{(\alpha)} \leq a'^{(\alpha)}_2$, **(2.25)**

2. $[a_1^{(\alpha)}, a'^{(\alpha)}_2]$ and $[a'^{(\alpha)}_1, a_2^{(\alpha)}]$, **(2.26)**

generate a membership function that is convex, and $[a_1^{(\alpha)}, a'^{(\alpha)}_2]$, generates a membership function that is normal.

3. $\forall\ \alpha, \alpha' \in [0, 1]$:

$$(\alpha' > \alpha) \Rightarrow ([a_1^{(\alpha)}, a_2^{(\alpha)}] \supset [a_1^{(\alpha')}, a_2^{(\alpha')}],$$
$$[a'^{(\alpha)}_1, a'^{(\alpha)}_2] \supset [a'^{(\alpha')}_1, a'^{(\alpha')}_2], \qquad \text{for } a_2^{(\alpha')} \geq a'^{(\alpha')}_1). \tag{2.27}$$

4. If the maximum of the membership function generated by $[a'^{(\alpha)}_1, a_2^{(\alpha)}]$ is the level α_m, that is, $[a'^{(\alpha_m)}_1, a_2^{(\alpha_m)}]$, then

$$[a'^{(\alpha_m)}_1, a_2^{(\alpha_m)}] \subset [a_1^{(\alpha = 1)}, a'^{(\alpha = 1)}_2]. \tag{2.28}$$

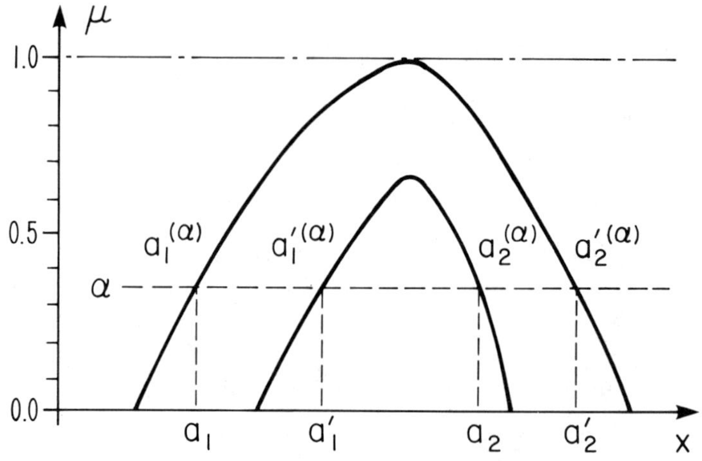

Figure 2.29 A fuzzy number of *type* 2.

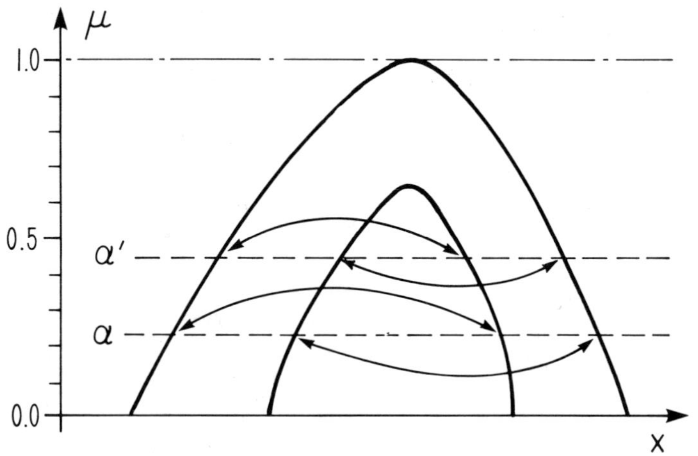

Figure 2.30 Property of Eq. (2.27) for fuzzy numbers of type 2.

In this case the sequence of intervals of confidence of *type* 2 describes a fuzzy number of *type* 2, as shown in Figure 2.29. Figure 2.30 illustrates the property (2.27).

EXAMPLE 2.15

Consider a fuzzy number **A** and a convex fuzzy subset **B**, as shown in Figure 2.31. We can use these two fuzzy subsets to construct a fuzzy number of *type* 2.

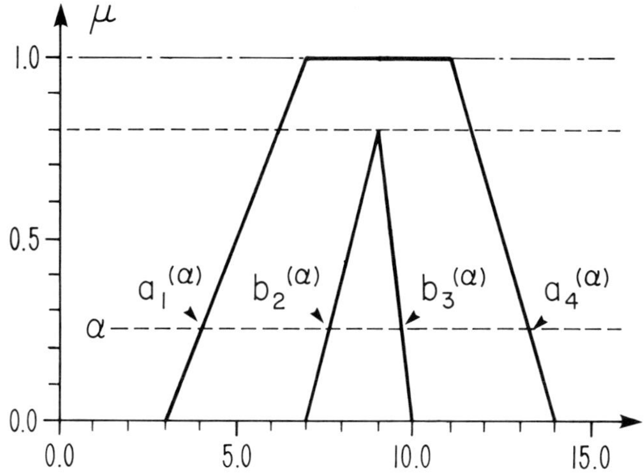

Figure 2.31 Construction of a fuzzy number of *type* 2, **A** and **B** (Example 2.15).

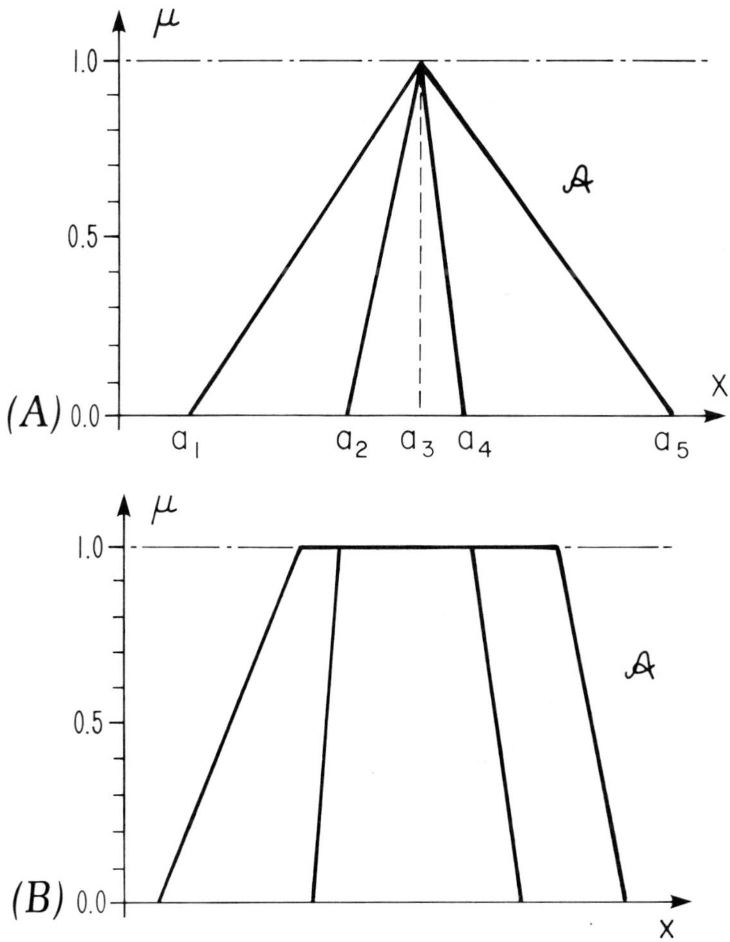

Figure 2.32 Examples of fuzzy number of *type* 2.

$\forall\, x \in R$:

$$\begin{aligned} \mu_{\mathbf{A}}(x) &= 0, & x &\le 3,\\ &= x/4 - 3/4, & 3 \le x &\le 7,\\ &= 1, & 7 \le x &\le 11,\\ &= -x/3 + 14/3, & 11 \le x &\le 14,\\ &= 0, & x &\ge 14. \end{aligned} \quad \textbf{(2.29)}$$

$$\begin{aligned} \mu_{\mathbf{B}}(x) &= 0, & x &\le 7,\\ &= 2x/5 - 14/5, & 7 \le x &\le 9,\\ &= -4x/5 + 8, & 9 \le x &\le 10,\\ &= 0, & x &\ge 10. \end{aligned} \quad \textbf{(2.30)}$$

The corresponding intervals of confidence will be, using the letters and indexes shown in Figure 2.31,

$$[a_1^{(\alpha)}, a_4^{(\alpha)}] = [3 + 4\alpha, 14 - 3\alpha], \qquad \alpha \in [0,1], \quad \textbf{(2.31)}$$

$$[b_2^{(\alpha)}, b_3^{(\alpha)}] = [7 + 5\alpha/2, 10 - 5\alpha/4], \qquad \alpha \in [0, 0.8]. \quad \textbf{(2.32)}$$

A fuzzy number of type 2 is then constructed as follows:

$$\mathcal{A}_\alpha = [[3 + 4\alpha, 7 + 5\alpha/2], [10 - \frac{5}{4}\alpha, 14 - 3\alpha]], \quad \text{for } \alpha \in [0, 0.8]$$

$$= [3 + 4\alpha, 14 - 3\alpha], \quad \alpha \in [0.8, 1].$$

It is easy to check constraints (2.25-2.28); this is left to the reader.

Figures 2.32(*A*) and 2.32(*B*) show some particular cases of fuzzy numbers of *type* 2.

EXAMPLE 2.16

In this example we examine the process of addition of two fuzzy numbers of *type* 2, as shown in Figure 2.33.

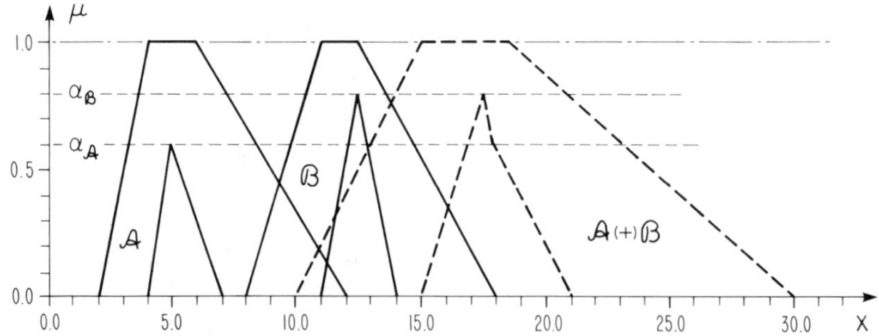

Figure 2.33 Addition of two fuzzy numbers of type 2 (Example 2.16).

$\forall\ \alpha \in [0, 1]$:

$$\mathcal{A}_\alpha\,(+)\,\mathcal{B}_\alpha = [[a_1^{(\alpha)}, a_2^{(\alpha)}], [a_3^{(\alpha)}, a_4^{(\alpha)}]]\,(+)\,[[b_1^{(\alpha)}, b_2^{(\alpha)}], [b_3^{(\alpha)}, b_4^{(\alpha)}]]$$

$$= [[a_1^{(\alpha)} + b_1^{(\alpha)}, a_2^{(\alpha)} + b_2^{(\alpha)}]\,[a_3^{(\alpha)} + b_3^{(\alpha)}, a_4^{(\alpha)} + b_4^{(\alpha)}]] \qquad 0 \le \alpha \le \alpha_{\mathcal{A}}$$

$$= [[a_1^{(\alpha)} + b_1^{(\alpha)}, a_2^{(\alpha_{\mathcal{A}})} + b_2^{(\alpha)}]\,[a_2^{(\alpha_{\mathcal{A}})}\,b_3^{(\alpha)}, a_4^{(\alpha)} + b_4^{(\alpha)}]], \qquad \alpha_{\mathcal{A}} \le \alpha \le \alpha_{\mathcal{B}}$$

$$= [[a_1^{(\alpha)} + b_1^{(\alpha)}, a_2^{(\alpha_{\mathcal{B}})} + b_2^{(\alpha_{\mathcal{B}})}]\,[a_2^{(\alpha_{\mathcal{B}})} + b_2^{(\alpha_{\mathcal{B}})}, a_4^{(\alpha)} + b_4^{(\alpha)}$$

$$= [a_1^{(\alpha)} + b_1^{(\alpha)}, a_4^{(\alpha)} + b_4^{(\alpha)}], \qquad \alpha_{\mathcal{B}} \le \alpha \le 1. \qquad \textbf{(2.33)}$$

Remarks

As was seen in Figure 2.33, the triangular shape may be lost after addition for the interval of $[a_2^{(\alpha)} + b_2^{(\alpha)}, a_3^{(\alpha)} + b_3^{(\alpha)}]$. This situation does not occur for the cases presented in Figures 2.32(A) and 2.32(B).

A fuzzy number of *type* 2, like the one in Figure 2.32(A) is said to be perfectly triangular, and one like that in Figure 2.32(B), perfectly trapezoidal. The first one is a special case of the second.

A fuzzy number of *type* 2 that is perfectly triangular can be written using a quintuplet, as follows.

$$\mathcal{A} = (a_1, a_2, a_3, a_4, a_5), \qquad a_1 \le a_2 \le a_3 \le a_4 \le a_5, \qquad \textbf{(2.34)}$$

where

$$a_1 = a_1^{(\alpha = 0)},$$
$$a_2 = a_2^{(\alpha = 0)},$$
$$a_3 = a_1^{(\alpha = 1)} = a_2^{(\alpha = 1)} = a_3^{(\alpha = 1)} = a_4^{(\alpha = 1)},$$
$$a_4 = a_3^{(\alpha = 0)},$$
$$a_5 = a_4^{(\alpha = 0)},$$

with

$$\mathcal{A}_\alpha = [[a_1^{(\alpha)}, a_2^{(\alpha)}], [a_3^{(\alpha)}, a_4^{(\alpha)}]]. \qquad \textbf{(2.35)}$$

The following discussion shows why we must be very careful when interchanging indexes. The trapezoidal case introduces a notation with eight numbers. For a perfectly triangular fuzzy number of *type* 2, it is necessary to have

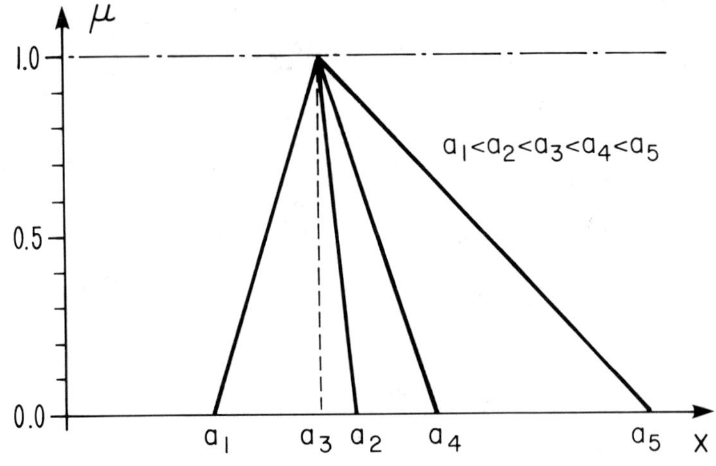

Figure 2.34 An example of a triangular number that is not a fuzzy number of type 2.

$$a_1 \leq a_2 \leq a_3 \leq a_4 \leq a_5, \tag{2.36}$$

since otherwise it will not be possible to satisfy constraints (2.24–2.28). Figure 2.34 shows an example of a fuzzy number that is not a fuzzy number of type 2. Let us consider the triangle with base $[a_2, a_4]$; since $a_3 \leq a_2$, this does not represent a fuzzy number, or even a fuzzy subset. For a TFN of type 2 of perfectly triangular nature to be specified, conditions (2.24–2.36) must be specified. For addition,

$$\mathcal{A}\ (+)\ \mathcal{B} = (a_1 + b_1, a_2 + b_2, a_3 + b_3, a_4 + b_4, a_5 + b_5).$$

Remark

Let us consider the interval of confidence of type 2 at level α:

$$A_\alpha = [[a_1^{(\alpha)}, a_2^{(\alpha)}], [a_3^{(\alpha)}, a_4^{(\alpha)}]].$$

Figure 2.35 shows five zones, namely,

1. $x \leq a_1^{(\alpha)}$.

This domain corresponds to a zero degree of uncertainty, while the following domain corresponds to one having a unity degree of uncertainty:

2. $a_1^{(\alpha)} \leq x \leq a_2^{(\alpha)}$.

Here the uncertainty is greater at its left but smaller at its right.

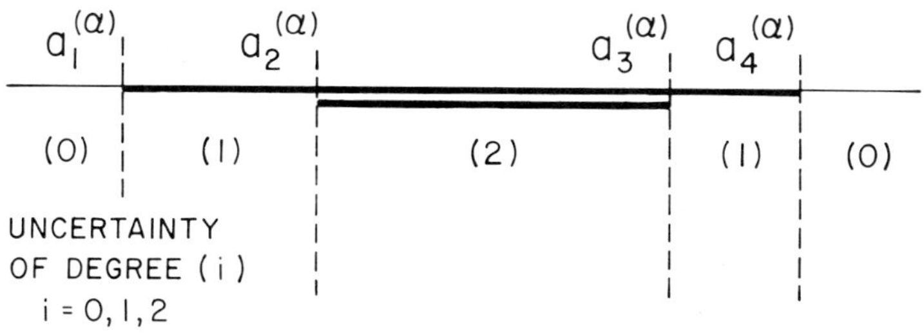

Figure 2.35 Type of uncertainties.

3. $a_2^{(\alpha)} \leq x \leq a_3^{(\alpha)}$.

This corresponds to a degree of uncertainty of two, which is the highest for this kind of concept.

4. $a_3^{(\alpha)} \leq x \leq a_4^{(\alpha)}$.

The degree of uncertainty is one here, while in

5. $a_4^{(\alpha)} \leq x$.

the degree is zero.

At each border, $a_1^{(\alpha)}$, $a_2^{(\alpha)}$, $a_3^{(\alpha)}$, and $a_4^{(\alpha)}$, the degree is ambiguous. When α increases, we pass from five zones to three.

Uncertainty Beyond Type 2

At this point we consider

$$[[[a_1^{(\alpha)}, a_2^{(\alpha)}], [a_3^{(\alpha)}, a_4^{(\alpha)}]], [[a_5^{(\alpha)}, a_6^{(\alpha)}] [a_7^{(\alpha)}, a_8^{(\alpha)}]]],$$

where, after stating some constraints, we are able to define a fuzzy number of *type* 3. In a similar way, we can postulate the concept of uncertainty for higher types of fuzzy numbers.

Φ-Fuzzy Subsets and Φ-Fuzzy Numbers

The concept of a fuzzy number of *type* 2 can be presented using an alternative approach. For this purpose let us recall the meaning of an Φ-fuzzy subset.[2] A fuzzy subset is characterized by the membership function

$\forall\, x \in E$:

$$\mu_{\mathbf{A}}(x) \in [0, 1];$$

which is an application of E in [0, 1].

A Φ-fuzzy subset is an application of E in $\mathcal{P}$ [0, 1], that is, a set of segments in [0, 1].

$\forall$ x $\in$ E:

$$\mu_{\mathbf{A}_\varphi}(x) = [a_1, a_2] \subset [0, 1].$$

In the above equation, the symbol φ has taken the place of the use of boldface so that we can distinguish between ordinary fuzzy subsets and Φ-subsets.

A general theory has been presented on Φ-fuzzy subsets[2] and in this book we consider only the special case of Φ-fuzzy numbers.

$\forall$ x $\in$ R:

$$\mu_{A_\varphi}(x) = [\mu_1(x), \mu_2(x)] \subset [0, 1]. \tag{2.37}$$

where $\mu_1(x)$ is the membership function of a fuzzy number in R and $\mu_2(x)$ is another fuzzy number in R.

For each Φ-fuzzy number (Fig. 2.36) we can write

$$A_\alpha = [[a_1^{(\alpha)}, a_2^{(\alpha)}], [a_3^{(\alpha)}, a_4^{(\alpha)}]].$$

If we take conditions (2.25-2.28), we again find the concept of a fuzzy number of *type* 2. Thus, from (2.37),

$$\mu_{A_\varphi}(x) = [\mu_{\mathbf{A}_1}(x), \mu_{\mathbf{A}_2}(x)],$$

which is shown in Figure 2.36, where $\mathbf{A}_1$ and $\mathbf{A}_2$ are fuzzy numbers in R with constraints (2.25-2.28). We are now able to write, for the operations that relate to A_φ and B_φ,

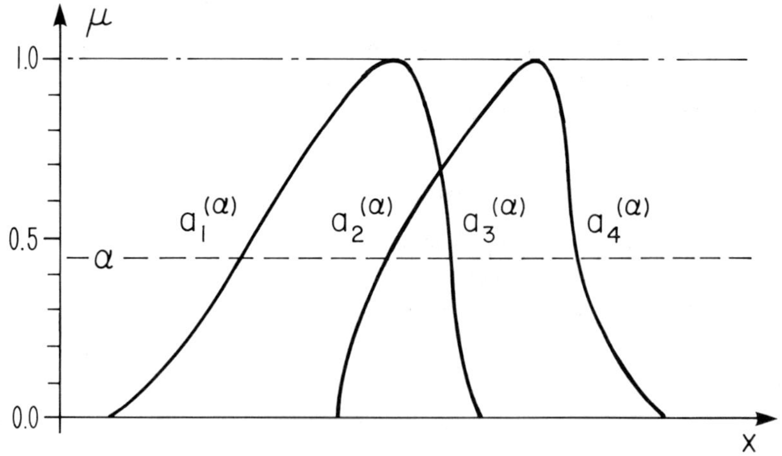

Figure 2.36 Φ-fuzzy subset.

$\forall$ x, y, z $\in$ R:

$$\mu_{A_\varphi (*) B_\varphi}(z) = [\mu_{\mathbf{A}_1 (*) \mathbf{B}_1}(z), \mu_{\mathbf{A}_2 (*) \mathbf{B}_2}(z)]$$

$$= [\bigvee_{z = x * y} (\mu_{\mathbf{A}_1}(x) \wedge \mu_{\mathbf{B}_1}(y), \bigvee_{z = x * y} (\mu_{\mathbf{A}_2}(x) \wedge \mu_{\mathbf{B}_2}(y)]$$

for every operation $(*)$ that satisfies constraints (2.25) to (2.28).

2.8 RANDOM FUZZY NUMBERS

Let us consider a fuzzy number of *type* 2.

$\forall\ \alpha \in [0, 1]$:

$$\mathcal{A}_\alpha = [[a_1^{(\alpha)}, a_2^{(\alpha)}], [a_3^{(\alpha)}, a_4^{(\alpha)}]]. \tag{2.38}$$

Now let us assign to each segment $[a_1^{(\alpha)}, a_2^{(\alpha)}]$ and $[a_3^{(\alpha)}, a_4^{(\alpha)}]$ a probability law $f_L(\alpha, x)$ and $f_R(\alpha, x)$, respectively. Therefore, in the interval of confidence (2.38), the lower and upper bounds become random variables. Figure 2.37 shows such a concept. Outside each interval the probability density is null. On the other hand, we can consider a fuzzy number of *type* 2 to be similar to a Φ-fuzzy number created by the two fuzzy numbers $\mathbf{A}_1$ and $\mathbf{A}_2$, which are unimodal with the same mode, as shown in Figure 2.37. In this case we can define a complementary cumulative law of probability at the right $H_R(\alpha, x)$, which has the value 1 from $a_2^{(\alpha)}$ to $a_3^{(\alpha)}$ and which is monotonically decreasing $a_3^{(\alpha)}$ to $a_4^{(\alpha)}$ and null after $a_4^{(\alpha)}$. In the same way we can define a complementary law of probability towards negative x at the left $H_L(\alpha, x)$, which has the value from $a_3^{(\alpha)}$ to $a_2^{(\alpha)}$ and is monotonically decreasing from $a_2^{(\alpha)}$ to $a_1^{(\alpha)}$ and null after $a_1^{(\alpha)}$. These hypotheses are shown in Figure 2.37.

In examining how the concept of convexity can be considered, we consider that all the bounds $a_1^{(\alpha)}$, $a_2^{(\alpha)}$, $a_3^{(\alpha)}$, $a_4^{(\alpha)}$ satisfy conditions (2.25–2.28). We then impose

$\forall\ \alpha', \alpha\ [0, 1]$, $\forall$ x $\in$ R:

$$(\alpha' > \alpha) \Rightarrow (H_R(\alpha', x) \le H_R(\alpha, x))$$
$$\Rightarrow (H_L(\alpha', -x) \le H_L(\alpha, -x)). \tag{2.39}$$

Equation (2.39) defines the convexity for random fuzzy numbers.

To make this clearer, we consider the following example.

EXAMPLE 2.17

Figure 2.38 shows a random fuzzy number. The probability that the lower bound or upper bound is located there is given in a square below

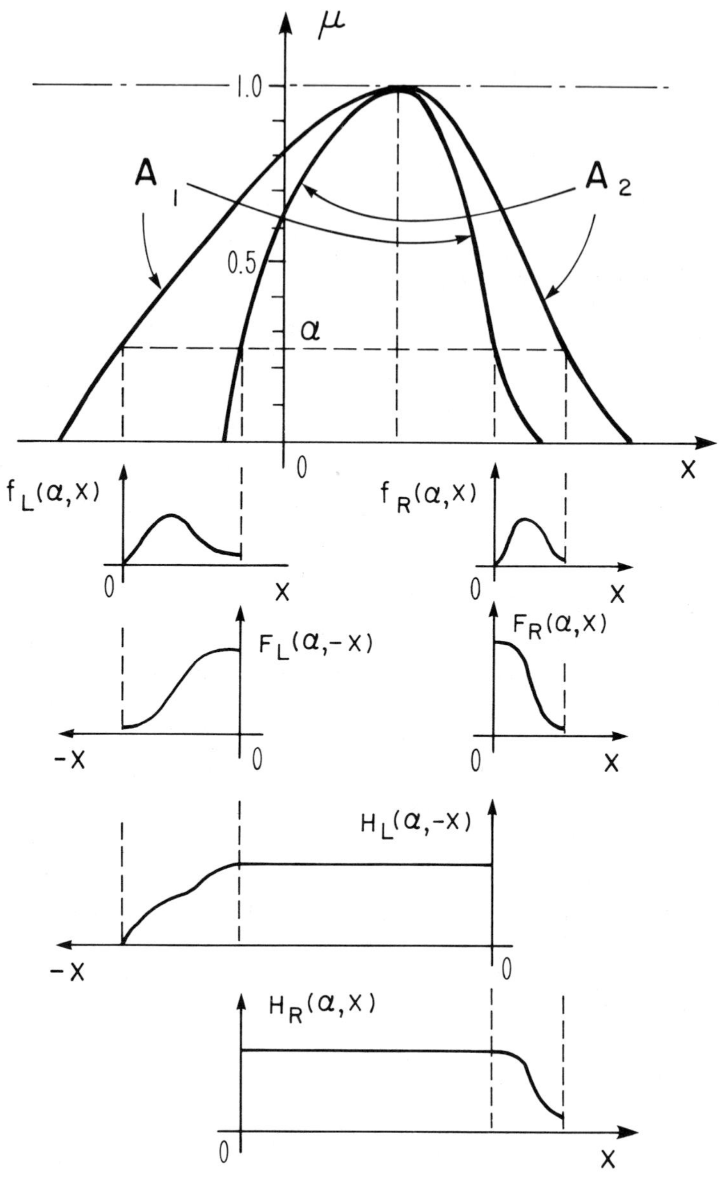

Figure 2.37 Concept of a random fuzzy number.

each individual diagonal. The cumulative probability is given above the small diagonal. Figure 2.38 shows that a probabilistic convexity according to 2.39 is satisfied, as is that of a probabilistic normality (for $x = 6$, $\mu = 1$ for $\alpha = 0, 0.1, 0.2, \ldots, 0.9, 1$). The left cumulative

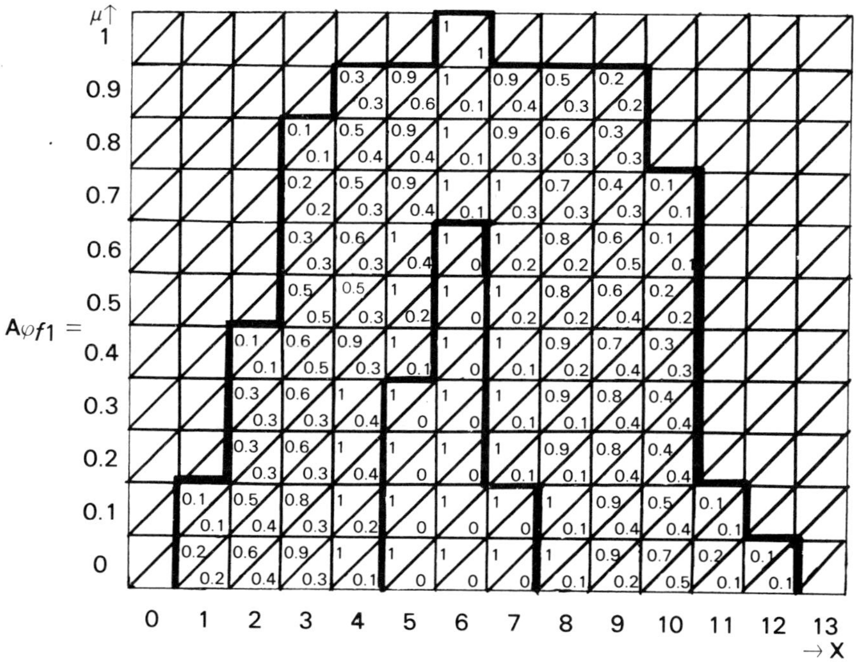

Figure 2.38 Random fuzzy number $\mathbf{A}_{\varphi f_1}$ (Example 2.17).

probability law is monotonically decreasing from right to left and the right cumulative probability law is monotonically decreasing from the left to right. For a given value of x, when the level of α increases the cumulative probability decreases.[3] For $\alpha = 1$, since we imposed an unimodial condition, both the cumulative probability and the noncumulative probability are equal to one.[4] The cumulative law at the left and at the right must be compatible; that is, in the central part where it is necessary to find a one for each level, the probability of x must be the same for the right law and the left law. For laws in R, densities for $\alpha < 1$ must be null. In our example, Figure 2.38, we are in N, and if we check, for example, at $x = 6$ and $\alpha = 0.9$, we find that pr(6) = 0.1 for the law at the right and pr(6) = 0.1 for the law at the left.

Let us now consider how to define an addition for two random fuzzy numbers. If we consider two segments (closed intervals), each having a probability law, the probability law of the sum of these segments for their bounds is obtained by the sum-product convolution and not by max-min convolution.

Let $A_{\varphi f_1}$ and $B_{\varphi f_2}$ be two random fuzzy number at the level α, where

$$A_{\varphi f_1}\alpha\ (+)\ B_{\varphi f_2}\alpha = [f_1 L_\alpha(x), f_1 R_\alpha(x)]\ (+)[f_2 L_\alpha(x), f_2 R_\alpha(x)]$$

$$= [\int_R f_1 L_\alpha(x-u) f_2 L_\alpha(u)\, du, \int_R f_1 R_\alpha(x-u) f_2 R_\alpha(u)\, du].$$

for a discrete case this becomes

$$\underset{pr_1}{A_\alpha}\ (+)\ \underset{pr_2}{B_\alpha} = [\sum_u p_{1L}(u-x)\, p_{2L}(u), \sum_u p_{1R}(u-x)\, p_{2R}(u)]. \quad \textbf{(2.40)}$$

We study addition by considering an example that uses (2.40).

EXAMPLE 2.18

Consider $A_{\varphi f_1}$ in Figure 2.38 and $B_{\varphi f_2}$ in Figure 2.39. The sum $A_{\varphi f_1}$ (+) $B_{\varphi f_2}$ is given in Figure 2.40.

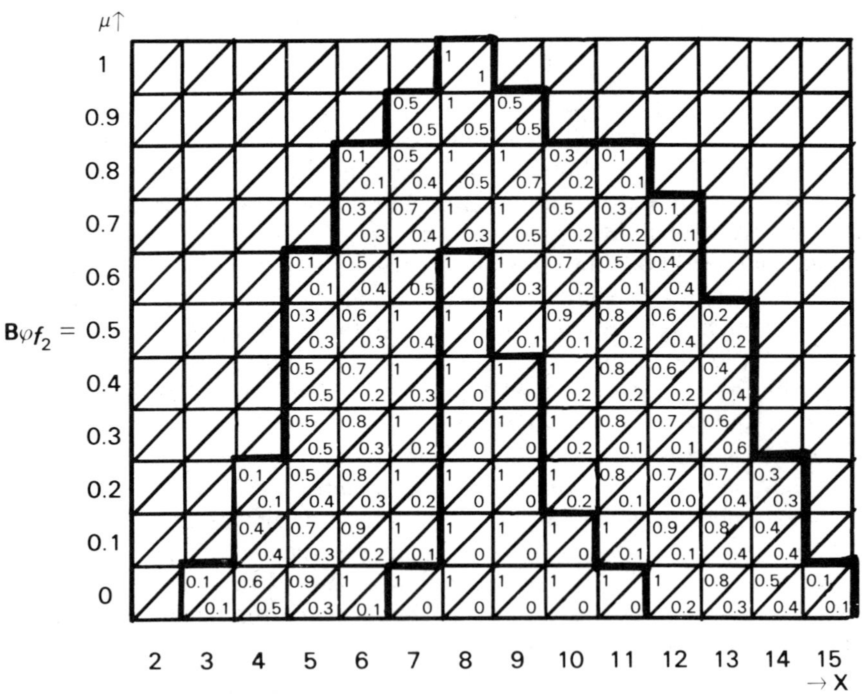

Figure 2.39 Random fuzzy number $\mathbf{B}_{\varphi f_2}$ (Example 2.18).

$A\varphi f_1 \, (+) \, B\varphi f_2 =$

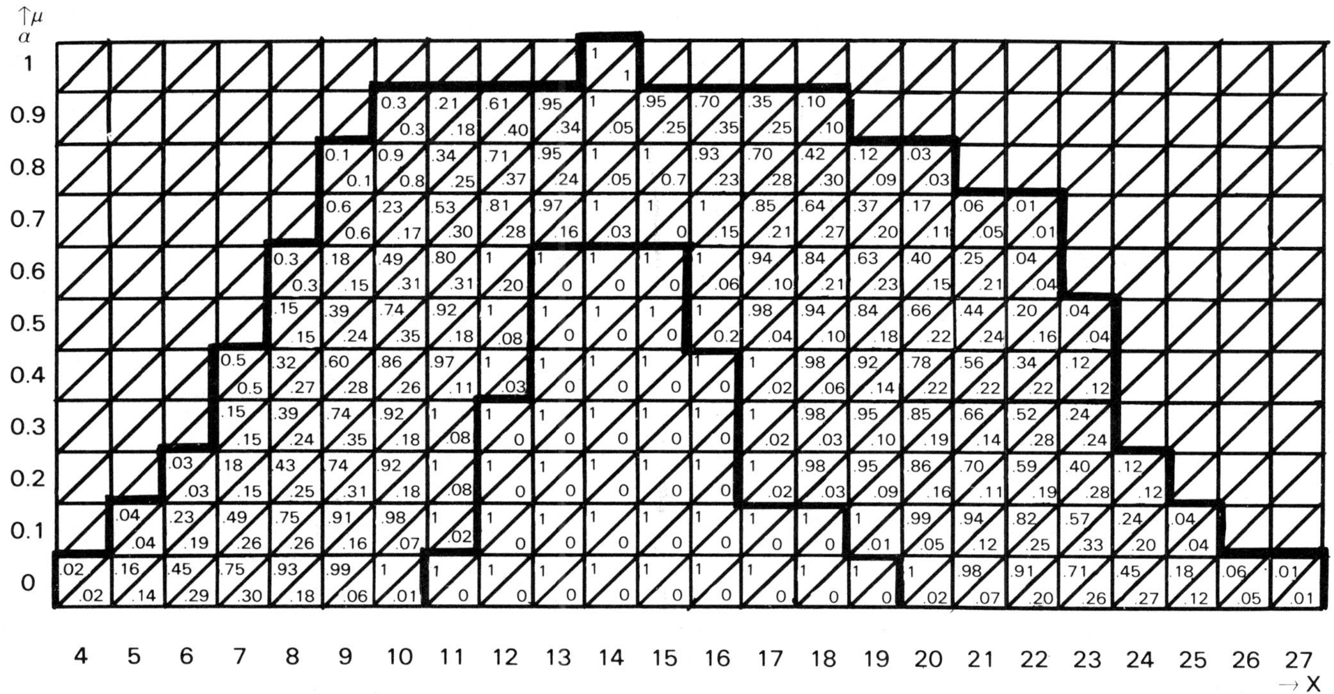

Figure 2.40 Sum-product of two random fuzzy numbers (Example 2.18).

For the two left laws at the level $\alpha = 0.2$, we find that

2	3	4
0.3	0.3	0.4

(+)

4	5	6	7
0.1	0.4	0.3	0.2

=

6	7	8	9	10	11
0.03	0.15	0.25	0.31	0.18	0.08

sum – product

And for the two right laws at the same level, $\alpha = 0.2$.

1	8	9	10
0.1	0.1	0.4	0.4

(+)

10	11	12	13	14
0.2	0.1	0.0	0.4	0.3

sum – product

=

17	18	19	20	21	22	23	24
0.02	0.03	0.09	0.16	0.11	0.19	0.28	0.12

Similarly, at the left for $\alpha = 0.7$,

3	4	5	6
0.2	0.3	0.4	0.1

(+)

6	7	8
0.3	0.4	0.3

=

9	10	11	12	13	14
0.06	0.17	0.30	0.28	0.16	0.03

sum – product

And at the right for $\alpha = 0.7$,

7	8	9	10
0.3	0.3	0.3	0.1

(+)

9	10	11	12
0.5	0.2	0.2	0.1

sum – product

=

16	17	18	19	20	21	22
0.15	0.21	0.27	0.20	0.11	0.05	0.01

Note that the probability for x = 15 is zero.

Remark

At the level $\alpha = 0$ we could assume that the cumulative probability is one everywhere and that the noncumulative probability is zero everywhere with value unity through to $-\infty$ for the left and value unity through to $+\infty$ for the right. This is an interesting situation and the one worth discussing. Several points of view can be considered.

Fuzzy Number of Type 2 Enclosing a Random Fuzzy Number

If we consider a random fuzzy number $A_{\varphi f}$, shown in Figure 2.41, which at the level α has the interval of confidence given by

$$[f_{L\alpha}(x),\quad f_{R\alpha}(x)],$$

we are able to associate with it a fuzzy number of *type* $2\mathcal{A}$ such that $\mathcal{A} = [[a_1{}^{(\alpha)}, a_2{}^{(\alpha)}], [a_3{}^{(\alpha)}, a_4{}^{(\alpha)}]]$, where

$$f_{L\alpha}(x < a_1{}^{(\alpha)}) = 0, \qquad f_{L\alpha}(x > a_2{}^{(\alpha)}) = 0,$$

$$f_{R\alpha}(x < a_3{}^{(\alpha)}) = 0, \qquad f_{R\alpha}(x > a_4{}^{(\alpha)}) = 0.$$

We call $\mathcal{A}$ the *envelope* of $A_{\varphi f}$. The random fuzzy numbers are added by the sum-product convolution, and their corresponding envelope is added by the max-min convolution.

Mean Fuzzy Number of a Random Fuzzy Number

If a probability law given by the density $f(x)$ is bounded at the left by m_1 and at the right by m_2, its mathematical expectation is located between m_1 and m_2. It is then possible to define a mean fuzzy number of a random fuzzy number as follows:

$$\mathbf{A}^m = [\xi(f_L(x)), \xi(f_R(x)]$$

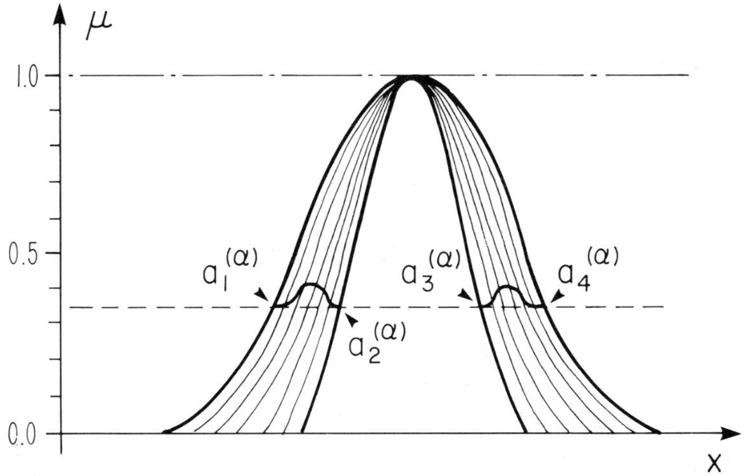

Figure 2.41 A fuzzy number of type 2 enclosing a random fuzzy number.

and $$\mathbf{A}^m{}_\alpha = [\xi(f_{L\alpha}(x)), \xi(f_{R\alpha}(x)],$$

with

$\forall\ \alpha \in [0, 1]$:

$$a_1^{(\alpha)} \leq \xi(f_{L\alpha}(x)) \leq a_2^{(\alpha)},$$
$$a_3^{(\alpha)} \leq \xi(f_{R\alpha}(x)) \leq a_4^{(\alpha)}.$$

This mean fuzzy number is shown in Figure 2.42.

Simulation of Random Fuzzy Numbers

We can simulate models with data that are random fuzzy numbers by using Monte Carlo methods with probability densities f_L and f_R or by using the equivalent laws for discrete cases.

Sheaf of Random Fuzzy Numbers

The concept of a sheaf of fuzzy numbers described in section 2.2 can be extended to random fuzzy numbers and fuzzy numbers of *type* 2. The concept of random fuzzy numbers also can be associated with that of hybrid numbers, as we explain next.

Fuzzy Random Numbers

Fuzzy random numbers are different from random fuzzy numbers. Recall the notion of a probabilistic set or a fuzzy random subset.[5,6]

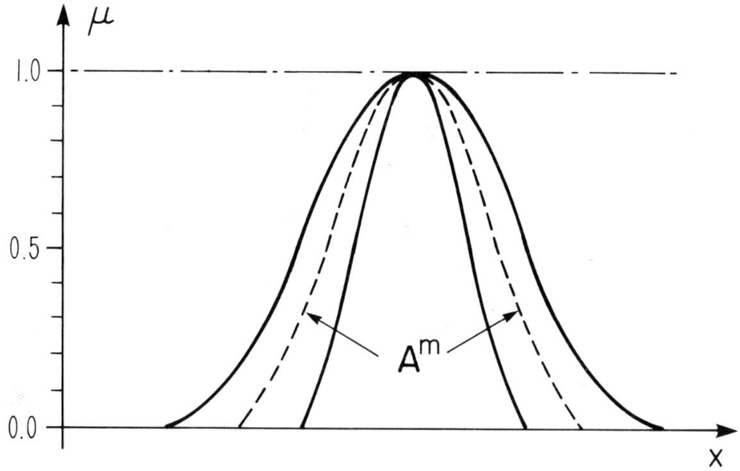

Figure 2.42 A mean fuzzy number of a random fuzzy number.

We consider a fuzzy subset where the membership function $\mu_{\mathbf{A}}(x)$ is a random variable on $[0, 1]$. $F(\mu, x)$ is the complementary cumulative function of the random variable such that

$\forall\ x \in E$:

$$F(\mu, x) = 1 \qquad \text{for } \mu = 0.$$

EXAMPLE 2.19

Consider the situation of a fuzzy random number as shown in Table 2.12. Let us assume that the referential set is R or Z, that for any level the fuzzy subset is convex, and, in addition, that all the nested fuzzy subsets of level α have the same normal value (or values). For these conditions the probabilistic set will be a fuzzy random number.

EXAMPLE 2.20

Let us consider a fuzzy random number in Z. We see that for each level of α the nesting property and the normality of each fuzzy subset of level α is as shown in Table 2.13.

Any operation that is applicable to fuzzy numbers can be extended, level by level, to fuzzy random numbers. To illustrate this, the Example 2.21 demonstrates the operation of addition by the max-min convolution.

Table 2.12 Fuzzy Random Number

α	a	b	c	d	e	f	g	h
1	0	0	0.2	0.1	0	0.1	1	0
0.9	0	0	0.2	0.1	0.5	0.2	1	0
0.8	0.3	0.1	0.3	0.1	0.5	0.2	1	0.3
0.7	0.6	0.1	0.4	0.2	0.5	0.5	1	0.5
0.6	0.8	0.2	0.4	0.3	1	0.6	1	0.5
A = 0.5	0.8	0.3	0.6	0.4	1	0.6	1	0.7
0.4	0.8	0.5	0.6	0.5	1	0.7	1	0.8
0.3	1	0.7	0.6	0.8	1	0.7	1	0.9
0.2	1	0.8	0.6	0.9	1	0.7	1	1
0.1	1	0.9	0.6	1	1	1	1	1
0	1	1	1	1	1	1	1	1

Table 2.13

α ↑						
1	0	0.1	0.1	1	0.1	0
0.9	0	0.1	0.3	1	0.1	0
0.8	0	0.1	0.4	1	0.2	0
0.7	0	0.2	0.4	1	0.3	0.2
0.6	0.1	0.4	0.6	1	0.5	0.2
0.5	0.2	0.5	0.7	1	0.6	0.3
0.4	0.3	0.6	0.8	1	0.7	0.4
0.3	0.5	0.7	0.8	1	0.7	0.6
0.2	0.5	0.8	0.9	1	0.8	0.6
0.1	0.7	0.9	0.9	1	0.8	0.6
0	1	1	1	1	1	1
	−1	0	1	2	3	4 → x

EXAMPLE 2.21

1	0.1	1	0	0
0.9	0.1	1	0.1	0
0.8	0.2	1	0.1	0
0.7	0.3	1	0.2	0.1
0.6	0.3	1	0.3	0.2
0.5	0.5	1	0.4	0.3
0.4	0.6	1	0.5	0.4
0.3	0.7	1	0.5	0.4
0.2	0.9	1	0.7	0.4
0.1	0.9	1	0.8	0.5
0	1	1	1	1
	−2	−1	0	1

(+)

1	0	1	0.2
0.9	0	1	0.2
0.8	0.1	1	0.2
0.7	0.2	1	0.2
0.6	0.3	1	0.3
0.5	0.4	1	0.5
0.4	0.6	1	0.6
0.3	0.6	1	0.6
0.2	0.6	1	0.7
0.1	0.8	1	0.9
0	1	1	1
	3	4	5

1	0	0.1	1	0.2	0	0
0.9	0	0.1	1	0.2	0.1	0
0.8	0.1	0.2	1	0.2	0.1	0
0.7	0.2	0.3	1	0.2	0.2	0.1
0.6	0.3	0.3	1	0.3	0.3	0.2
= 0.5	0.4	0.5	1	0.5	0.4	0.3
0.4	0.6	0.6	1	0.6	0.5	0.4
0.3	0.6	0.7	1	0.6	0.5	0.4
0.2	0.6	0.9	1	0.7	0.7	0.4
0.1	0.8	0.9	1	0.9	0.8	0.5
	1	1	1	1	1	1
	1	2	3	4	5	6

A fuzzy number is a particular case of a fuzzy random number; that is,

1				1	
0.9				1	
0.8			1	1	
0.7			1	1	1
0.6			1	1	1
0.5			1	1	1
0.4		1	1	1	1
0.3		1	1	1	1
0.2	1	1	1	1	1
0.1	1	1	1	1	1
0	1	1	1	1	1
	−1	0	1	2	3

A Fuzzy Number

1		1			
0.9		1			
0.8		1			
0.7		1			
0.6		1			
0.5		1			
0.4		1			
0.3		1			
0.2		1			
0.1		1			
0	1	1	1	1	1
	−2	−1	0	1	2

An Ordinary Number

2.9 FUZZY NUMBERS OF DIMENSION 2 AND MORE

We consider a fuzzy graph $\mathbf{G} \subset R \times R$, that is, a functional fuzzy relation in R^2 such that its membership function $\mu_{\mathbf{G}}(x, y)$, $(x, y) \in R^2$, $\mu_{\mathbf{G}}(x, y) \in [0, 1]$, has the following properties:

1. $\forall\ x_0 \in R$:

$$\mu_{\mathbf{G}}(x_0, y) \in [0, 1], \tag{2.41}$$

is a convex membership function.

2. $\forall\ y_0 \in R$:

$$\mu_{\mathbf{G}}(x, y_0) \in [0, 1], \tag{2.42}$$

is a convex membership function.

3. $\forall\ \alpha \in [0, 1]$:

$$\mu_{\mathbf{G}}(x, y) = \alpha, \tag{2.43}$$

is a convex surface.

4. $\exists (x_2, y_2) \in R^2$:

$$\mu_{\mathbf{G}}(x_2, y_2) = 1. \tag{2.44}$$

If conditions (2.41-2.44) are satisfied, the fuzzy subset $\mathbf{G} \subset R^2$ is called a *fuzzy number of dimension 2*. Figure 2.43(*A*) represents such a fuzzy number. Figure 2.43(*B*) shows a cut of the surface that defines the membership function for a level $\alpha \in [0, 1]$. If (x_2, y_2) is unique, the fuzzy number is unimodal. Otherwise, it belongs to a plate or flat.

Figure 2.44(*A*) shows a fuzzy number in R^2 with a semipyramidal shape. Figure 2.44(*B*) shows the same fuzzy number of dimension 2 from a top view.

A fuzzy member of dimension 2 can also be considered in Z^2, where the convexity takes into account $x = \text{constant}$ or $y = \text{constant}$. Example 2.22 shows a fuzzy number in Z^2.

EXAMPLE 2.22

We have a fuzzy number in Z^2 shown in Table 2.14. Let us now check to see if this is indeed a fuzzy number of dimension 2 (continued on page **149**):

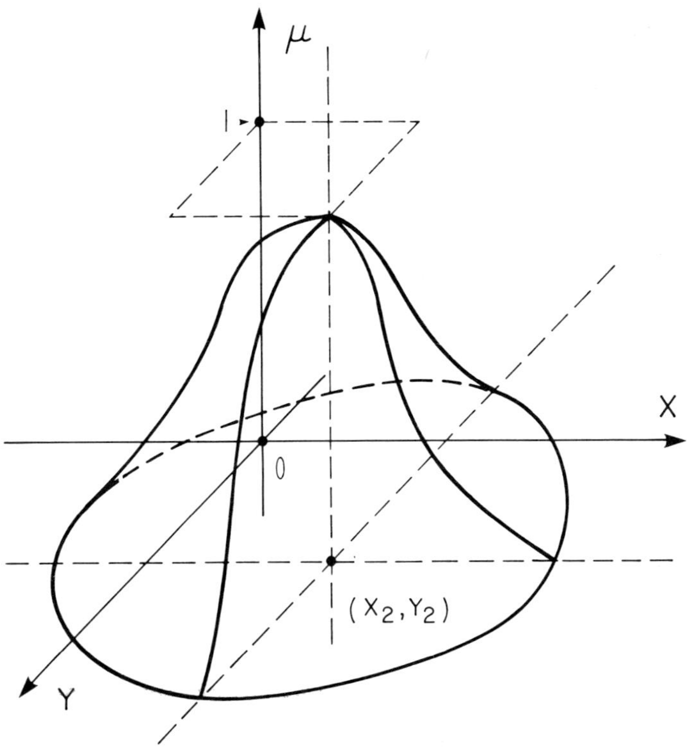

Figure 2.43(A) A fuzzy number of dimension 2.

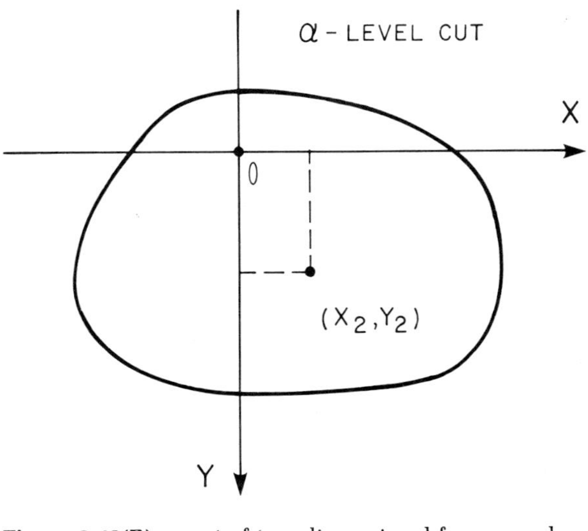

Figure 2.43(B) α-cut of two-dimensional fuzzy number.

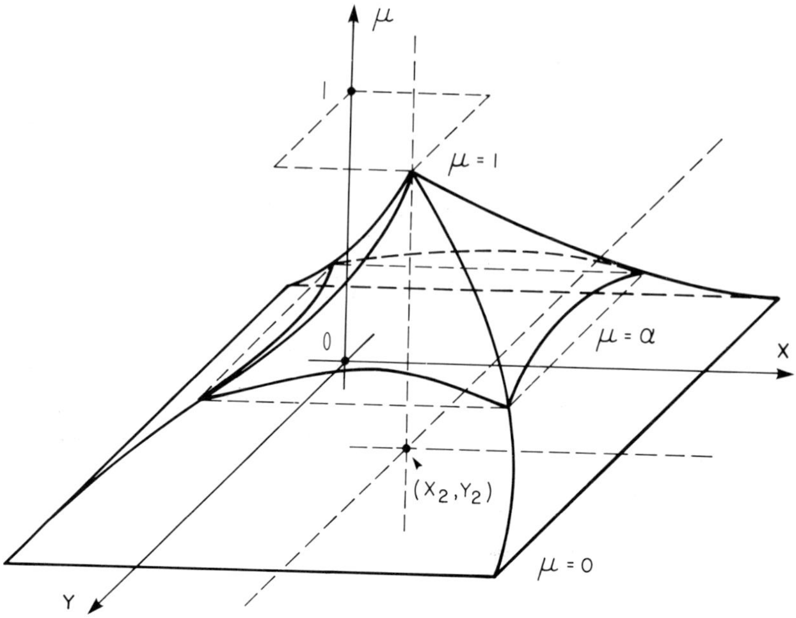

Figure 2.44(A) A semipyramidal-shape fuzzy number.

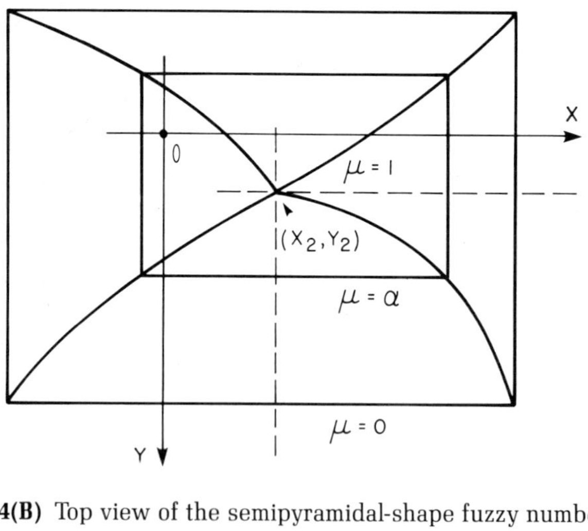

Figure 2.44(B) Top view of the semipyramidal-shape fuzzy number.

Table 2.14

G: x \ y	−3	−2	−1	0	1	2	3	4
−4	0	0	0	0	0	0	0	0
−3	0	0.1	0.3	0.4	0.2	0.1	0.1	0
−2	0	0.2	0.5	0.9	0.8	0.4	0.2	0
−1	0	0.2	0.6	1	0.8	0.5	0.3	0
0	0	0.2	0.6	0.7	0.6	0.4	0.3	0
1	0	0.1	0.2	0.3	0.2	0.1	0.1	0
2	0	0	0.1	0.1	0.1	0.1	0.1	0
3	0	0	0.1	0.1	0.1	0.1	0.1	0
4	0	0	0	0	0	0	0	0

	−3	−2	−1	0	1	2	3	4
G(−4, y) =	0	0	0	0	0	0	0	0
G(−3, y) =	0	0.1	0.3	0.4	0.2	0.1	0.1	0
G(−2, y) =	0	0.2	0.5	0.9	0.8	0.4	0.2	0
G(−1, y) =	0	0.2	0.6	1	0.8	0.5	0.3	0
G(0, y) =	0	0.2	0.6	0.7	0.6	0.4	0.3	0
G(1, y) =	0	0.1	0.2	0.3	0.2	0.1	0.1	0
G(2, y) =	0	0	0.1	0.1	0.1	0.1	0.1	0
G(3, y) =	0	0	0.1	0.1	0.1	0.1	0.1	0
G(4, y) =	0	0	0	0	0	0	0	0

	−4	−3	−2	−1	0	1	2	3	4
G(x, −3) =	0	0	0	0	0	0	0	0	0
G(x, −2) =	0	0.1	0.2	0.2	0.2	0.1	0	0	0
G(x, −1) =	0	0.3	0.5	0.6	0.6	0.2	0.1	0.1	0
G(x, 0) =	0	0.4	0.9	1	0.7	0.3	0.1	0.1	0
G(x, 1) =	0	0.2	0.8	0.8	0.6	0.2	0.1	0.1	0
G(x, 2) =	0	0.1	0.4	0.5	0.4	0.1	0.1	0.1	0
G(x, 3) =	0	0.1	0.2	0.3	0.1	0.1	0.1	0.1	0
G(x, 4) =	0	0	0	0	0	0	0	0	0

(2.45)

All these fuzzy subsets are convex, and **G** $(-1, y)$ and **G** $(x, 0)$ are fuzzy numbers. The maximum value is at $\mu_{\mathbf{G}}(-1, 0) = 1$, which is unique in this case.

When we examine the α cuts, level by level, we have the results in Tables 2.15-2.25.

Table 2.15 $\alpha = 0$

	−3	−2	−1	0	1	2	3	4
−4	1	1	1	1	1	1	1	1
−3	1	1	1	1	1	1	1	1
−2	1	1	1	1	1	1	1	1
−1	1	1	1	1	1	1	1	1
0	1	1	1	1	1	1	1	1
1	1	1	1	1	1	1	1	1
2	1	1	1	1	1	1	1	1
3	1	1	1	1	1	1	1	1
4	1	1	1	1	1	1	1	1

Table 2.16 $\alpha = 0.1$

	−3	−2	−1	0	1	2	3	4
−4								
−3		1	1	1	1	1	1	
−2		1	1	1	1	1	1	
−1		1	1	1	1	1	1	
0		1	1	1	1	1	1	
1		1	1	1	1	1	1	
2			1	1	1	1	1	
3			1	1	1	1	1	
4								

Table 2.17 $\alpha = 0.2$

	−3	−2	−1	0	1	2	3	4
−4								
−3			1	1	1	1		
−2		1	1	1	1	1	1	
−1		1	1	1	1	1	1	
0		1	1	1	1	1	1	
1			1	1	1			
2								
3								
4								

Table 2.18 $\alpha = 0.3$

	−3	−2	−1	0	1	2	3	4
−4								
−3			1	1	1	1		
−2			1	1	1	1	1	
−1			1	1	1	1	1	
0			1	1	1	1		
1				1				
2								
3								
4								

Table 2.19 $\alpha = 0.4$

	−3	−2	−1	0	1	2	3	4
−4								
−3				1				
−2			1	1	1			
−1			1	1	1	1		
0			1	1	1	1		
1								
2								
3								
4								

Table 2.20 $\alpha = 0.5$

	−3	−2	−1	0	1	2	3	4
−4								
−3								
−2			1	1	1			
−1			1	1	1	1		
0			1	1	1			
1								
2								
3								
4								

Table 2.21 $\alpha = 0.6$

	−3	−2	−1	0	1	2	3	4
−4								
−3								
−2				1	1			
−1			1	1	1			
0			1	1	1			
1								
2								
3								
4								

Table 2.22 $\alpha = 0.7$

	−3	−2	−1	0	1	2	3	4
−4								
−3								
−2				1	1			
−1				1	1			
0				1				
1								
2								
3								
4								

Table 2.23 $\alpha = 0.8$

	−3	−2	−1	0	1	2	3	4
−4								
−3								
−2				1	1			
−1				1	1			
0								
1								
2								
3								
4								

Table 2.24 $\alpha = 0.9$

	−3	−2	−1	0	1	2	3	4
−4								
−3								
−2				1				
−1				1				
0								
1								
2								
3								
4								

Table 2.25 $\alpha = 1$

	−3	−2	−1	0	1	2	3	4
−4								
−3								
−2								
−1				1				
0								
1								
2								
3								
4								

Domain of Confidence

The concept of the *domain of confidence* in R^2 is a generalization of that of the interval of confidence in R. It is a convex surface in R^2. Recall the conditions under which a surface is convex in R^2. An ordinary subset in R^2 is convex if and only if, considering any two points of this subset, $(x_1) = (x_1^{(1)}, x_1^{(2)})$ and $(x_2) = (x_2^{(1)}, x_2^{(2)})$. Then any point $(x_3) = (x_3^{(1)}, x_3^{(2)})$ located in the segment between (x_1) and (x_2) belongs to the subset. For this reason, the subset or surface S in Figure 2.45(A) is convex and that in Figure 2.45(B) is not convex, since we can see in the figure that (x_3) does not belong to S'.

Convexity can be presented in a more general way in R^n, $n = 1, 2, 3, 4, \ldots$. Consider an ordinary subset D in R^n such as $D \subset R^n$. If a point $(x) = (x^{(1)}, x^{(2)}, \ldots, x^{(n)}) \in D$; and we put $\mu_D(x) = 1$ if $(x) \in D$; and $\mu_D(x) = 0$ if $(x) \notin D$, then D is convex if and only if

$\forall\ (x_1), (x_2) \in D$ and $\lambda\ [0, 1]$:

$$\mu_D[\lambda(x_1) + (1 - \lambda)(x_2)] \geq \mu_D(x_1) \wedge \mu_D(x_2). \qquad \textbf{(2.46)}$$

This equation means that if (x_1) and (x_2) belong to D, that is, $\mu_D(x_1) = 1$ and $\mu_D(x_2) = 1$, then

$$\mu_D[\lambda(x_1) + (1 - \lambda)(x_2)] = 1$$

implies that

$$(x_3) = \lambda(x_1) + (1 - \lambda)(x_2) \in D.$$

The notion of convexity in a domain D in Z^2 is extended as follows. We call a subset of Z^2 that contains each element of the row i according to $x^{(i)}$, which belongs to D, a *cut for* $x_1^{(i)}$. We call a subset of Z^2 that contains each element of the column j according to $x^{(j)}$, which belongs at D, a *cut for* $x_2^{(j)}$. A cut for $x^{(i)}$ will be represented by $C_1(x_1^{(i)})$ and a cut for $x_2^{(j)}$ by $C_2(x_2^{(j)})$. Then D will be convex if and only if the two following conditions are satisfied.

1. $\exists\ x_1^{(i)} \in Z$ such that

$$\ldots \subset C_1(x_1^{(i-2)}) \subset C_1(x_1^{(i-1)}) \subset C_1(x_1^{(i)}) \supset C_1(x_1^{(i+1)}) \supset C_1(x_1^{(i+2)}) \supset \ldots. \qquad \textbf{(2.47)}$$

2. $\exists\ x_2^{(j)} \in Z$ such that

$$\ldots \subset C_2(x_2^{(j-2)}) \subset C_2(x_2^{(j-1)}) \subset C_2(x_2^{(j)}) \supset C_2(x_2^{(j+1)}) \supset C_2(x_2^{(j+2)}) \supset \ldots. \qquad \textbf{(2.48)}$$

$\subset$ is the symbol of nonstrict inclusion.

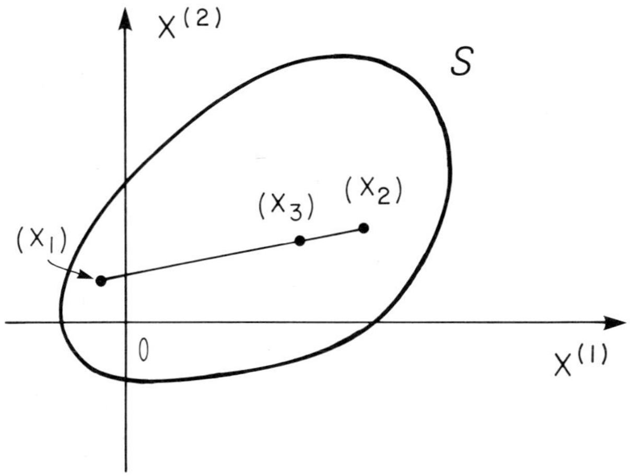

Figure 2.45(A) Definition of a convex surface.

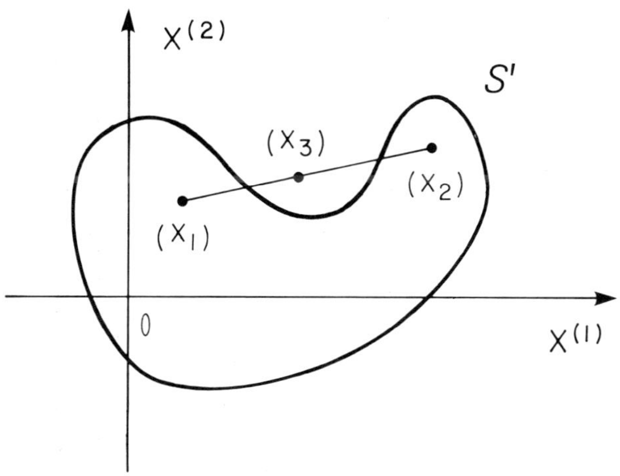

Figure 2.45(B) Definition of a nonconvex surface.

From Table 2.26 we have

$$C_1(0) \subset C_1(1) \subset C_1(2) \subset C_1(3) = C_1(4) \supset C_1(5) \supset C_1(6),$$

$$C_2(0) \subset C_2(1) \subset C_2(2) \supset C_2(3) \supset C_2(4) \supset C_2(5) \supset C_2(6) \supset C_2(7) \supset C_2(8).$$

Table 2.26

x_1 \ x_2	−2	−1	0	1	2	3	4	5	6	7	8	9
−2												
−1												
0					1							
1					1	1	1					
2				1	1	1	1	1				
3			1	1	1	1	1	1	1	1	1	
4			1	1	1	1	1	1	1	1	1	
5				1	1	1	1	1				
6					1	1	1					
7												
8												
9												

Table 2.27

x_1 \ x_2	−2	−1	0	1	2	3	4	5	6	7	8	9
−2												
−1												
0												
1						1	1		1			
2				1	1	1	1	1	1			
3					1	1	1	1	1	1		
4					1	1	1	1	1			
5						1	1					
6												
7												
8												
9												

Since conditions (2.47) and (2.48) are satisfied in the domain, the domain given in Table 2.26 is convex.

From Table 2.27 we see that $C_1(2) \not\supset C_1(3)$, and $C_1(3) \not\supset C_1(2)$, so the condition (2.47) is not satisfied. Therefore the domain in Table 2.27 is not convex.

From Table 2.28 we see that

Table 2.28

$x_1 \backslash x_2$	−2	−1	0	1	2	3	4	5	6	7	8	9
−2												
−1												
0												
1				1								
2				1		1						
3				1	1	1						
4				1	1	1						
5				1	1	1	1	1	1			
6												
7						1						
8												
9												

Table 2.29

$x_1 \backslash x_2$	−2	−1	0	1	2	3	4	5	6	7	8	9
−2												
−1												
0												
1				1	1	1	1	1	1			
2				1					1			
3				1					1			
4				1	1	1	1	1	1			
5												
6												
7												
8												
9												

$$C_1(1) \subset C_1(2) \subset C_1(3) \subset C_1(4) \subset C_1(5) \supset C_1(6) \subset C_1(7) = \varnothing .$$

Since condition (2.23) is not satisfied, therefore the domain is not convex.

Finally, in Table 2.29 we see that

$$C_1(1) \supset C_1(2) = C_1(3) \subset C_1(4).$$

Again, condition (2.23) is not satisfied, and the domain is not convex.

Addition of Two Domains of Confidence

Initially we do not consider convexity.

Let D_1 be a domain or subset of R^2. We consider a point (x_1) of D_1, which we denote as follows:

$$(x_1) = (x_1^{(1)}, x_1^{(2)}).$$

Similarly, we let D_2 be a domain or subset of R^2 and consider a point of D_2 as follows:

$$(x_2) = (x_2^{(1)}, x_2^{(2)}).$$

The addition of D_1 and D_2 is $D = D_1 \,(+)\, D_2$ and is defined by

$$\begin{aligned} D = \{(x^{(1)}, x^{(2)}) \mid (x^{(1)}, x^{(2)}) &= (x_1^{(1)} + x_2^{(1)}, x_1^{(2)} + x_2^{(2)}), \quad (x_1^{(1)}, x_1^{(2)}) \in D_1, \\ &(x_2^{(\alpha)}, x_2^{(1)}) \in D_2\}, \end{aligned} \qquad \textbf{(2.49)}$$

or, in short,

$$D = \{(x) \mid (x) = (x_1) \,(+)\, (x_2), \quad (x_1) \in D_1, (x_2) \in D_2\}. \qquad \textbf{(2.50)}$$

Equation (2.50) can be extended to represent an addition in R^n, $n = 1, 2, 3, 4, \ldots$, where (x) is a point in R^n similar to (x_1) and (x_2) in R^2. Definitions (2.46) and (2.47) are also valid in Z^2 and Z^n.

THEOREM 2.4

Let $D_1 \subset R^2$ and $D_2 \subset R^2$ be convex domains, $(x_1) \in D_1$ and $(x_2) \in D_2$. If we build a domain D using addition, we have

$$(x) = (x_1) \,(+)\, (x_2);$$

then the domain D is a convex domain in R^2.

Proof:

If D_1 is convex, then any point $(x_{1,3})$ located on the segment between $(x_{1,1}) \in D_1$ and $(x_{1,2}) \in D_1$ also belongs to D_1. That is,

$$(x_{1,3}) = \lambda\,(x_{1,1}) + (1 - \lambda)\,(x_{1,2}) \in D_1, \qquad \lambda \in [0, 1]. \qquad \textbf{(2.51)}$$

Similarly, if D_2 is convex, then any point $(x_{2,3})$ located on the segment between $(x_{2,1}) \in D_2$ and $(x_{2,2}) \in D_2$ also belongs to D_2. That is,

$$(x_{2,3}) = \lambda'\,(x_{2,1}) + (1 - \lambda')\,(x_{2,2}) \in D_2, \qquad \lambda' \in [0, 1]. \qquad \textbf{(2.52)}$$

Before we go any further, let us recall that property (2.51) or (2.52) can be rewritten as follows:

$$(y_4) = \lambda_1(y_1) + \lambda_2(y_2) + \lambda_3(y_3),$$

$$\lambda_1, \lambda_2, \lambda_3 \geq 0, \quad \text{and} \quad \lambda_1 + \lambda_2 + \lambda_3 = 1, \qquad \textbf{(2.53)}$$

which means that any (y_4) located inside or on the edge of the triangle $((y_1), (y_2), (y_3))$ also belongs to the domain of the triangle. Now, let us add[7] (2.51) and (2.52).

$$(x_{1,3})\ (+)\ (x_{2,3}) = \lambda(x_{1,1}) + (1-\lambda)\ (x_{1,2}) + \lambda'(x_{2,1}) + (1-\lambda')\ (x_{2,2}).$$

Let us suppose that $\lambda \leq \lambda'$ (or the contrary). We then have

$$\begin{aligned}(x_{1,3})\ (+)\ (x_{2,3}) = \lambda(x_{1,1}) + \lambda(x_{2,1}) + (\lambda' - \lambda)\ (x_{2,1}) \\ + (1-\lambda')\ (x_{1,2}) + (1-\lambda')\ (x_{2,2}) \\ + (\lambda' - \lambda)x_{1,2},\end{aligned}$$

or

$$(x_{1,3}) + (x_{2,3}) = \lambda((x_{1,1})\ (+)\ (x_{2,1})) + (1-\lambda')\ ((x_{1,2})\ (+)\ (x_{2,2})) + (\lambda' - \lambda)\ ((x_{2,1})\ (+)\ (x_{1,2})).$$

But

$$\lambda + (1-\lambda') + (\lambda' - \lambda) = 1$$

and

$$\lambda,\ (1-\lambda'),\ (\lambda - \lambda') \geq 0.$$

Hence, using the property (2.52), we prove that $\{(x_{1,3})\ (+)\ (x_{2,3})\}$ belongs to the triangular domains defined by

$$\{((x_{1,1})\ (+)\ (x_{2,1})),\ ((x_{1,2})\ (+)\ (x_{2,2})),\ ((x_{2,1})\ (+)\ (x_{1,2}))\}.$$

Therefore, if D is the domain obtained by $D_1\ (+)\ D_2$, where D_1, and D_2 are convex, then D is also convex.

A proof based on (2.47) and (2.48) having this property is also true in Z^2. It is easy to prove that this is also true in R^n and Z^n, $n = 1, 2, 3, 4, \ldots$.

EXAMPLE 2.23

Let us consider two rectangular domains D_1 and D_2 shown in Figure 2.46, defined by their respective vertices.

$$\begin{aligned}D_1 &= ((a_1), (b_1), (c_1), (d_1)) \\ &= ((2, -4), (7, -4), (7, -2), (2, -2))\end{aligned}$$

and

$$\begin{aligned}D_2 &= ((a_2), (b_2), (c_2), (d_2)) \\ &= ((9, 2), (12, 2), (12, 6)\ (9, 6)).\end{aligned}$$

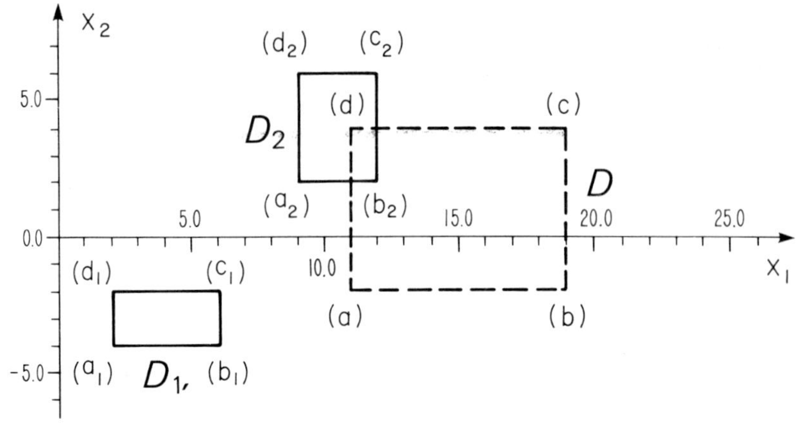

Figure 2.46 Sum of two fuzzy rectangular domains, $D = D_1 + D_2$ (Example 2.23).

It is trivial to show that the addition of these two rectangular domains gives a rectangular domain that can immediately be obtained by adding of points that are vertices of the rectangles. Of course, if the edges are parallel to the axis we have $0x_1$ and $0x_2$, respectively. We then obtain

$$
\begin{aligned}
D = D_1 (+) D_2 &= ((a_1 (+) (a_2), (b_1) (+) (b_2), \\
&\quad (c_1) (+) (c_2), (d_1) (+) (d_2)) \\
&= ((2, -4) (+) (9, 2), (7, -4) (+) (12, 2), (7, -2) \\
&\quad (+) (12, 6), (2, -2) (+) (9, 6)) \\
&= ((11, -2), (19, -2), (19, 4), (11, 4)).
\end{aligned}
$$

This result is shown in Figure 2.46.

EXAMPLE 2.24

Let $D_a \subset R^2$ be a circle with center $(a) = (a_1, a_2)$ and radius r_a, and $D_b \subset R^2$ be another circle with center $(b) = (b_1, b_2)$ and radius r_b. We construct the sum D_c as shown in Figure 2.47.

$$D_c = D_a (+) D_b,$$

where

$$D_a = (a_1, a_2) (+) (r_a \cos\theta, r_a \sin\theta),$$[8]

and

$$D_b = (b_1, b_2) (+) (r_b \cos\theta, r_b \sin\theta),$$

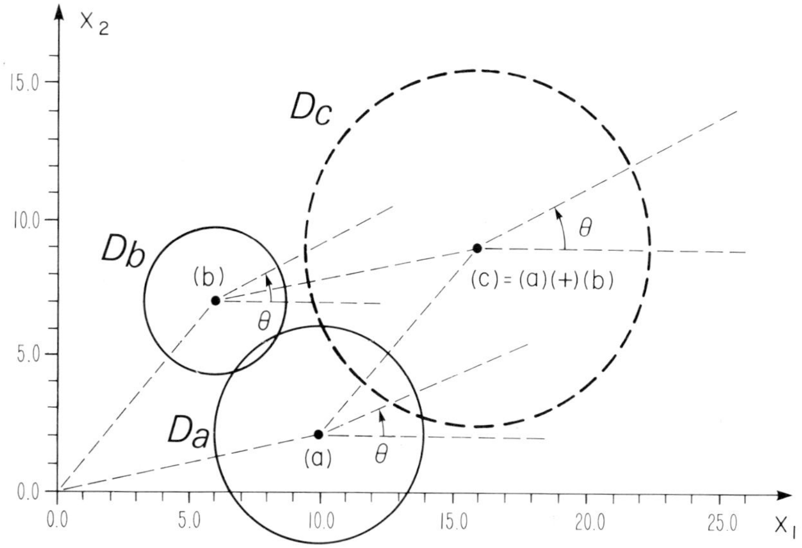

Figure 2.47 Sum of two fuzzy circular domains (Example 2.24).

and

$$\begin{aligned} D_c &= (a_1, a_2)\,(+)\,(r_a \cos\theta, r_a \sin\theta) \\ &\quad (+)\,(b_1, b_2)\,(+)\,(r_b \cos\theta, r_b \sin\theta) \\ &= (a_1 + b_1, a_2 + b_2)\,(+)\,((r_a + r_b)\cos\theta, (r_a + r_b)\sin\theta) \\ &= (c_1, c_2)\,(+)\,(r_c \cos\theta, r_c \sin\theta). \end{aligned}$$

Therefore,

$$\begin{aligned} (c) &= (a)\,(+)\,(b), \\ r_c &= r_a + r_b. \end{aligned}$$

EXAMPLE 2.25

This example in Z^2 is given in Table 2.30 (on the following page).

Max-Min Convolution Formulas in R^2 and R^n

Consider a max-min convolution problem in R^2.

$\forall\ (x_1, x_2), (y_1, y_2), (z_1, z_2) \in R^2$:

$$\mu_{D_c}(z_1, z_2) = \bigvee (\mu_{D_a}(x_1, x_2) \wedge \mu_{D_b}(y_1, y_2)),$$

$$(z_1, z_2) = (x_1, x_2)\,(+)\,(y_1, y_2) \qquad \textbf{(2.54)}$$

Table 2.30

	9	10	11	12	13	14	15
15							
14			1	1	1		
13		1	1	1	1	1	
12			1	1	1		
11				1	1		
10							

D_1

(+)

	17	18	19	20	21	22	23
17							
16					1		
15				1	1	1	
14		1	1	1	1	1	
13			1	1	1		
12							

D_2

=

	27	28	29	30	31	32	33	34	35	36	37
31											
30						1	1	1			
29					1	1	1	1	1		
28			1	1	1	1	1	1	1	1	
27		1	1	1	1	1	1	1	1	1	
26			1	1	1	1	1	1	1		
25				1	1	1	1	1	1		
24					1	1	1	1			
23											

$D = D_1 \,(+)\, D_2$

where

$$\mu_{D_a}(x_1, x_2) = 1, \quad \text{iff } (x_1, x_2) \in D_a,$$
$$= 0, \quad \text{iff } (x_1, x_2) \notin D_a;$$
$$\mu_{D_b}(y_1, y_2) = 1, \quad \text{iff } (y_1, y_2) \in D_b,$$
$$= 0, \quad \text{iff } (y_1, y_2) \notin D_b;$$

and,

$$\mu_{D_c}(z_1, z_2) = 1, \quad \text{iff } (z_1, z_2) \in D_c,$$
$$= 0, \quad \text{iff } (z_1, z_2) \notin D_c. \tag{2.55}$$

Equation (2.54) may now be extended to R^n:

$\forall$ $(x), (y), (z) \in R^n$, $n = 1, 2, 3, 4, \ldots$

$$\mu_{D_c}(z) = \bigvee_{(z) = (x)\,(+)\,(y)} (\mu_{D_a}(x) \wedge \mu_{D_b}(y)),$$

where

$$\begin{aligned}
\mu_{D_a}(x) &= 1 && \text{iff } (x) \in D_a, \\
&= 0 && \text{iff } (x) \notin D_a; \\
\mu_{D_b}(y) &= 1 && \text{iff } (y) \in D_b, \\
&= 0 && \text{iff } (y) \notin D_b; \\
\mu_{D_c}(z) &= 1 && \text{iff } (z) \in D_c, \\
&= 0 && \text{iff } (z) \notin D_c.
\end{aligned}$$

Equation (2.54) has been used both explicitly and implicitly in Examples 23, 24, and 25. These equations can be adapted to subtraction, multiplication, division, and maximum and minimum operations, as shown below.[9]

$\forall\ (x), (y), (z) \in R^n, \qquad n = 1, 2, 3, 4, \ldots$

$$\mu_{D_c}(z) = \bigvee_{(z) = (x)(-)(y)} (\mu_{D_a}(x) \wedge \mu_{D_b}(y));$$

$\forall\ (x), (y), (z) \in R^{+n}, \qquad n = 1, 2, 3, 4, \ldots$

$$\mu_{D_c}(z) = \bigvee_{(z) = (x)(\cdot)(y)} (\mu_{D_a}(x) \wedge \mu_{D_b}(y));$$

$\forall\ (x), (y), (z) \in R^{+n}, \qquad n = 1, 2, 3, 4, \ldots$

$$\mu_{D_c}(z) = \bigvee_{(z) = (x)(:)(y)} (\mu_{D_a}(x) \wedge \mu_{D_b}(y));$$

$\forall\ (x), (y), (z) \in R^n, \qquad n = 1, 2, 3, 4, \ldots$

$$\mu_{D_c}(z) = \bigvee_{(z) = (x)(\wedge)(y)} (\mu_{D_a}(x) \wedge \mu_{D_b}(y));$$

where

$$\begin{aligned}
&(x'_1, x'_2, \ldots, x'_n) \wedge (x''_1, x''_2, \ldots, x''_n) \\
&= (x'_1 \wedge x''_1, x'_2 \wedge x''_2, \ldots, x'_n \wedge x''_n);
\end{aligned}$$

$\forall\ (x), (y), (z) \in R^n$:

$$\mu_{D_c}(z) = \bigvee_{(z) = (x) \vee (y)} (\mu_{D_a}(x) \wedge \mu_{D_b}(y)),$$

where

$$\begin{aligned}
&(x'_1, x'_2, \ldots, x'_n) \vee (x''_1, x''_2, \ldots, x''_n) \\
&= (x'_1 \vee x''_1, x'_1 \vee x''_2, \ldots, x'_n \vee x''_n).
\end{aligned}$$

More generally, we can carry out some operations on domains by the convolution max-min with an operator $*$ for $(z) = (x) * (y)$ provided that the operation is significant and gives useful results.

Addition of Fuzzy Numbers of Dimension 2, and More by Max-Min Convolution

To pass from domains of confidence in R^2 (and in R^n) to fuzzy numbers in R^2 (and in R^n), it is sufficient to introduce the concept of the level of presumption. At the beginning of this section, we gave the definition of a fuzzy number in R^2 (and in R^n). The addition by convolution max-min of two fuzzy numbers in R^2 (and in R^n) will be an extension of equations (2.54) and (2.55). A fuzzy graph can also be defined by using a domain of confidence that depends on a level α of presumption, $\alpha \in [0, 1]$. The words "domain of confidence" are sometimes used with the corresponding symbols.

The equation for addition is

$$\forall\ (x_1, x_2), (y_1, y_2), (z_1, z_2) \in R^2:$$

$$\mu_{\mathbf{D}_C}(z_1, z_2) = \bigvee_{(z_1, z_2) = (x_1, x_2)\ (+)\ (y_1, y_2)} \{\mu_{\mathbf{D}_a}(x_2) \wedge \mu_{\mathbf{D}_b}(y_1, y_2)\}; \tag{2.56}$$

$$\mathbf{D}_a, \mathbf{D}_b, \mathbf{D}_C, \subset R^2,$$

$$\mu_{\mathbf{D}_a}(x_1, x_2), \mu_{\mathbf{D}_b}(y_1, y_2), \mu_{\mathbf{D}_C}(z_1, z_2) \in [0, 1].$$

For each level α, $\alpha \in [0, 1]$, we have, therefore,

$$\mathbf{D}_{C,\,\alpha} = \mathbf{D}_{a,\,\alpha}\ (+)\ \mathbf{D}_{b,\,\alpha}.$$

Equation (2.56) may be extended directly to R^n.

EXAMPLE 2.26

We now consider fuzzy pyramidal numbers in R^2 with edges having rectangular bases parallel to the axis $0x_1$ and $0x_2$. Three coordinates are required for the descriptions of triangular fuzzy numbers in R and five coordinates are required for the description of the pyramidal numbers in R^2, as shown in Figure 2.48.

$$\begin{aligned}
\mathbf{D}_a &= ((a_1), (a_2), (a_3), (a_4), (a_5)) \\
&= ((4, 5), (4, 9), (2, 9), (2, 5), (3, 7)), \\
\mathbf{D}_b &= ((b_1), (b_2), (b_3), (b_4), (b_5)) \\
&= ((1, 11), (1, 14), (-2, 14), (-2, 11), (0, 12)), \\
\mathbf{D}_C &= ((a_1), (a_2), (a_3), (a_4), (a_5)) \\
&\quad (+)\ ((b_1), (b_2), (b_3), (b_4), (b_5))
\end{aligned}$$

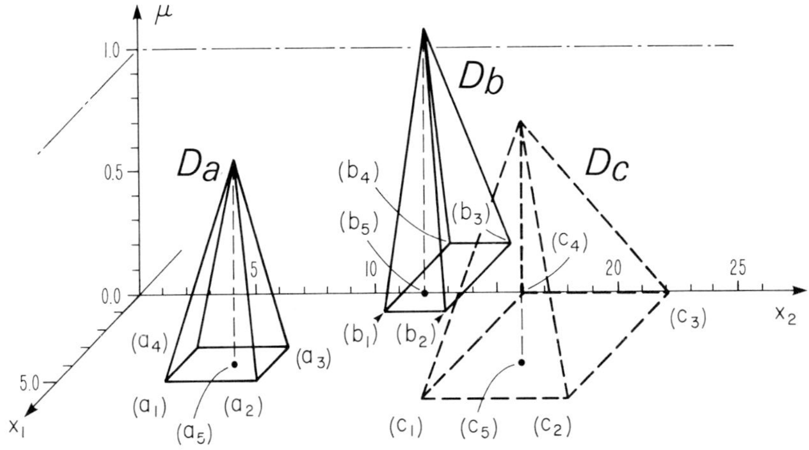

Figure 2.48 Sum of two fuzzy pyramidal numbers (Example 2.26).

$$= ((a_1)\ (+)\ (b_1),\ (a_2)\ (+)\ (b_2),\ (a_3)\ (+)\ (b_3)$$
$$(a_4)\ (+)\ (b_4),\ (a_5)\ (+)\ (b_5))$$
$$= ((4 + 1, 5 + 11), (4 + 1, 9 + 14), (2 - 2, 9 + 14)$$
$$(2 - 2, 5 + 11), (3 + 0, 7 + 12)$$
$$= ((5, 16), (5, 23), (0, 23), (0, 16), (3, 19)).$$

EXAMPLE 2.27

We now describe cone-shaped fuzzy numbers in R^2 by their level at $\alpha = 0$. Their respective bases are circles and their points are at level $\alpha = 1$, with projections on the base. The projection is not necessarily the center of the circle, but lies inside the circle. We use the following representation for this purpose.

D = (center of the circle, radius of the circle, projection of the top on the base)

Using Figure 2.47 for the base and projecting the top exactly at the center for each circle, we have

$$\mathbf{D}_a = ((10, 2), 4, (12, 3)),$$
$$\mathbf{D}_b = ((6, 7), 3, (6, 5)).$$

Therefore the sum is

$$\mathbf{D}_C = ((10 + 6, 2 + 7), 4 + 3, (12 + 6, 3 + 5))$$
$$= ((16, 9), 7\ (18, 8)).$$

EXAMPLE 2.28

For fuzzy numbers in Z^2, consider Tables 2.31 and 2.32.

Table 2.31

x_1 \ x_2	−3	−2	−1	0	1	2	3
2	0.1	0.2	0.3	0.2	0.2	0.1	0
3	0.2	0.5	0.9	0.7	0.5	0.1	0
$\mathbf{D}_a$ = 4	0.2	0.7	**1**	0.8	0.6	0.2	0.1
5	0.1	0.4	0.5	0.3	0.2	0.1	0
6	0	0.2	0.5	0.3	0.2	0	0

Table 2.32

y_1 \ y_2	2	3	4	5
−1	0	0.1	0.1	0.1
0	0.1	0.4	0.5	0.2
1	0.1	0.7	0.9	0.3
$\mathbf{D}_b$ = 2	0.2	0.8	**1**	0.4
3	0.2	0.7	0.8	0.3
4	0.1	0.6	0.6	0.3
5	0.1	0.2	0.3	0

To review in detail the computations that must be carried out for this case, we use Equation (2.56) and the max-min convolution, adding the rows of $\mathbf{D}_a$ with the rows of $\mathbf{D}_b$ as given in Table 2.33.

Table 2.33

Row 1 of $\mathbf{D}_a$ (+) $\mathbf{D}_b$

	−1	0	1	2	3	4	5	6	7	8
2 + (−1):	0	0.1	0.1	0.1	0.1	0.1	0.1	0.1	0.1	0

Row 2		−1	0	1	2	3	4	5	6	7	8
	2 + 0:	0.1	0.1	0.2	0.3	0.3	0.2	0.2	0.2	0.1	0.1
	3 + (−1):	0	0.1	0.1	0.1	0.1	0.1	0.1	0.1	0.1	0
	∨	0.1	0.1	0.2	0.3	0.3	0.2	0.2	0.2	0.1	0.1

Row 3		−1	0	1	2	3	4	5	6	7	8
	2 + 1:	0.1	0.1	0.2	0.3	0.3	0.3	0.3	0.2	0.1	0
	3 + 0:	0.1	0.2	0.4	0.5	0.5	0.5	0.5	0.2	0.1	0
	4 + (−1):	0	0.1	0.1	0.1	0.1	0.1	0.1	0.1	0.1	0.1
	∨	0.1	0.2	0.4	0.5	0.5	0.5	0.5	0.2	0.1	0.1

Row 4		−1	0	1	2	3	4	5	6	7	8
	2 + 2:	0.1	0.2	0.3	0.3	0.3	0.3	0.2	0.2	0.1	0.1
	3 + 1:	0.1	0.2	0.5	0.7	0.9	0.7	0.5	0.3	0.1	0
	4 + 0:	0.1	0.2	0.4	0.5	0.5	0.5	0.5	0.2	0.2	0.1
	5 + (−1):	0	0.1	0.1	0.1	0.1	0.1	0.1	0.1	0.1	0
	∨	0.1	0.2	0.5	0.7	0.9	0.7	0.5	0.3	0.2	0.1

Row 5		−1	0	1	2	3	4	5	6	7	8
	2 + 3:	0.1	0.2	0.2	0.3	0.3	0.3	0.2	0.2	0.1	0
	3 + 2:	0.2	0.2	0.5	0.8	0.9	0.7	0.5	0.4	0.1	0
	4 + 1:	0.1	0.2	0.7	0.7	0.9	0.8	0.6	0.3	0.2	0.1
	5 + 0:	0.1	0.2	0.4	0.4	0.5	0.5	0.5	0.2	0.2	0.1
	6 + (−1):	0	0	0.1	0.1	0.1	0.1	0.1	0.1	0	0
	∨	0.2	0.2	0.7	0.8	0.9	0.8	0.6	0.4	0.2	0.1

Table 2.33 (*continued*)

		−1	0	1	2	3	4	5	6	7	8
	2 + 4:	0.1	0.2	0.2	0.3	0.3	0.3	0.2	0.2	0.1	0
	3 + 3:	0.2	0.2	0.5	0.7	0.8	0.8	0.6	0.3	0.2	0.1
Row 6	4 + 2:	0.2	0.2	0.7	0.8	1	0.8	0.6	0.4	0.2	0.1
	5 + 1:	0.1	0.1	0.4	0.5	0.5	0.3	0.3	0.2	0.1	0
	6 + 0:	0	0.1	0.2	0.4	0.5	0.3	0.2	0.2	0	0
	$\vee$	0.2	0.2	0.7	0.8	1	0.8	0.6	0.4	0.2	0.1

		−1	0	1	2	3	4	5	6	7	8
	2 + 5:	0.1	0.1	0.2	0.2	0.3	0.2	0.2	0.1	0	0
	3 + 4:	0.1	0.2	0.5	0.6	0.6	0.6	0.5	0.3	0.1	0
Row 7	4 + 3:	0.2	0.2	0.7	0.7	0.8	0.8	0.6	0.3	0.2	0.1
	5 + 2:	0.1	0.2	0.4	0.5	0.5	0.4	0.3	0.2	0.1	0
	6 + 1:	0	0.1	0.2	0.5	0.5	0.3	0.3	0.3	0.2	0.1
	$\vee$	0.2	0.2	0.7	0.7	0.8	0.8	0.6	0.3	0.2	0.1

		−1	0	1	2	3	4	5	6	7	8
	3 + 5:	0.1	0.2	0.2	0.3	0.3	0.3	0.3	0.1	0	0
	4 + 4:	0.1	0.2	0.6	0.6	0.6	0.6	0.6	0.3	0.2	0.1
Row 8	5 + 3:	0.1	0.2	0.4	0.5	0.5	0.3	0.3	0.2	0.1	0
	6 + 2:	0	0.2	0.2	0.5	0.5	0.4	0.3	0.2	0	0
	$\vee$	0.1	0.2	0.6	0.6	0.6	0.6	0.6	0.3	0.2	0.1

		−1	0	1	2	3	4	5	6	7	8
	4 + 5:	0.1	0.2	0.2	0.3	0.3	0.3	0.3	0.2	0.1	0
Row 9	5 + 4:	0.1	0.2	0.4	0.5	0.5	0.3	0.3	0.2	0.1	0
	6 + 3:	0	0.2	0.2	0.5	0.5	0.3	0.3	0.2	0	0
	$\vee$	0.1	0.2	0.4	0.5	0.5	0.3	0.3	0.2	0.1	0

Row 10

	−1	0	1	2	3	4	5	6	7	8
5 + 5:	0.1	0.1	0.2	0.3	0.3	0.3	0.2	0.1	0	0
6 + 4:	0	0.1	0.2	0.5	0.5	0.3	0.3	0.2	0	0
∨	0.1	0.1	0.2	0.5	0.5	0.3	0.3	0.2	0	0

Row 11

	−1	0	1	2	3	4	5	6	7	8
6 + 5	0	0.1	0.2	0.2	0.3	0.3	0.2	0	0	0

We find, therefore, that we have the results shown in Table 2.34. This table can be computed in many other ways, for instance, level by level as done in Table 1.1 in z. It may also be computed by considering the Cartesian products of the intervals of confidence, which are dependent. The method presented here appears to be the simplest, however.

Table 2.34

$\mathbf{D}_c = \mathbf{D}_a \ (+) \ \mathbf{D}_b =$

$z_1 \backslash z_2$	−1	0	1	2	3	4	5	6	7	8
1	0	0.1	0.1	0.1	0.1	0.1	0.1	0.1	0.1	0
2	0.1	0.1	0.2	0.3	0.3	0.2	0.2	0.2	0.1	0.1
3	0.1	0.2	0.4	0.5	0.5	0.5	0.5	0.2	0.1	0.1
4	0.1	0.2	0.5	0.7	0.9	0.7	0.5	0.3	0.2	0.1
5	0.2	0.2	0.7	0.8	0.9	0.8	0.6	0.4	0.2	0.1
6	0.2	0.2	0.7	0.8	**1**	0.8	0.6	0.4	0.2	0.1
7	0.2	0.2	0.7	0.7	0.8	0.8	0.6	0.3	0.2	0.1
8	0.1	0.2	0.6	0.6	0.6	0.6	0.6	0.3	0.2	0.1
9	0.1	0.2	0.4	0.5	0.5	0.3	0.3	0.2	0.1	0
10	0.1	0.1	0.2	0.5	0.5	0.3	0.3	0.2	0	0
11	0	0.1	0.2	0.2	0.3	0.3	0.2	0	0	0

Other Operations with Fuzzy Numbers of Dimension 2, and More

Consider the following in R^n:

$\forall\ (x), (y), (z) \in R^n$, n = 1, 2, 3, 4 . . .

$$\mu_{\mathbf{D}_c}(z) = \bigvee_{(z) = (x)\,(*)\,(y)} (\mu_{\mathbf{D}_a}(x) \wedge \mu_{\mathbf{D}_b}(y))$$

$\mathbf{D}_a, \mathbf{D}_b, \mathbf{D}_c \subset R^n$,

where

$$(*) = (+), (-), (\wedge) \text{ or } (\vee) \text{ in } R^n$$
$$= (\cdot), (:) \text{ in } R^{+n} \text{ with possible extensions to } R^n.$$

For $(+)$, $(\cdot)$, $(\wedge)$, $(\vee)$, the structures are monoids. Hence, except for the complexity of the computations, the rules in R^n are the same as in R.

Noninteractive Fuzzy Numbers

Let us consider a fuzzy number $\mathbf{D} \subset R^2$ such that

$\forall\ (x, y) \in R^2$:

$$\mu_{\mathbf{D}}(x, y) = \mu_{\mathbf{A}}(x) \wedge \mu_{\mathbf{B}}(y). \qquad \textbf{(2.57)}$$

We see that the fuzzy number $\mathbf{D} \subset R^2$ is the direct product of two fuzzy numbers $\mathbf{A} \subset R$ and $\mathbf{B} \subset R$. If condition (2.57) is satisfied, then the numbers **A** and **B** are called noninteractive.

The first projection (on x O μ) of D is

$$\bigvee_y \mu_{\mathbf{D}}(x, y) = \bigvee_y (\mu_{\mathbf{A}}(x) \wedge \mu_{\mathbf{B}}(y)) = \mu_{\mathbf{A}}(x). \qquad \textbf{(2.58)}$$

and the second projection is

$$\bigvee_x \mu_{\mathbf{D}}(x, y) = \bigvee_x (\mu_{\mathbf{A}}(x) \wedge \mu_{\mathbf{B}}(y)) = \mu_{\mathbf{B}}(y) \qquad \textbf{(2.59)}$$

It is obvious that conditions (2.58) and (2.59), when taken together, are equivalent to condition (2.57). When these conditions are not satisfied, fuzzy numbers **A** and **B** are said to be *interactive*.

EXAMPLE 2.29

Let us return to Figure 2.48 and consider the pyramidal fuzzy number $\mathbf{D}_a$. We write

$$(a_1) = (a_{11}, a_{12}),$$
$$(a_2) = (a_{21}, a_{22}),$$
$$(a_3) = (a_{31}, a_{32}),$$
$$(a_4) = (a_{41}, a_{42}),$$
$$(a_5) = (a_{51}, a_{52}),$$

By construction,

$$a_{21} = a_{11},\ a_{12} = a_{22},\ a_{41} = a_{31},\ a_{12} = a_{42}.$$

The first projection of the pyramid D_a in $x_1 0\ \mu$ is thc fuzzy triangular number (a_{41}, a_{51}, a_{11}), and the second projection of D_a on $x_2 0\ u$ is the fuzzy triangular number (a_{42}, a_{52}, a_{32}). Thus, the triangular fuzzy numbers (a_{41}, a_{51}, a_{11}) and (a_{42}, a_{52}, a_{32}) are noninteractive.

EXAMPLE 2.30

Consider a fuzzy number **D** $(x, y) \subset Z^2$ and its projections **A**(x) and **B**(y) as shown in Table 2.35.

Table 2.35

	x \ y	2	3	4	5	6	7	**A**(x)
	−3	0.1	0.1	0.1	0.1	0.1	0.1	0.1
	−2	0.1	0.4	0.4	0.4	0.4	0.1	0.4
	−1	0.1	0.7	0.7	0.7	0.6	0.1	0.7
D(x, y) =	0	0.1	0.8	0.8	0.8	0.6	0.1	0.8
	1	0.1	0.8	1	0.9	0.6	0.1	1
	2	0.1	0.5	0.5	0.5	0.5	0.1	0.5
	3	0.1	0.2	0.2	0.2	0.2	0.1	0.2
B(y) =		0.1	0.8	1	0.9	0.6	0.1	1

Table 2.36

$\mathbf{A}(x) \times \mathbf{B}(y) =$

x	A(x)
−3	0.1
−2	0.4
−1	0.7
0	0.8
1	1
2	0.5
3	0.2

$\wedge$

y	2	3	4	5	6	7
B(y)	0.1	0.8	1	0.9	0.6	0.1

If we take the Cartesian or direct product we have Table 2.36. We thus find **D** (x, y), where the fuzzy number **D**(x, y) is realized by the direct product of **A**(x) and **B**(y), which are noninteractive.

EXAMPLE 2.31

If we consider a fuzzy number **D**(x, y) $\subset Z^2$ (Table 2.37), we then have Table 2.38. Projections **A**(x) and **B**(y) are therefore interactive.

Table 2.37

D(x, y) =

x \ y	−1	0	1	2	3	4	5	6	**A**(x)
−3	0	0	0.2	0.2	0.2	0.2	0.1	0.1	0.2
−2	0.1	0.3	0.3	0.6	0.6	0.5	0.2	0.1	0.6
−1	0.2	0.4	0.4	0.8	1	0.7	0.5	0.1	1
0	0.1	0.1	0.3	0.6	0.8	0.7	0.3	0	0.8
1	0.1	0.1	0.2	0.5	0.5	0.4	0.1	0	0.5
2	0	0	0.1	0.1	0.1	0.1	0.1	0	0.1

B(y) =

0.2	0.4	0.4	0.8	1	0.7	0.5	0.1

Table 2.48

$\mathbf{A}(x) \times \mathbf{B}(y) =$

x	$\mu_{\mathbf{A}}$
−3	0.2
−2	0.6
−1	1
0	0.8
1	0.5
2	0.1

$\wedge$

y	−1	0	1	2	3	4	5	6
$\mu_{\mathbf{B}}$	0.2	0.4	0.4	0.8	1	0.8	0.5	0.1

=

x \ y	−1	0	1	2	3	4	5	6
−3	0.2	0.2	0.2	0.2	0.2	0.2	0.2	0.1
−2	0.2	0.4	0.4	0.6	0.6	0.6	0.5	0.1
−1	0.2	0.4	0.4	0.8	1	0.8	0.5	0.1
0	0.2	0.4	0.4	0.8	0.8	0.8	0.5	0.1
1	0.2	0.4	0.4	0.5	0.5	0.5	0.5	0.1
2	0.1	0.1	0.1	0.1	0.1	0.1	0.1	0.1

$\neq \mathbf{D}(x, y)$

Remark

If $\mathbf{A}(x)$ and $\mathbf{B}(y)$ are projections of $\mathbf{D}(x, y)$, then we have

$$\mathbf{A}(x) \times \mathbf{B}(y) \supset \mathbf{D}(x, y).$$

Indeed,

$\forall\ x \in R$:

$$\bigvee_{y} \mu_{\mathbf{D}}(x, y) \geq \mu_{\mathbf{D}}(x, y),$$

$\forall\ y \in R$:

$$\bigvee_{x} \mu_{\mathbf{D}}(x, y) \geq \mu_{\mathbf{D}}(x, y),$$

and

$$(\bigvee_{y} \mu_{\mathbf{D}}(x, y) \wedge (\bigvee_{x} \mu_{\mathbf{D}}(x, y)) \geq \mu_{\mathbf{D}}(x, y).$$

Hybrid Addition in R^2 and in R^n

The theory of hybrid numbers discussed in section 2.2 can be extended from R to R^n, $n = 1, 2, 3, 4, \ldots$, fairly easily. Thus, in R^2, for example, we can do the hybrid addition of a fuzzy number $\mathbf{D}(x, y)$ with membership function $\mu_{\mathbf{D}}(x, y)$ and a random variable (X, Y) of density $f(x, y)$. One may extend the operations of hybrid numbers in R to operations of hybrid number in R^n with some very interesting results.

NOTES

1. See A. Kaufmann, *Introduction to the Theory of Fuzzy Subsets*, vol. 1, Academic Press, New York, Sec. 26, p. 139.
2. See A. Sambuc, *Applications Φ-Fuzzy Sets in Diagnostic Patholigic Thyrohoid*, Ph.D in Medicine, Faculty de Medicine Marseilles, December 1975.
3. In this book the word "increasing" or "decreasing" is used with a non-strict meaning.
4. In Figure 2.38, thick lines give contour for bounds. The inferior thick line does not reach level 1 because we are considering numbers in N. If the contour is considered with cummulative probabilities, the thick-line contours will reach level 1.
5. See R. Feron, Ensembles Aliatoires Flous, *Acad. Sci. (Paris)* C.R. ser. A, **282**:903-906.
6. See K. Hirota, Concepts of probabilistic sets, *Fuzzy Sets and Systems* 5:31-46(1981).
7. The symbol $(+)$ means a generalized addition.
8. Using the same angle θ in this equation and the previous one, we obtain extremal points; then the domain D corresponds to addition of two domains D_a and D_b.
9. We consider here the domain R^{+n} but the proof is also good for R^n.

Chapter 3

ARITHMETIC AND COMBINATORICS WITH UNCERTAIN NUMBERS

3.1 INTRODUCTION

Even though every operation that can be carried out using ordinary numbers can also be carried with fuzzy numbers, many properties change as we pass from one family to the other. Although this is important, it should not prevent us from using the concepts and theory of fuzzy numbers. The world of perception does not have sharp edges. It is full of ambiguity and uncertainty, and it is only reasonable to promote the use of this fuzzy field.

In this chapter we examine and study certain properties pertaining to arithmetic and combinatoric operations with uncertain numbers.

3.2 REVIEW OF RELATIVE INTEGERS MODULO *n*

In a ring we consider a set and two compositive laws $(*)$ and $(*')$ such that

1. $\forall\ x, y, z \in E$:

$$(x * y) * z = x * (y * z)$$

is the associativity for $*$;

$$x * e = e * x = x,$$

is the existence of a neutral for $*$;

$$x * x^{-1} = x^{-1} * x = e,$$

where each element has a single symmetric for $*$;

$$x * y = y * x,$$

is the commutativity for $*$;

2. $\forall\ x, y, z \in E$:

$$(x *' y) *' z = x *' (y *' z),$$

is the associativity for $*'$; and

3. $\forall\ x, y, z \in E$:

$$(x * y) *' z = (x *' z) * (y *' z),$$
$$z *' (x * y) = (z *' x) * (z *' y),$$

are the left distributivity and rights distributivity of $*$ to $*'$.

The ring is commutative if, in addition,

$$x *' y = y *' x.$$

Since the properties of $(*)$ and $(*')$ are similar to the properties of $(+)$ and $(\cdot)$ respectively in R, it is common to put $(* = +)$ and $(*' = \cdot)$.

Every ring structure has the following properties:

(a) In a ring, a difference or a subtraction exists that is the inverse of $* = +$ and is denoted by $(x - y)$. This inverse law is neither associative nor commutative.

(b) Multiplication $*' = \cdot$ is distributive for the difference operation and is expressed by

$$x \cdot (y - z) = (x \cdot y) - (x \cdot z)$$

and

$$(x - y) \cdot z = (x \cdot z) - (y \cdot z).$$

(c) From the properties of a ring, we can write

$$(-x) \cdot (-y) = x \cdot y,$$
$$(-x) \cdot y = -(x \cdot y),$$

and

$$x \cdot (-y) = -(x \cdot y).$$

(d) Exponential properties are also present in a ring.

$$x^n = x \cdot x \cdots x \qquad (n \text{ times } x),$$
$$x^p \cdot x^q = x^q \cdot x^p = x^{p+q}.$$

(e) In a ring, we also note

$$n \cdot x, \qquad \text{where } n \in Z \text{ and } x \in R.$$

The sum of n elements equal to x is

$$n \cdot x = \underbrace{x + x + \cdots + x}_{n \text{ times}}, \qquad \text{if } x > 0,$$
$$= 0, \qquad \text{if } x = 0,$$
$$= \underbrace{(-x) + (-x) + \cdots + (-x)}_{n \text{ times}}, \qquad \text{if } x < 0.$$

(f) If e is the neutral for the law $*' = \cdot$, we may use it to avoid confusing $n \cdot e$ and n. It is possible that $n \cdot e = 0$ when $n \neq 0$. The smallest such positive integer, $n \cdot e = 0$, is the *characteristic* of the ring. If this characteristic does not exist, the ring is said to have an infinite characteristic.

(g) In commutative rings with a nonnull characteristic n (n prime), we have

$$C_n^p = n \cdot q \qquad (q \in N, p \neq 0, p \neq n),$$

$$C_n^n = C_n^0 = 1.$$

If a unit e exists, then

$$C_n^p = n \cdot (q \cdot e) = 0$$

and

$$(a + b)^n = a^n + b^n,$$
$$(a - b)^n = a^n - b^n.$$

(h) An important case is concerned with residual classes modulo n (n integer). Consider the equivalence relation or congruence on Z defined as follows:

$$n \in N_0, \qquad r \in N, \qquad a, b, q, q' \in Z.$$

Two numbers a and b are said to be *equivalent modulo* n or *congruent modulo* n if their difference is divisible by n.

Let

$$a = n \cdot q + r$$

and

$$b = n \cdot q' + r;$$

then

$$a - b = n(q - q').$$

We may then write $a = b$ modulo n. In addition, every member having the form $a + kn$, where $a, k \in Z$, and $n \in N$, is put into the same class, which we call the *residual class*. For this reason, there are n different classes, and the *quotient ring* is represented by Z/n. In Table 3.1 we present rings Z/n for $n = 2, 3, 4, 5, 6, 7, 8$ and 9.

Table 3.1 Rings z/n for $n = 2, 3, \ldots, 9$

Modulo 2

+	0	1
0	0	1
1	1	0

·	0	1
0	0	0
1	0	1

Modulo 3

+	0	1	2
0	0	1	2
1	1	2	0
2	2	0	1

·	0	1	2
0	0	0	0
1	0	1	2
2	0	2	1

Modulo 4

+	0	1	2	3
0	0	1	2	3
1	1	2	3	0
2	2	3	0	1
3	3	0	1	2

·	0	1	2	3
0	0	0	0	0
1	0	1	2	3
2	0	2	0	2
3	0	3	2	1

Modulo 5

+	0	1	2	3	4
0	0	1	2	3	4
1	1	2	3	4	0
2	2	3	4	0	1
3	3	4	0	1	2
4	4	0	1	2	3

·	0	1	2	3	4
0	0	0	0	0	0
1	0	1	2	3	4
2	0	2	4	1	3
3	0	3	1	4	2
4	0	4	3	2	1

+	0	1	2	3	4	5
0	0	1	2	3	4	5
1	1	2	3	4	5	0
2	2	3	4	5	0	1
3	3	4	5	0	1	2
4	4	5	0	1	2	3
5	5	0	1	2	3	4

·	0	1	2	3	4	5
0	0	0	0	0	0	0
1	0	1	2	3	4	5
2	0	2	4	0	2	4
3	0	3	0	3	0	3
4	0	4	2	0	4	2
5	0	5	4	3	2	1

Modulo 6

+	0	1	2	3	4	5	6
0	0	1	2	3	4	5	6
1	1	2	3	4	5	6	0
2	2	3	4	5	6	0	1
3	3	4	5	6	0	1	2
4	4	5	6	0	1	2	3
5	5	6	0	1	2	3	4
6	6	0	1	2	3	4	5

·	0	1	2	3	4	5	6
0	0	0	0	0	0	0	0
1	0	1	2	3	4	5	6
2	0	2	4	6	1	3	5
3	0	3	6	2	5	1	4
4	0	4	1	5	2	6	3
5	0	5	3	1	6	4	2
6	0	6	5	4	3	2	1

Modulo 7

+	0	1	2	3	4	5	6	7
0	0	1	2	3	4	5	6	7
1	1	2	3	4	5	6	7	0
2	2	3	4	5	6	7	0	1
3	3	4	5	6	7	0	1	2
4	4	5	6	7	0	1	2	3
5	5	6	7	0	1	2	3	4
6	6	7	0	1	2	3	4	5
7	7	0	1	2	3	4	5	6

·	0	1	2	3	4	5	6	7
0	0	0	0	0	0	0	0	0
1	0	1	2	3	4	5	6	7
2	0	2	4	6	0	2	4	6
3	0	3	6	1	4	7	2	5
4	0	4	0	4	0	4	0	4
5	0	5	2	7	4	1	6	3
6	0	6	4	2	0	6	4	2
7	0	7	6	5	4	3	2	1

Modulo 8

+	0	1	2	3	4	5	6	7	8
0	0	1	2	3	4	5	6	7	8
1	1	2	3	4	5	6	7	8	0
2	2	3	4	5	6	7	8	0	1
3	3	4	5	6	7	8	0	1	2
4	4	5	6	7	8	0	1	2	3
5	5	6	7	8	0	1	2	3	4
6	6	7	8	0	1	2	3	4	5
7	7	8	0	1	2	3	4	5	6
8	8	θ	1	2	3	4	5	6	7

·	0	1	2	3	4	5	6	7	8
0	0	0	0	0	0	0	0	0	0
1	0	1	2	3	4	5	6	7	8
2	0	2	4	6	8	1	3	5	7
3	0	3	6	0	3	6	0	3	6
4	0	4	8	3	7	2	6	1	5
5	0	5	1	6	2	7	3	8	4
6	0	6	3	0	6	3	0	6	3
7	0	7	5	3	1	8	6	4	2
8	0	8	7	6	5	4	3	2	1

Modulo 9

Some well-known properties of rings Z/n are as follows.

1. In Z/n the characteristic is n.
2. In Z/n, if n is prime, the neutral of multiplication is 1.
3. In Z/n, an *inversible element* is one that has a symmetric or inverse for the law $*' = \cdot$ such that

$$a \cdot a^{-1} = a^{-1} \cdot a = 1.$$

Consequently, in $Z/2$ the inversible element is 1,
in $Z/3$ the inversible elements are 1 and 3,
in $Z/4$ the inversible elements are 1 and 4,
in $Z/5$ the inversible elements are 1, 2, 3, and 4,
in $Z/6$ the inversible elements are 1 and 5.

3.3 FUZZY RELATIVE INTEGERS MODULO n

We now examine the representation of the residual classes modulo n in a nonfuzzy case.

	−7	−6	−5	−4	−3	−2	−1	0	1	2	3	4	5	6	7	
...	0	1	0	1	0	1	0	1	0	1	0	1	0	1	0	...

a = 0, residual class $n = 2$

	−7	−6	−5	−4	−3	−2	−1	0	1	2	3	4	5	6	7	
...	1	0	1	0	1	0	1	0	1	0	1	0	1	0	1	...

$a = 1$, residual class $n = 2$

	−7	−6	−5	−4	−3	−2	−1	0	1	2	3	4	5	6	7	
...	0	1	0	0	1	0	0	1	0	0	1	0	0	1	0	...

$a = 0$, residual class $n = 3$

	−7	−6	−5	−4	−3	−2	−1	0	1	2	3	4	5	6	7	
...	0	0	1	0	0	1	0	0	1	0	0	1	0	0	1	...

$a = 1$, residual class $n = 3$

	−7	−6	−5	−4	−3	−2	−1	0	1	2	3	4	5	6	7	
...	1	0	0	1	0	0	1	0	0	1	0	0	1	0	0	...

$a = 2$, residual class $n = 3$

	−7	−6	−5	−4	−3	−2	−1	0	1	2	3	4	5	6	7	
...	0	0	0	1	0	0	0	1	0	0	0	1	0	0	0	...

$a = 0$, residual class $n = 4$ **(3.1)**

Equation 3.1 shows that residual classes are ordinary subsets of Z. We now examine how to use such relationships.

Max-Min Convolution of Fuzzy Relative Integers Modulo *n*

In the following the symbols $+_n$ and $\cdot_n$ are used for addition modulo n and multiplication modulo n ($n \in N_0$), respectively. Consider the following convolutions of $\mathbf{A}, \mathbf{B} \subset Z$:

$(\mathbf{A}\,(+)_n\,\mathbf{B}) \Leftrightarrow (\forall\, x, y, z \in Z, n \in N_0$:

$$\mu_{\mathbf{A}(+)_n\mathbf{B}}(z) = \bigvee_{z = x +_n y} (\mu_{\mathbf{A}}(x) \wedge \mu_{\mathbf{B}}(y))). \tag{3.2}$$

$(\mathbf{A}\,(\cdot)_n\,\mathbf{B}) \Leftrightarrow (\forall\, x, y, z \in Z, n \in N_0$:

$$\mu_{\mathbf{A}(\cdot)_n\mathbf{B}}(z) = \bigvee_{z = x \cdot_n y} (\mu_{\mathbf{A}}(x) \wedge \mu_{\mathbf{B}}(y))). \tag{3.3}$$

We now define a fuzzy subset for which convolutions (3.2) and (3.3) will give the following.

Fuzzy Relative Integers Modulo *n* Cyclomonotonic

Let $\mathbf{A} \subset Z$; then

1. With $\mu_{\mathbf{A}}(x) = 1, x \in Z$, we impose

$$\mu_{\mathbf{A}}(x) \geq \mu_{\mathbf{A}}(x+1) \geq \mu_{\mathbf{A}}(x+2) \geq \ldots \geq \mu_{\mathbf{A}}(x+n+1), \qquad n \in N_0. \tag{3.4}$$

2. Let us also impose

$\forall\, x \in Z,\ \forall\, n \in N_0,\ \forall\, k \in Z$:

$$\mu_{\mathbf{A}}(x+kn) = \mu_{\mathbf{A}}(x). \tag{3.5}$$

If conditions (3.4) and (3.5) are both satisfied, then we call **A** a class of fuzzy relative integers modulo *n* cyclomonotonic. Condition (3.5) indicates how these cycles are created. Let us study one such example.

EXAMPLE 3.1

The following is a class of fuzzy relative integers modulo 6 cyclomonotonic.

	−2	−1	0	1	2	3	4	5	6	7	8	9	10	11	12	
...	0.8	0.5	0.1	0	1	0.9	0.8	0.5	0.1	0	1	0.9	0.8	0.5	0.1	...
					↑						↑					

(3.6)

From (3.1) we see that (3.6) is really a fuzzy subset of Z.

We now simplify the process.

A =

0	1	2	3	4	5
0.1	0	1	0.9	0.8	0.5
		↑			

↻ 6

or

2	3	4	5	6	7
1	0.9	0.8	0.5	0.1	0
↑					

↻ 6 (3.7)

When presented as (3.7), the fuzzy subset **A** is not, properly speaking, a fuzzy number but rather a representative of a class of fuzzy relative integers modulo n. Of course, (3.7) is normal, but its normality is qualified as cyclic.

THEOREM 3.1

If **A** and **B** are fuzzy relative integers modulo n cyclomonotonic, then their max-min convolution for addition $(+)_n$ gives a fuzzy relative integer modulo n cyclomonotonic.

Proof:

The max-min convolution for + of two fuzzy numbers such as (3.7) gives a fuzzy number. Here **A** and **B** are extra monotically decreasing and so is their sum. Assume that **A** is a representative of a class of fuzzy relative integers modulo n cyclomonotonic, and that **B** is a representative of a class of fuzzy relative integers modulo n cyclomonotonic. The max-min convolution of these two classes will give another class of the same type modulo n and cyclomonotonic fuzzy relative integer.

Example 3.2 involves such a convolution.

EXAMPLE 3.2

Let

A =

2	3	4	5	6	
1	0.8	0.5	0.1	0	↺ 5

B =

1	2	3	4	5	
1	0.9	0.6	0.3	0.2	↺ 5

(3.6)

We use two tables, Table 3.2 for addition modulo 5, which is concerned with ordinary numbers, and Table 3.3 to compare the membership function in the fuzzy case.

Table 3.2

$+_5$	1	2	3	4	5
2	3	4	0	1	2
3	4	0	1	2	3
4	0	1	2	3	4
5	1	2	3	4	0
6	2	3	4	0	1

Table 3.3

A \ B ($+_5$)	1	2	3	4	5
2	1 / 1	0.9 / 1	0.6 / 1	0.3 / 1	0.2 / 1
3	1 / 0.8	0.9 / 0.8	0.6 / 0.8	0.3 / 0.8	0.2 / 0.8
4	1 / 0.5	0.9 / 0.5	0.6 / 0.5	0.3 / 0.5	0.2 / 0.5
5	1 / 0.1	0.9 / 0.1	0.6 / 0.1	0.3 / 0.1	0.2 / 0.1
	1 / 0	0.9 / 0	0.6 / 0	0.3 / 0	0.2 / 0

To obtain the convolution it is sufficient to consider each diagonal case in Table 3.2 and to compute the max-min in Table 3.3. We thus find

$$
\begin{aligned}
\mu_{\mathbf{A}(+)_5\mathbf{B}}(3) &= (1 \wedge 1) \vee (0.2 \wedge 0.8) \vee (0.5 \wedge 0.3) \vee (0.1 \wedge 0.6) \vee \\
&\quad (0 \wedge 0.9) = 1, \\
(4) &= (1 \wedge 0.9) \vee (0.8 \wedge 1) \vee (0.5 \wedge 0.2) \vee (0.1 \wedge 0.3) \vee \\
&\quad (0 \wedge 0.6) = 0.9, \\
(0) &= (1 \wedge 0.6) \vee (0.8 \wedge 0.9) \vee (0.5 \wedge 1) \vee (0.1 \wedge 0.2) \vee \\
&\quad (0 \wedge 0.3) = 0.8, \\
(1) &= (1 \wedge 0.3) \vee (0.8 \wedge 0.6) \vee (0.5 \wedge 0.9) \vee (0.1 \wedge 1) \vee \\
&\quad (0 \wedge 0.2) = 0.6, \\
(2) &= (1 \wedge 0.2) \vee (0.8 \wedge 0.3) \vee (0.5 \wedge 0.6) \vee (0.1 \wedge 0.9) \vee \\
&\quad (0 \wedge 1) = 0.5,
\end{aligned}
$$

or

$\mathbf{A}\,(+)_5\,\mathbf{B} =$

0	1	2	3	4
0.8	0.6	0.5	1	0.9

(cyclic, 5)

but also

	−5	−4	−3	−2	−1	0	1	2	3
...	0.8	0.6	0.5	1	0.9	0.8	0.6	0.5	1
				↑					↑

4	5	6	7	8	9	10	
0.9	0.8	0.6	0.5	1	0.9	0.8	...
				↑			

is a fuzzy subset representing the whole class.

It should be noted that Theorem 3.1 is also valid when the normal fuzzy subset contains several 1's for the value of the membership function. Also, the monotonicity that has been considered is not strict. A specification will be given when it is strict.

Fuzzy Relative Integers Modulo *n* Cycloconvex

Let **A** $\subset$ Z. Then

1. With $\mu_{\mathbf{A}}(x+r)=1$, $x, r \in Z$, we impose the condition

$$\mu_{\mathbf{A}}(x) \leq \mu_{\mathbf{A}}(x+1) \leq \ldots \leq \mu_{\mathbf{A}}(x+r) \geq \ldots \geq \mu_{\mathbf{A}}(x+n-1) \geq \mu_{\mathbf{A}}(x+n), \qquad n \in N_0. \tag{3.8}$$

2. We also impose the condition

$\forall\, x \in Z,\ \forall\, n \in N_0,\ \forall\, k \in Z:$

$$\mu_{\mathbf{A}}(x+kn)=\mu_{\mathbf{A}}(x). \tag{3.9}$$

If conditions (3.8) and (3.9) are both satisfied, we call **A** the class of fuzzy relative integer modulo *n* cycloconvex.

EXAMPLE 3.3

	−5	−4	−3	−2	−1	0	1	2	3	4	5	6
...	0.5	1	0.9	0.7	0.1	0.5	1	0.9	0.7	0.1	0.5	1
		↑					↑					↑

7	8	9	10	
0.9	0.7	0.1	0.5	...

In shorter notation we have

	0	1	2	3	4	(mod)
A =	0.5	1	0.9	0.7	0.1	5

or

	1	2	3	4	5	(mod)
	1	0.9	0.7	0.1	0.5	5

As we can see, cyclic convexity is defined for the number n of modulo n.

THEOREM 3.2

If **A** and **B** are fuzzy relative integers modulo n cycloconvex, then their max-min convolution gives a fuzzy relative integer modulo n cycloconvex.

Proof:

It is sufficient to represent **A** and **B** by their respective symbols. We can then shift their representatives in such a way so as to obtain the fuzzy numbers.

EXAMPLE 3.4

	0	1	2	3	4	5	(mod)
A =	0.7	0.8	1	0.9	0.1	0.6	6

(3.10)

	0	1	2	3	4	5	(mod)
B =	0.5	0	0.3	0.6	1	0.8	6

(3.11)

To transform these representatives to fuzzy numbers, we have to shift appropriately.

	−1	0	1	2	3	4	(mod)
A =	0.6	0.7	0.8	1	0.9	0.1	6

	1	2	3	4	5	6	(mod)
B =	0	0.3	0.6	1	0.8	0.5	6

Now we use a modulo 6 table (Tables 3.4 and 3.5).

Table 3.4

	1	2	3	4	5	6
−1	0	1	2	3	4	5
0	1	2	3	4	5	0
1	2	3	4	5	0	1
2	3	4	5	0	1	2
3	4	5	0	1	2	3
4	5	0	1	2	3	4

Table 3.5

	1	2	3	4	5	6
−1	0 / 0.6	0.3 / 0.6	0.6 / 0.6	1 / 0.6	0.8 / 0.6	0.5 / 0.6
0	0 / 0.7	0.3 / 0.7	0.6 / 0.7	1 / 0.7	0.8 / 0.7	0.5 / 0.7
1	0 / 0.8	0.3 / 0.8	0.6 / 0.8	1 / 0.8	0.8 / 0.8	0.5 / 0.8
2	0 / 1	0.3 / 1	0.6 / 1	1 / 1	0.8 / 1	0.5 / 1
3	0 / 0.9	0.3 / 0.9	0.6 / 0.9	1 / 0.9	0.8 / 0.9	0.5 / 0.9
4	0 / 0.1	0.3 / 0.1	0.6 / 0.1	1 / 0.1	0.8 / 0.1	0.5 / 0.1

We find that

$$
\begin{aligned}
\mu_{\mathbf{A}(+)_6\mathbf{B}}(0) &= (0.6 \wedge 0) \vee (0.1 \wedge 0.3) \vee (0.9 \wedge 0.6) \vee (1 \wedge 1) \\
&\quad \vee (0.8 \wedge 0.8) \vee (0.7 \wedge 0.5) = 1, \\
(1) &= (0.7 \wedge 0) \vee (0.6 \wedge 0.3) \vee (0.1 \wedge 0.6) \vee (0.9 \wedge 1) \\
&\quad \vee (1 \wedge 0.8) \vee (0.8 \wedge 0.5) = 0.9, \\
(2) &= (0.8 \wedge 0) \vee (0.7 \wedge 0.3) \vee (0.6 \wedge 0.6) \vee (0.1 \wedge 1) \\
&\quad \vee (0.9 \wedge 0.8) \vee (1 \wedge 0.5) = 0.8,
\end{aligned}
$$

$$\begin{aligned}(3) &= (1 \wedge 0) \vee (0.8 \wedge 0.3) \vee (0.7 \wedge 0.6) \vee (0.6 \wedge 1)\\ &\quad \vee (0.1 \wedge 0.8) \vee (0.9 \wedge 0.5) = 0.6,\\ (4) &= (0.9 \wedge 0) \vee (1 \wedge 0.3) \vee (0.8 \wedge 0.6) \vee (0.7 \wedge 1)\\ &\quad \vee (0.6 \wedge 0.8) \vee (0.1 \wedge 0.5) = 0.7,\\ (5) &= (0.1 \wedge 0) \vee (0.9 \wedge 0.3) \vee (1 \wedge 0.6) \vee (0.8 \wedge 1)\\ &\quad \vee (0.7 \wedge 0.8) \vee (0.6 \wedge 0.5) = 0.8.\end{aligned}$$

That is,

$\mathbf{A}\,(+)_6\,\mathbf{B} =$

0	1	2	3	4	5	
1	0.9	0.8	0.6	0.7	0.8	↻ 6

and by shifting

$\mathbf{A}\,(+)_6\,\mathbf{B} =$

−2	−1	0	1	2	3	
0.7	0.8	1	0.9	0.8	0.6	↻ 6

	−6	−5	−4	−3	−2	−1	0	1
...	1	0.9	0.8	0.6	0.7	0.8	1	0.9
	↑						↑	

2	3	4	5	6	7	8	9	
0.8	0.6	0.7	0.8	1	0.9	0.8	0.6	...
				↑				

Let us now consider the algebraic properties of fuzzy relative integers modulo n cycloconvex, which are as follows.

$\forall\ \mathbf{A}, \mathbf{B}, \mathbf{C} \subset Z$ modulo n cycloconvex:

$$\mathbf{A}\,(+)_n\,(\mathbf{B}\,(+)_n\,\mathbf{C}) = (\mathbf{A}\,(+)_n\,\mathbf{B})\,(+)_n\,\mathbf{C}, \qquad \text{associativity,}$$

$$\mathbf{A}\,(+)_n\,\mathbf{B} = \mathbf{B}\,(+)_n\,\mathbf{A}, \qquad \text{commutativity.}$$

A unique neutral exists for each n and for every $\mathbf{A}$ cycloconvex.

$$\begin{aligned}\mu O_n(kn) &= 1, \qquad k \in N_0,\\ &= 0, \qquad x \neq kn,\end{aligned}$$

such that

$$\mathbf{A}\,(+)_n\,O = O\,(+)_n\,\mathbf{A} = \mathbf{A}.$$

For instance, with $n = 4$, we have

$O_4 =$

	−2	−1	0	1	2	3	4	5	6	7	8	9	10	
...	0	0	1	0	0	0	1	0	0	0	1	0	0	...

We must conclude, therefore, that fuzzy relative integers modulo *n* cycloconvex have a commutative monoid structure. This conclusion can be verified by making certain that a symmetry does not exist in this monoid. By comparison with ordinary relative integers modulo *n*, we here see a structure that is not a group.

We note also that fuzzy relative integers modulo *n* cycloconvex can belong to *pseudoneutrals*; that is, to neutrals for some but not for all in the same class.

EXAMPLE 3.5

Consider the neutral modulo 5,

$O_5 =$

0	1	2	3	4
1	0	0	0	0

↻ 5

two fuzzy relative integers modulo 5 cycloconvex,

$A =$

0	1	2	3	4
0.3	0.2	0.1	1	0.7

↻ 5

$B =$

0	1	2	3	4
0.3	0.2	0	1	0.7

↻ 5

and a pseudo-neutral

$C =$

0	1	2	3	4
1	0.1	0	0	0

↻ 5

Indeed,

$\mathbf{A}\ (+)_5\ \mathbf{C} =$

0	1	2	3	4
0.3	0.2	0.1	1	0.7

$(+)_5$

0	1	2	3	4
1	0.1	0	0	0

$=$

0	1	2	3	4
0.3	0.2	0.1	1	0.7

$= \mathbf{A}.$

$\mathbf{B}\ (+)_5\ \mathbf{C} =$

0	1	2	3	4
0.3	0.2	0	1	0.7

$(+)_5$

0	1	2	3	4
1	0.1	0	0	0

$=$

0	1	2	3	4
0.3	0.2	0.1	1	0.7

$\neq \mathbf{B}.$

Now let us consider multiplication modulo n by max-min convolution. First we ask whether this convolution is conservative for cycloconvexity. The answer is no.

EXAMPLE 3.6

Consider (3.10) and (3.11) after a convenient shifting.

$\mathbf{A} =$

−1	0	1	2	3	4
0.6	0.7	0.8	1	0.9	0.1

↻ 6

$\mathbf{B} =$

1	2	3	4	5	6
0	0.3	0.6	1	0.8	0.5

↻ 6 **(3.12)**

Now we use a multiplication table modulo 6 (Tables 3.6 and 3.7).

We compute the max-min convolution in Equation (3.12) row by row. The method of computation used is different from the one used in section 1.5 for ordinary multiplication. Indeed we cannot pass from Z to R; we must stay in Z, in fact in Z/n.

$$\begin{aligned}\mu_{\mathbf{A}\,(\cdot)_6\mathbf{B}}(0) = {} & (0.6 \wedge 0.5) \vee (0.7 \wedge 0) \vee (0.7 \wedge 0.3) \vee (0.7 \wedge 0.6)\\ & \vee (0.7 \wedge 0.1) \vee (0.7 \wedge 0.8) \vee (0.7 \wedge 0.5) \vee (0.8 \wedge 0.5)\\ & \vee (1 \wedge 0.6) \vee (1 \wedge 0.5) \vee (0.9 \wedge 0.3) \vee (0.9 \wedge 0.1)\\ & \vee (0.9 \wedge 0.5) \vee (0.1 \wedge 0.6) \vee > (0.1 \wedge .5) = 0.9,\end{aligned}$$

$$\begin{aligned}(1) = {} & (0.6 \wedge 0.8) \vee (0.8 \wedge 0) = 0.6,\\ (2) = {} & (0.6 \wedge 0.8) \vee (0.3 \wedge 0.8) \vee (1 \wedge 0) \vee (1 \wedge 1) \vee (0.1 \wedge 0.3)\\ & \vee (0.1 \wedge 0.8) = 1,\\ (3) = {} & (0.6 \wedge 0.6) \vee (0.6 \wedge 0.8) \vee (0.9 \wedge 0) \vee (0.6 \wedge 0.8)\\ & \vee (0.8 \wedge 0.9) = 0.8,\\ (4) = {} & (0.6 \wedge 0.3) \vee (0.8 \wedge 1) \vee (1 \wedge 0.3) \vee (1 \wedge 0.8) \vee (0.1 \wedge 0)\\ & \vee (0.1 \wedge 1) = 0.8,\\ (5) = {} & (0.6 \wedge 0) \vee (0.8 \wedge 0.8) = 0.8,\end{aligned}$$

Table 3.6

$A\ (\cdot)_6\ B$	1	2	3	4	5	6
−1	5	4	3	2	1	0
0	0	0	0	0	0	0
1	1	2	3	4	5	0
2	2	4	0	2	4	0
3	3	0	3	0	3	0
4	4	2	0	4	2	0

Table 3.7

A \ B	1	2	3	4	5	6
−1	0 / 0.6	0.3 / 0.6	0.6 / 0.6	1 / 0.6	0.8 / 0.6	0.5 / 0.6
0	0 / 0.7	0.3 / 0.7	0.6 / 0.7	1 / 0.7	0.8 / 0.7	0.5 / 0.7
1	0 / 0.8	0.3 / 0.8	0.6 / 0.8	1 / 0.8	0.8 / 0.8	0.5 / 0.8
2	0 / 1	0.3 / 1	0.6 / 1	1 / 1	0.8 / 1	0.5 / 1
3	0 / 0.9	0.3 / 0.9	0.6 / 0.9	1 / 0.9	0.8 / 0.9	0.5 / 0.9
4	0 / 0.1	0.3 / 0.1	0.6 / 0.1	1 / 0.1	0.8 / 0.1	0.5 / 0.1

which gives

$\mathbf{A}\,(+)_6\,\mathbf{B} =$

0	1	2	3	4	5	
0.9	0.6	1	0.8	0.8	0.8	↺ 6

In this multiplication modulo n operation cycloconvexity of **A** and **B** is not preserved. We thus cannot generalize the theory of relative integers modulo n. It does not have the structure of a ring for $(+)_n$ and $(\cdot)_n$, but only the structure of a monoid for $(+)_n$.

Special Case with Ring Structure

Consider the following fuzzy relative integer of modulo n cycloconvex.

$\mathbf{1} =$

0	1	2	3	4	---	$n-1$		
a	1	a	a	a	---	a	↺ n,	$0 \le a < 1$

(3.13)

From this fuzzy number modulo we construct

$$\mathbf{2} = \mathbf{1}\,(+)_n\,\mathbf{1}$$

as

$\mathbf{2} =$

0	1	2	3	4	---	$n-1$		
a	a	1	a	a	---	a	↺ n,	$0 < a < 1$

$\mathbf{3} =$

0	1	2	3	4	---	$n-1$		
a	a	a	1	a	---	a	↺ n,	$0 \le a < 1$

$\mathbf{N} - \mathbf{i} =$

0	1	2	3	4	---	$n-1$		
a	a	a	a	a	---	1	↺ n,	$0 \le a < 1$

0	1	2	3	4	---	$n-1$		
1	a	a	a	a	---	a	↺ n,	$0 \le a \le 1$

Between these fuzzy relative integers modulo n cycloconvex and relative integers modulo n Z/n we see an isomorphism that relates to addition and multiplication. Thus, for these special and almost trivial forms, these fuzzy relative integers modulo n cycloconvex have a ring structure. An interesting question arises from this study: Will fuzzy relative integers modulo n cycloconvex other than those of (3.13) also have a ring structure? We suspect that they do not.

3.4 FUZZY REAL NUMBERS MODULO 1

The ordinary concept of real numbers modulo n is denoted by R/n. Two real numbers a, $b \in R$ will be the equivalent modulo n, $n \in N_0$, if and only if a and b have the same result when divided by n. We write

$$a = b, \quad \text{modulo } n.$$

For instance,

$$\cdots = -2.63 = -0.63 = 1.37 = 3.37 = \cdots \qquad \text{modulo } 2,$$
$$\cdots = -1.518 = -0.518 = 0.482 = 1.482 = \cdots \qquad \text{modulo } 1.$$

The properties of the modulo arithmetic are as follows:

1. $(a +_n b) +_n c = a +_n (b +_n c)$, associative.

2. If a neutral exists, the ordinary 0 is

$$a +_n 0 = 0 +_n a = a.$$

3. A symmetric exits for each a

$$a +_n (-a) = (-a) +_n a = 0. \tag{3.14}$$

4. $a +_n c = c +_n a$, commutative.

In this case a group structure exists.

Multiplication modulo n is not associative. As an example, consider a modulo 2 multiplication:

$$(1.8 \cdot_2 1.7) \cdot_2 1.6 = 1.06 \cdot_2 1.6 = 1.696,$$
$$(1.8 \cdot_2 (1.7 \cdot_2 1.6) = 1.8 \cdot_2 0.72 = 1.296,$$

which is not associative.

We must be satisfied with only the property of the group for addition, since a ring structure is not available for the couplet $(+_n, \cdot_n)$.

We now introduce the concept of modulo 1.

1. The properties of associativity for $\cdot_n$ for $n = 1$ are given below:

$$a \cdot_1 b \in [0, 1), \text{ then } a \cdot_1 b = a \cdot b.$$
$$(a \cdot_1 b) \cdot_1 c \in [0, 1), \text{ then } (a \cdot_1 b) \cdot_1 c = (a \cdot b) \cdot c.$$
$$b \cdot_1 c \in [0, 1), \text{ then } b \cdot_1 c = b \cdot c.$$
$$a \cdot_1 (b \cdot_1 c) \in [0, 1), \text{ then } a \cdot_1 (b \cdot_1 c) = a \cdot (b \cdot c).$$

Therefore, we have $(a \cdot_1 b) \cdot_1 c = a \cdot_1 (b \cdot_1 c)$.

2. The neutral of multiplication modulo 1 cannot be one because $1 = 0$ modulo 1.

3. Commutativity is satisfied if $a \cdot_1 b = b \cdot_1 a$.

Because a neutral does not exist for $\cdot_1$ does not allow us to specify a monoid. That is, $+_1$ and $\cdot_1$ are not distributive. For example,

$$(0.9 +_1 0.8) \cdot_1 0.4 = 0.7 \cdot_1 0.4 = 0.28.$$
$$(0.9 \cdot_1 0.4) +_1 (0.8 \cdot_1 0.4) = 0.36 +_1 0.32 = 0.68.$$

Even if fuzzy real modulo 1 numbers do not have a ring structure and their use is limited to addition, nevertheless they are of interest in several applications such as the Theory of Gomory for programming with integer values in economic models. Let us now study fuzzy real modulo 1.

Fuzzy Reals Modulo 1 Cycloconvex

This concept is defined as follows:

$$\mu_{\mathbf{A}}(h) = 1, \qquad h \in [a, a+1), \qquad a \in R,$$
$$\mu_{\mathbf{A}}(x) \text{ is monotonically increasing for } x \in [a, h]$$
$$\mu_{\mathbf{A}}(x) \text{ is monotonically decreasing for } x \in [h, a+1),$$
$$\mu_{\mathbf{A}}(kx) = \mu(x), \qquad k \in Z.$$

Let us study the following example.

EXAMPLE 3.7

This example is shown in Figure 3.1. The fuzzy real modulo 1 cycloconvex number is represented by the abscissa h; that is, **h**. Later, we use Table 3.8 of addition modulo 1 from 0 to 1.

The max-min convolution for addition modulo 1 is given by

$(\mathbf{A}\,(+)_1 \mathbf{B}) \Leftrightarrow (\forall\ x, y, z \in R$:

$$\mu_{\mathbf{A}\,(+)_1 \mathbf{B}}(z) = \bigvee_{z = x +_1 y} (\mu_{\mathbf{A}}(x) \wedge \mu_{\mathbf{B}}(y)). \qquad \textbf{(3.15)}$$

Let us examine how we may use (3.15).

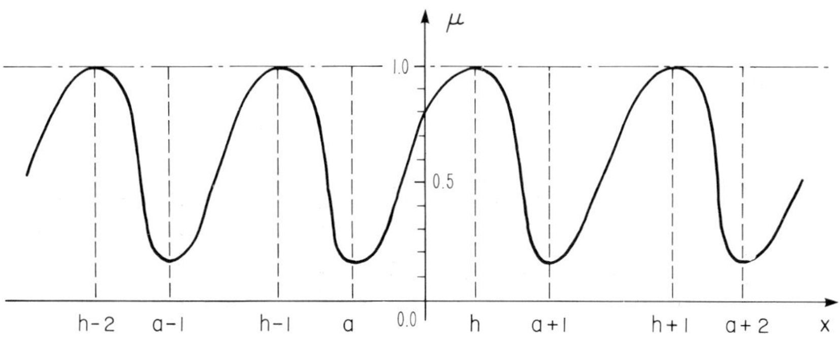

Figure 3.1 A cycloconvex real number modulo 1 (Example 3.7).

Table 3.8

$+_1$	0	0.1	0.2	0.3	0.4	0.5	0.6	0.7	0.8	0.9	
0	0	0.1	0.2	0.3	0.4	0.5	0.6	0.7	0.8	0.9	0.9
0.1	0.1	0.2	0.3	0.4	0.5	0.6	0.7	0.8	0.9	0	0
0.2	0.2	0.3	0.4	0.5	0.6	0.7	0.8	0.9	0	0.1	0.1
0.3	0.3	0.4	0.5	0.6	0.7	0.8	0.9	0	0.1	0.2	0.2
0.4	0.4	0.5	0.6	0.7	0.8	0.9	0	0.1	0.2	0.3	0.3
0.5	0.5	0.6	0.7	0.8	0.9	0	0.1	0.2	0.3	0.4	0.4
0.6	0.6	0.7	0.8	0.9	0	0.1	0.2	0.3	0.4	0.5	0.5
0.7	0.7	0.8	0.9	0	0.1	0.2	0.3	0.4	0.5	0.6	0.6
0.8	0.8	0.9	0	0.1	0.2	0.3	0.4	0.5	0.6	0.7	0.7
0.9	0.9	0	0.1	0.2	0.3	0.4	0.5	0.6	0.7	0.8	0.8
	0.9	0	0.1	0.2	0.3	0.4	0.5	0.6	0.7	0.8	

EXAMPLE 3.8

In Tables 3.2 and 3.3, let **A** and **B** be two fuzzy reals modulo 1 cycloconvex given at each tenth value. To add these numbers, we use Table 3.8 and Equation (3.15), taking the diagonals in Table 3.8 as shown by Equations (3.16) and (3.17) and Table 3.9.

	0	0.1	0.2	0.3	0.4	0.5	0.6	0.7	0.8	0.9		
A =	0.3	0.3	0.1	0.2	0.7	0.9	1	0.8	0.6	0.5	↰ 1	**(3.16)**

	0	0.1	0.2	0.3	0.4	0.5	0.6	0.7	0.8	0.9		
B =	0.4	0.2	0	0.1	0.4	0.7	0.7	1	0.9	0.7	↰ 1	**(3.17)**

Table 3.9

A \ B	0	0.1	0.2	0.3	0.4	0.5	0.6	0.7	0.8	0.9	
0	0.4 / 0.3	0.2 / 0.3	0 / 0.3	0.1 / 0.3	0.4 / 0.3	0.7 / 0.3	0.7 / 0.3	1 / 0.3	0.9 / 0.3	0.7 / 0.3	0.9
0.1	0.4 / 0.3	0.2 / 0.3	0 / 0.3	0.1 / 0.3	0.4 / 0.3	0.7 / 0.3	0.7 / 0.3	1 / 0.3	0.9 / 0.3	0.7 / 0.3	0
0.2	0.4 / 0.1	0.2 / 0.1	0 / 0.1	0.1 / 0.1	0.4 / 0.1	0.7 / 0.1	0.7 / 0.1	1 / 0.1	0.9 / 0.1	0.7 / 0.1	0.1
0.3	0.4 / 0.2	0.2 / 0.2	0 / 0.2	0.1 / 0.2	0.4 / 0.2	0.7 / 0.2	0.7 / 0.2	1 / 0.2	0.9 / 0.2	0.7 / 0.2	0.2
0.4	0.4 / 0.7	0.2 / 0.7	0 / 0.7	0.1 / 0.7	0.4 / 0.7	0.7 / 0.7	0.7 / 0.7	1 / 0.7	0.9 / 0.7	0.7 / 0.7	0.3
0.5	0.4 / 0.9	0.2 / 0.9	0 / 0.9	0.1 / 0.9	0.4 / 0.9	0.7 / 0.9	0.7 / 0.9	1 / 0.9	0.9 / 0.9	0.7 / 0.9	0.4
0.6	0.4 / 1	0.2 / 1	0 / 1	0.1 / 1	0.4 / 1	0.7 / 1	0.7 / 1	1 / 1	0.9 / 1	0.7 / 1	0.5
0.7	0.4 / 0.8	0.2 / 0.8	0 / 0.8	0.1 / 0.8	0.4 / 0.8	0.7 / 0.8	0.7 / 0.8	1 / 0.8	0.9 / 0.8	0.7 / 0.8	0.6
0.8	0.4 / 0.6	0.2 / 0.6	0 / 0.6	0.1 / 0.6	0.4 / 0.6	0.7 / 0.6	0.7 / 0.6	1 / 0.6	0.9 / 0.6	0.7 / 0.6	0.7
0.9	0.4 / 0.5	0.2 / 0.5	0 / 0.5	0.1 / 0.5	0.4 / 0.5	0.7 / 0.5	0.7 / 0.5	1 / 0.5	0.9 / 0.5	0.7 / 0.5	0.8
	0.9	0	0.1	0.2	0.3	0.4	0.5	0.6	0.7	0.8	

Diagonals

By searching in the diagonals, we find

	0	0.1	0.2	0.3	0.4	0.5	0.6	0.7	0.8	0.9	
$\mathbf{A}\ (+)_1\ \mathbf{B} =$	0.7	0.7	0.9	1	0.9	0.8	0.7	0.6	0.5	0.7	↺ 1

This process remains unchanged no matter what values we take for the subdivisions. It is interesting, therefore, to give a theorem for the property of associativity.

THEOREM 3.3

The addition modulo 1 is associative for fuzzy reals modulo 1. That is, $\forall$ **A**, **B**, **C** fuzzy reals modulo 1 cycloconvex:

$$(\mathbf{A}\,(+)_1\,\mathbf{B})\,(+)_1\,\mathbf{C} = \mathbf{A}\,(+)_1\,(\mathbf{B}\,(+)_1\,\mathbf{C}). \tag{3.18}$$

Proof:

We can modify a fuzzy real modulo 1 cycloconvex number in such a way that each representative is shown as a convex and normal fuzzy subset by a convenient shifting.

$$\mathbf{A}' = \mathbf{A} \text{ modulo } 1,$$
$$\mathbf{B}' = \mathbf{B} \text{ modulo } 1,$$
$$\mathbf{C}' = \mathbf{C} \text{ modulo } 1.$$

The max-min convolution for the addition of fuzzy numbers being associative, we can write

$$(\mathbf{A}'\,(+)_1\,\mathbf{B}')\,(+)_1\,\mathbf{C}' = \mathbf{A}'\,(+)_1\,(\mathbf{B}'\,(+)_1\,\mathbf{C}').$$

Since the result at the left is the same as that at the right, another shifting will give a result in [0, 1), for example. Equation (3.18) must always be checked at the end.

EXAMPLE 3.9

Taking (3.16) and (3.17) again we have

C = (cyclic, period 1) (3.19)

0	0.1	0.2	0.3	0.4	0.5	0.6	0.7	0.8	0.9
0.4	0.4	0.7	1	0.8	0.6	0.3	0	0.1	0.3

First let us shift **A** in (3.16) and **B** in (3.17) so that they are convex. Thus

A′ = (cyclic, period 1) (3.20)

0.2	0.3	0.4	0.5	0.6	0.7	0.8	0.9	1	1.1
0.1	0.2	0.7	0.9	1	0.8	0.6	0.5	0.3	0.3

B′ = (cyclic, period 1) (3.21)

0.2	0.3	0.4	0.5	0.6	0.7	0.8	0.9	1	1.1
0	0.1	0.4	0.7	0.7	1	0.9	0.7	0.4	0.2

We find that

A′ (+) **B′** =

0.4	0.5	0.6	0.7	0.8	0.9	1	1.1	1.2	1.3
0	0.1	0.1	0.2	0.4	0.7	0.7	0.7	0.9	1

1.4	1.5	1.6	1.7	1.8	1.9	2.0	2.1	2.2
0.9	0.8	0.7	0.6	0.5	0.4	0.3	0.3	0.2

In this result we have to take the largest value of μ to respect the max-min convolution. We thus have

$\mathbf{A}\ (+)_1\ \mathbf{B} =$

0	0.1	0.2	0.3	0.4	0.5	0.6	0.7	0.8	0.9
0.7	0.7	0.9	1	0.9	0.8	0.7	0.6	0.5	0.7

↻ 1 **(3.22)**

which shows that we have computed (3.14) once again, but in another way.

Now let us shift (3.22) to show its convexity.

$(\mathbf{A}\ (+)_1\ \mathbf{B})' =$

0.9	0	0.1	0.2	0.3	0.4	0.5	0.6	0.7	0.8
0.7	0.7	0.7	0.9	1	0.9	0.8	0.7	0.6	0.5

↻ 1 **(3.23)**

Let us also shift (3.19).

$\mathbf{C}' =$

0.7	0.8	0.9	0	0.1	0.2	0.3	0.4	0.5	0.6
0	0.1	0.3	0.4	0.4	0.7	1	0.8	0.6	0.3

↻ 1 **(3.24)**

Now let us take the max-min convolution of (3.23) and (3.24) for ordinary addition.

$(\mathbf{A}\ (+)\ \mathbf{B})'\ (+)\ \mathbf{C}' =$

1.6	1.7	1.8	1.9	2	2.1	2.2	2.3	2.4	2.5	2.6
0	0.1	0.3	0.4	0.4	0.4	0.7	0.7	0.7	0.9	1

2.7	2.8	2.9	3.0	3.1	3.2	3.3	3.4
0.9	0.8	0.8	0.7	0.6	0.6	0.5	0.3

(3.25)

We can now compute $\mathbf{B}'\ (+)\ \mathbf{C}'$ using (3.21) and (3.24)

$\mathbf{B}'\ (+)\ \mathbf{C}' =$

0.9	1	1.1	1.2	1.3	1.4	1.5	1.6	1.7	1.8	1.9	2.0
0	0	0.1	0.1	0.3	0.4	0.4	0.4	0.7	0.7	0.7	1
							0.6	0.7	0.8	0.9	0

2.1	2.2	2.3	2.4	2.5	2.6	2.7
0.9	0.8	0.7	0.6	0.4	0.3	0.2
0.1	0.2	0.3	0.4	0.5		

This gives

$(\mathbf{B}\ (+)_1\ \mathbf{C})' =$ **(3.26)**

0.6	0.7	0.8	0.9	1	0.1	0.2	0.3	0.4	0.5
0.4	0.7	0.7	0.7	1	0.9	0.8	0.7	0.6	0.4

↺ 1

We will now compute $\mathbf{A}'\ (+)_1\ (\mathbf{B}\ (+)_1\ \mathbf{C})'$ using (3.20) and (3.26):

$\mathbf{A}'\ (+)_1\ (\mathbf{B}\ (+)_1\ \mathbf{C})' =$

0.8	0.9	1	1.1	1.2	1.3	1.4	1.5	1.6
0.1	0.2	0.4	0.7	0.7	0.7	0.7	0.9	1
				0.2	0.3	0.4	0.5	0.6

1.7	1.8	1.9	2.0	2.1	2.2	2.3	2.4	2.5	2.6
0.9	0.8	0.8	0.7	0.6	0.6	0.5	0.4	0.3	0.3
0.7	0.8	0.9	0	0.1					

(3.27)

Comparing (3.25) and (3.27) we find the same result:

$$(\mathbf{A}\ (+)_1\ \mathbf{B})\ (+)_1\ \mathbf{C} = \mathbf{A}\ (+)_1\ (\mathbf{B}\ (+)_1\ \mathbf{C}) =$$

0	0.1	0.2	0.3	0.4	0.5	0.6	0.7	0.8	0.9
0.7	0.6	0.7	0.7	0.7	0.9	1	0.9	0.8	0.8

↺ 1

We have thus shown that addition modulo 1 by max-min convolution is associative and commutative. This kind of addition has a neutral. It is the usual 0 of real numbers.

$$\mathbf{A}\,(+)_1\,0 = 0\,(+)_1\,\mathbf{A} = \mathbf{A}.$$

Hence, fuzzy real numbers modulo 1 cycloconvex have a commutative monoid structure for addition modulo 1 by max-min convolution.

Addition of a Real Number (Nonfuzzy) with a Fuzzy Real Modulo 1 Cycloconvex

We consider the following real nonfuzzy number:

$\forall\ x \in R$:

$$\mu_{\mathbf{B}}(x) = 1, \qquad x = b \in R,$$
$$= 0, \qquad x \neq b.$$

To add such a real to a fuzzy real modulo 1 cycloconvex number we must use ordinary addition modulo 1 followed by a shift.

EXAMPLE 3.10

Let

$\mathbf{A} =$

0	0.1	0.2	0.3	0.4	0.5	0.6	0.7	0.8	0.9
0.3	0	0.1	0.5	1	1	0.8	0.5	0.5	0.4

↻ 1

and then add 3.2 to **A**. The result is 3.2 = 0.2 modulo 1. We then have

$\mathbf{A}\,(+)_1\,\mathbf{3.2} =$

0.2	0.3	0.4	0.5	0.6	0.7	0.8	0.9	1	1.1
0.3	0	0.1	0.5	1	1	0.8	0.5	0.5	0.4

↻ 1

or

$\mathbf{A}\,(+)_1\,\mathbf{0.2} =$

0	0.1	0.2	0.3	0.4	0.5	0.6	0.7	0.8	0.9
0.5	0.4	0.3	0	0.1	0.5	1	1	0.8	0.5

↻ 1

Now let us add −1.3 to **A**. The result is (−1.3) = 0.7 modulo 1.

$\mathbf{A}\,(+)_1\,\mathbf{0.7} =$

0.7	0.8	0.9	0	0.1	0.2	0.3	0.4	0.5	0.6
0.3	0	0.1	0.5	1	1	0.8	0.5	0.5	0.4

↻ 1

$\mathbf{A}\,(+)_1\,\mathbf{0.7} =$

0	0.1	0.2	0.3	0.4	0.5	0.6	0.7	0.8	0.9
0.5	1	1	0.8	0.5	0.5	0.4	0.3	0	0.1

↻ 1

Multiplication of a Fuzzy Real Modulo 1 Cycloconvex by a Nonfuzzy Integer

Consider an integer, $n \in N_0^+$, and, **A**, a fuzzy real number modulo 1 cycloconvex. We then have

$$n \cdot \mathbf{A} = \underbrace{\mathbf{A}\,(+)_1\,\mathbf{A}\,(+)_1 \cdots (+)_1\,\mathbf{A}}_{n \text{ times}}.$$

EXAMPLE 3.11

$\mathbf{A} =$

0	0.1	0.2	0.3	0.4	0.5	0.6	0.7	0.8	0.9
0.4	0.5	0.8	1	0.7	0.2	0.1	0	0.1	0.3

↻ 1

Let us transpose **A** to a fuzzy number:

$\mathbf{A}' =$

0.6	0.7	0.8	0.9	1	1.1	1.2	1.3	1.4	1.5	
0.1	0.3	0.4	0.5	0.8	1	0.7	0.2	0.1	0	↑ 1

then

$\mathbf{A}' \, (+)_1 \, \mathbf{A}' =$

1.2	1.3	1.4	1.5	1.6	1.7	1.8	1.9
0.1	0.1	0.3	0.3	0.4	0.4	0.5	0.5

2.0	2.1	2.2	2.3	2.4	2.5	2.6	2.7	2.8	2.9	3.0
0.8	0.8	1	0.7	0.7	0.2	0.2	0.1	0.1	0	0
0	0.1	0.2	0.3	0.4	0.5	0.6	0.7	0.8	0.9	

That is,

$\mathbf{2} \cdot \mathbf{A} =$

0	0.1	0.2	0.3	0.4	0.5	0.6	0.7	0.8	0.9	
0.8	0.8	1	0.7	0.7	0.3	0.4	0.4	0.5	0.5	↑ 1

$(\mathbf{2} \cdot \mathbf{A})' =$

0.6	0.7	0.8	0.9	1	1.1	1.2	1.3	1.4	1.5
0.4	0.4	0.5	0.5	0.8	0.8	1	0.7	0.7	0.3

$\mathbf{A}' \, (+)_1 \, (\mathbf{2} \cdot \mathbf{A})' =$

1.2	1.3	1.4	1.5	1.6	1.7	1.8	1.9
0.1	0.3	0.4	0.4	0.4	0.5	0.5	0.5

2.0	2.1	2.2	2.3	2.4	2.5	2.6	2.7	2.8	2.9	3.0
0.8	0.8	0.8	1	0.7	0.7	0.7	0.3	0.2	0.1	0
0	0.1	0.2	0.3	0.4	0.5	0.6	0.7	0.8	0.9	

$\mathbf{3} \cdot \mathbf{A} =$

0	0.1	0.2	0.3	0.4	0.5	0.6	0.7	0.8	0.9	
0.8	0.8	0.8	1	0.7	0.7	0.7	0.5	0.5	0.5	↑ 1

$(\mathbf{3} \cdot \mathbf{A})' =$

0	0.1	0.2	0.3	0.4	0.5	0.6	0.7	0.8	0.9
0.8	0.8	0.8	1	0.7	0.7	0.7	0.5	0.5	0.5

$(\mathbf{3} \cdot \mathbf{A})' \, (+)_1 \, (\mathbf{3} \cdot \mathbf{A})' =$

0	0.1	0.2	0.3	0.4	0.5	0.6
0.8	0.8	0.8	0.8	0.8	0.8	1

0 0.1 0.2 0.3 0.4 0.5 0.6

0.7	0.8	0.9	1	1.1	1.2	1.3	1.4	1.5	1.6	1.7	1.8
0.7	0.7	0.7	0.7	0.7	0.7	0.5	0.5	0.5	0.5	0.5	0.5

0.7 0.8 0.9

$\mathbf{6} \cdot \mathbf{A} =$

0	0.1	0.2	0.3	0.4	0.5	0.6	0.7	0.8	0.9
0.8	0.8	0.8	0.8	0.8	0.8	1	0.7	0.7	0.7

↻ 1

We see from these equations that the property of convergence appears. You may check the corresponding theorem for

$$\lim_{n \to \infty} n \cdot \mathbf{A}, n \in N_0^+.$$

EXAMPLE 3.12

Let us now consider this example. Figure 3.2 shows the characteristic of a **O** fuzzy real modulo 1 cycloconvex. If we study Figure 3.2 we see that **O**, which is not the neutral 0 but a fuzzy real modulo 1 among others, becomes asymptotic to the straight line $\mu = 1$ as $k \to \infty$. The same phenomenon appears for any fuzzy real modulo 1 cycloconvex in R because each time such a number is added to itself the spaces, as can be seen from Figure 3.2, become smaller and finally disappear when $k \to \infty$, where k is the number of such repetitive additions. This is possible, of course, only when the fuzzy real modulo 1 does not have a discontinuity. If, in this example, we have

$$\begin{aligned} \mu_{\mathbf{A}}(x) &= 1, \qquad x = 0, \\ &= \beta, \qquad x = [1 - a, 0) \cup (0, a], \end{aligned}$$

the number will stretch over all noninteger values up to $\alpha = \beta$. We leave it to the reader to study this phenomenon.

Concept of Interval of Confidence Modulo 1

We now define a new concept by using the following symbols: $(a —— b)$, $a, b \in [0, 1)$, such that

$$\begin{aligned} (a —— b) &= \{x \in [0, 1) \,|\, a \leq x \leq b, \text{ if } a \leq b\} \\ &= \{x \in [0, 1) \,|\, 0 \leq x \leq b, a \leq x < 1, \text{ if } a \geq b\}. \end{aligned}$$

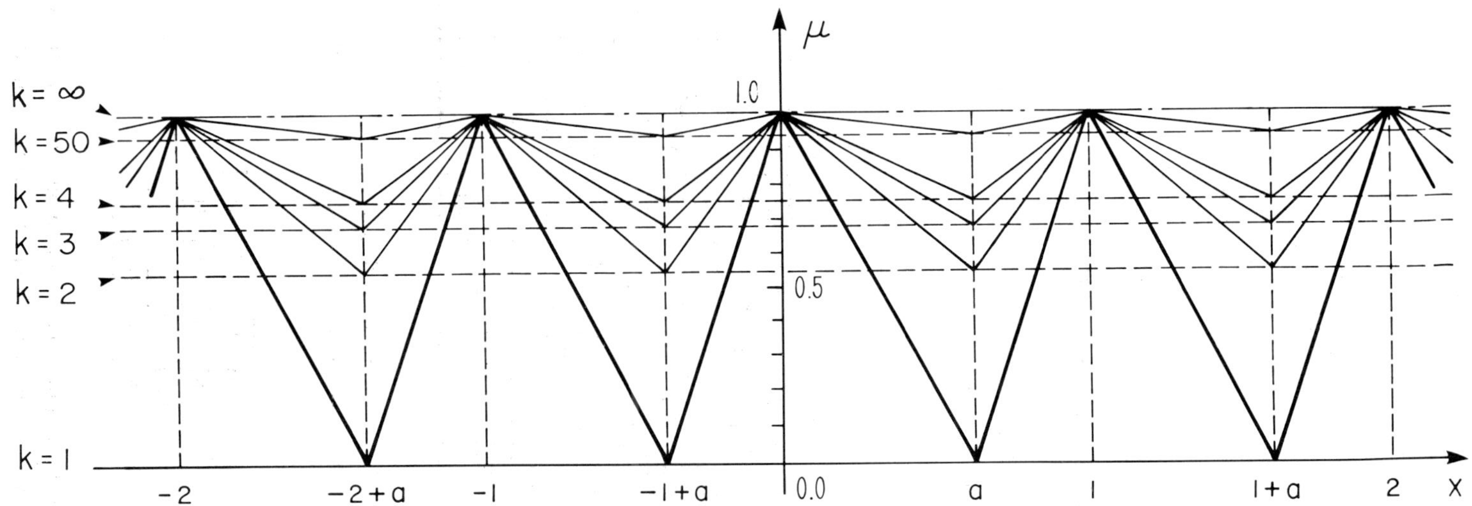

Figure 3.2 A convergence property of $\lim_{n \to \infty} n \cdot \mathbf{A}$, $n \in N_0^+$ for a fuzzy real number modulo 1 cycloconvex **A** (Example 3.8).

This concept is called the *interval of confidence modulo 1*.[2] Hence, (0.2—0.7) means the continum of reals between 0.2 to 0.7, and (0.8—0.1) means the continum from 0.8 to $1-\epsilon$ and its union with the continum from 0 to 0.1. Alternatively, we have

$$\begin{aligned} \mu(0.2-0.7)(x) &= 1, && 0.2 \le x \le 0.7, \\ &= 0, && 0 \le x \le 0.2, 0.7 \le x \le 1. \\ \mu(0.8-0.1)(x) &= 1, && 0.8 \le x \le 1, 0 \le x \le 0.1, \\ &= 0, && 0 < x < 0.8. \end{aligned}$$

Addition of Two Intervals of Confidence Modulo 1

To illustrate the addition of two intervals of confidence modulo 1 in a direct form for a given level α, we again use Table 3.8 and obtain Table 3.10.

Table 3.10

$(+)_1$	0	0.1	0.2	0.3	0.4	0.5	0.6	0.7	0.8	0.9
0	0	0.1	0.2	0.3	0.4	0.5	0.6	0.7	0.8	0.9
0.1	0.1	0.2	0.3	0.4	0.5	0.6	0.7	0.8	0.9	0
0.2	0.2	0.3	0.4	0.5	0.6	0.7	0.8	0.9	0	0.1
0.3	0.3	0.4	0.5	0.6	0.7	0.8	0.9	0	0.1	0.2
0.4	0.4	0.5	0.6	0.7	0.8	0.9	0	0.1	0.2	0.3
0.5	0.5	0.6	0.7	0.8	0.9	0	0.1	0.2	0.3	0.4
0.6	0.6	0.7	0.8	0.9	0	0.1	0.2	0.3	0.4	0.5
0.7	0.7	0.8	0.9	0	0.1	0.2	0.3	0.4	0.5	0.6
0.8	0.8	0.9	0	0.1	0.2	0.3	0.4	0.5	0.6	0.7
0.9	0.9	0	0.1	0.2	0.3	0.4	0.5	0.6	0.7	0.8

Let us suppose that we have to add (0.5——0.8) (+) (0.2——0.4). We then have Table 3.11. As we see, the result gives (0.7——0.2).

Table 3.11

+	0.2	0.3	0.4
0.5	0.7	0.8	0.9
0.6	0.8	0.9	0
0.7	0.9	0	0.1
0.8	0	0.1	0.2

Table 3.11 can be decomposed, obviously, into hundredths, thousandths, or whatever values we wish to use. However, we see that

$$(a_1 \text{——} b_1)\,(+)_1\,(a_2 \text{——} b_2) = (a_1 +_1 a_2 \text{——} b_1 +_1 b_2).$$

Using the notations and concepts introduced previously, we are able to decompose (3.15) level by level.

Let for this purpose,

$$A_\alpha = (a_1^{(\alpha)} — a_2^{(\alpha)}),$$
$$B_\alpha = (b_1^{(\alpha)} — b_2^{(\alpha)}),$$
$$A_\alpha (+)_1 B_\alpha = (a_1^{(\alpha)} — a_2^{(\alpha)})\,(+)_1\,(b_1^{(\alpha)} — b_2^{(\alpha)})$$
$$= (a_1^{(\alpha)} +_1 b_1^{(\alpha)} — a_2^{(\alpha)} +_1 b_2^{(\alpha)}). \qquad \textbf{(3.28)}$$

To compute using (3.15) or (3.28) level by level gives the same result. This can be seen from Table 3.12.

Table 3.12

	0.2	0.3	0.4	0.5	0.6	0.7	0.8	0.9	0	0.1
α	0.1	0.2	0.7	0.9	1	.8	.6	.5	.3	.3
0	1	1	1	1	1	1	1	1	1	1
0.1	1	1	1	1	1	1	1	1	1	1
0.2		1	1	1	1	1	1	1	1	1
0.3			1	1	1	1	1	1	1	1
0.4			1	1	1	1	1	1		
0.5			1	1	1	1	1	1		
0.6			1	1	1	1	1			
0.7			1	1	1	1				
0.8				1	1	1				
0.9				1	1					
					1					

	0.3	0.4	0.5	0.6	0.7	0.8	0.9	0	0.1	0.2
α	0.1	0.4	0.7	0.7	1	0.9	0.7	0.1	0.2	0
	1	1	1	1	1	1	1	1	1	1
	1	1	1	1	1	1	1	1	1	
		1	1	1	1	1	1	1	1	
		1	1	1	1	1	1	1		
$(+)_1$		1	1	1	1	1	1	1		
			1	1	1	1	1			
			1	1	1	1	1			
			1	1	1	1	1			
					1	1				
					1	1				
					1					

Table 3.12 (*continued*)

	0.5	0.6	0.7	0.8	0.9	0	0.1	0.2	0.3	0.4
	1	1	1	1	1	1	1	1	1	1
	1	1	1	1	1	1	1	1	1	1
	1	1	1	1	1	1	1	1	1	1
	1	1	1	1	1	1	1	1	1	1
	1	1	1	1	1	1	1	1	1	1
=	1	1	1	1	1	1	1	1	1	1
	1	1	1		1	1	1	1	1	1
	1	1			1	1	1	1	1	1
	1							1	1	1
								1	1	1
									1	
	0.8	0.7	0.6	0.5	0.7	0.7	0.7	0.9	1	0.9

Restoring the order from 0 to 0.9 we obtain

	0	0.1	0.2	0.3	0.4	0.5	0.6	0.7	0.8	0.9	
$\mathbf{A}\ (+)_1\ \mathbf{B} =$	0.7	0.7	0.9	1	0.9	0.8	0.7	0.6	0.5	0.7	↶ 1

It is easy to show now that for addition the interval of confidence modulo 1 is associative and commutative and has a neutral (0). This structure is, therefore, a commutative monoid.

3.5 FACTORIALS, SEQUENCES, AND SERIES WITH FUZZY NUMBERS

Factorial with Fuzzy Numbers

The concept of the factorial

$$n \in N, \qquad n! = 1 \cdot 2 \cdot 3 \cdot 4 \ldots (n-1) \cdot n,$$

with $0! = 1$, is easily extended to fuzzy case. Before we do this let us consider multiplication by max-min convolution. Here, as explained in section 1.5, when we pass from R^+ to N, and if we consider, only integers in N, then we may lose convexity. To avoid this loss, we must proceed by means of nesting. Let **1** be a fuzzy number in R^+ with the following membership function.

$\forall\, x \in R^+$:

$$\mu_{\mathbf{1}}(x) \in [0, 1], \quad \text{and} \quad \mu_{\mathbf{1}}(1) = 1.$$

We now proceed with the successive construction of fuzzy numbers as follows:

$$\mathbf{2} = \mathbf{1}\,(+)\,\mathbf{1},\ \mathbf{3} = \mathbf{2}\,(+)\,\mathbf{1},\ \mathbf{4} = \mathbf{3}\,(+)\,\mathbf{1} \cdots,$$

and from this,

$$\mathbf{1}! = \mathbf{1},\ \mathbf{2}! = \mathbf{1}\,(\cdot)\,\mathbf{2},\ \mathbf{3}! = \mathbf{1}\,(\cdot)\,\mathbf{2}\,(\cdot)\,\mathbf{3},\ \mathbf{4}! = \mathbf{1}\,(\cdot)\,\mathbf{2}\,(\cdot)\,\mathbf{3}\,(\cdot)\,\mathbf{4} \cdots, \text{ etc.}$$

The above shows how fuzzy numbers, **1**, **2**, **3**, **4** . . ., and their fuzzy factorials **1**!, **2**!, **3**!, **4**!, . . . are constructed. Another point to note is that what is found using ordinary numbers is not necessarily found using fuzzy numbers. For example,

$$\mathbf{2}\,(+)\,\mathbf{2} \neq \mathbf{2}\,(\cdot)\,\mathbf{2}.$$

Conversely, it is possible to define as many fuzzy numbers as we wish since there is an infinite number of membership functions. Only the number 1 is arbitrary, however, since all the others are constructed from it.

In the following study we use these symbols for fuzzy numbers:

$\mathbf{A}_1$ for **1**, $\mathbf{A}_2$ for **2**, $\mathbf{A}_3$ for **3**, . . ., etc.
$\mathbf{A}_1!$ for **1**!, $\mathbf{A}_2!$ for **2**!, $\mathbf{A}_3!$ for **3**!, . . ., etc.

We also use $\mathbf{0}! = 1$ (the ordinary 1 of the real numbers), or

$$\begin{aligned} \mu_{\mathbf{0}!}(x) &= 1, \qquad x = 1, \\ &= 0, \qquad x \neq 1. \end{aligned}$$

$\mathbf{A}_0!$ is universal and independent of **A**.

Let us now define each factorial as it is constructed from $\mathbf{A}_1$ by the use of the intervals of confidence of level α.

$$\begin{aligned} \mathbf{A}_1{}^{(\alpha)} &= [a^{(\alpha)}, b^{(\alpha)}], \\ \mathbf{A}_2{}^{(\alpha)} &= [a^{(\alpha)} + a^{(\alpha)}, b^{(\alpha)} + b^{(\alpha)}] = [2a^{(\alpha)}, 2b^{(\alpha)}], \\ \mathbf{A}_3{}^{(\alpha)} &= [2a^{(\alpha)} + a^{(\alpha)}, 2b^{(\alpha)} + b^{(\alpha)}] = [3a^{(\alpha)}, 3b^{(\alpha)}], \\ \ldots &= \ldots \\ \ldots &= \ldots \\ A_n{}^{(\alpha)} &= [na^{(\alpha)}, nb^{(\alpha)}]. \end{aligned} \tag{3.29}$$

Now

$$A_1^{(\alpha)}(\cdot)A_2^{(\alpha)} = [a^{(\alpha)} \cdot 2a^{(\alpha)}, b^{(\alpha)} \cdot 2b^{(\alpha)}]$$
$$= [2a^{(\alpha)2}, 2b^{(\alpha)2}],$$
$$A_1^{(\alpha)} \cdot A_2^{(\alpha)} \cdot A_3^{(\alpha)} = [2a^{(\alpha)2} \cdot 3a^{(\alpha)}, 2b^{(\alpha)2} \cdot 3b^{(\alpha)}]$$
$$= [1 \cdot 2 \cdot 3a^{(\alpha)3}, 1 \cdot 2 \cdot 3b^{(\alpha)3}],$$
$$\ldots = \ldots$$
$$\ldots = \ldots$$
$$A_1^{(\alpha)}(\cdot)A_2^{(\alpha)}(\cdot)A^{(\alpha)}(\cdot)\cdots(\cdot)A_n^{(\alpha)} = [n!a^{(\alpha)n}, n!b^{(\alpha)n}],$$
$$\ldots = \ldots.$$

That is,

$$\mathbf{A}_n^{(\alpha)}! = [n!\, a^{(\alpha)n}, n!b^{(\alpha)n}]. \tag{3.30}$$

From now on we use $\mathbf{A}_n^{(\alpha)}$ for the fuzzy numbers and $\mathbf{A}_\mathbf{n}^{(\alpha)}!$ for the fuzzy factorials. For purposes of clarity, consider Example 3.13.

EXAMPLE 3.13

Consider the fuzzy number defined by Table 3.13, from which we have Table 3.14 and Table 3.15. These tables show that the fuzzy numbers $\mathbf{A_n}!$ move away from zero as if they were of the form $n!x^n$ and not as if they were of the form $n!$.

Table 3.13

α	$A_1(\alpha)$
0	[0, 6]
0.1	[0.10, 5]
0.2	[0.25, 4]
0.3	[0.40, 3]
0.4	[0.55, 2.50]
0.5	[0.65, 2.05]
0.6	[0.75, 1.80]
0.7	[0.85, 1.40]
0.8	[0.90, 1.25]
0.9	[0.95, 1.20]
1	1

Table 3.14

α	$A_2(\alpha)!$
0	[0, 72]
0.1	[0.020, 50]
0.2	[0.125, 32]
0.3	[0.320, 18]
0.4	[0.605, 12.5]
0.5	[0.845, 8.405]
0.6	[1.125, 5.480]
0.7	[1.445, 3.920]
0.8	[1.620, 3.125]
0.9	[1.805, 2.880]
1	2

Table 3.15

α	$A_3(\alpha)!$
0	[0, 1296]
0.1	[0.00600, 750]
0.2	[0.09375, 384]
0.3	[0.38400, 162]
0.4	[0.99825, 93.75]
0.5	[1.64775, 51.69075]
0.6	[2.53125, 34.99200]
0.7	[3.68485, 16.46400]
0.8	[4.37400, 11.71875]
0.9	[5.14425, 10.36800]
1	6

Sequences of Fuzzy Numbers

In the field of ordinary numbers, a sequence σ_n is a numerical set that is linerarly ordered and is generated by a unique algebraic expression that gives an element of this set for each value $n \in N$. That is, if $\sigma_n = 1/n$, then the sequence is

$$1, 1/2, 1/3, 1/4, 1/5, \ldots \quad \textbf{(3.31)}$$

Or, if $\sigma_n = [3n^2]/[4n^2+1]$, then the sequence is

$$0, 3/5, 12/17, 27/37, 48/65, \ldots. \quad \textbf{(3.32)}$$

An infinite sequence is said to converge iff

$$\lim_{n \to \infty} \sigma_n = \sigma^*.$$

Thus, (3.31) converges to

$$\lim_{n \to \infty} 1/n = 0$$

and (3.32) converges to

$$\lim_{n \to \infty} [3n^2]/[4n^2+1] = 3/4.$$

Conversely, the sequence $\sigma_n = n^2$ does not converge. Also, the sequence $\sigma_n = (-1)^n$ tends to 1 for n even, and -1 for odd n.

This concept is easily extended to fuzzy numbers if we consider them with their α-cut.

Let $\sigma_n^{(\alpha)} = [\sigma_a^{(\alpha)}(n), \sigma_b^{(\alpha)}(n)]$, where $\sigma_a^{(\alpha)}(n)$ and $\sigma_b^{(\alpha)}(n)$ are sequences such that

1. $\forall\ \alpha \in [0,1],\ \forall\ n \in N_0$:

$$\sigma_a^{(\alpha)}(n), \sigma_b^{(\alpha)}(n) \in R$$

and

$$\sigma_a^{(\alpha)}(n) \le \sigma_b^{(\alpha)}(n)$$

2. $(\alpha' > \alpha) \Rightarrow ([\sigma_a^{(\alpha)}(n), \sigma_b^{(\alpha)}(n)] \subset ([\sigma_a^{(\alpha')}(n), \sigma_b^{(\alpha')}(n)])$.

Using this definition, we can write a sequence of fuzzy numbers as follows:

$$\sigma(n) = \bigcap_{\alpha} \alpha [\sigma_a^{(\alpha)}(n), \sigma_b^{(\alpha)}(n)], \qquad \alpha \in [0,1],$$

where the fuzzy sequence $\sigma(n)$ converges iff both $\sigma_a{}^{(\alpha)}(n)$ and $\sigma_b{}^{(\alpha)}(n)$ are converged for every $\alpha \in [0, 1]$.

EXAMPLE 3.14

We now show how to extend the sequence $\sigma_n = 1/n, \quad n \in N,$ to the fuzzy case. For this purpose let us define the α-cut for the fuzzy number 1 as follows.

$$A_1{}^{(\alpha)} = [a^{(\alpha)}, b^{(\alpha)}], \qquad \alpha \in [0, 1].$$

Using (3.30), we find that

$$\begin{aligned} A_2{}^{(\alpha)} &= [2a^{(\alpha)}, 2b^{(\alpha)}], \\ \ldots &= \ldots, \\ \ldots &= \ldots, \\ A_n{}^{(\alpha)} &= [na^{(\alpha)}, nb^{(\alpha)}]. \end{aligned}$$

Also, using the inversion formula (1.50), we obtain

$$\begin{aligned} A_1{}^{-1(\alpha)} &= [(1/b^{(\alpha)}), (1/a^{(\alpha)}], \\ A_2{}^{-1(\alpha)} &= [(1/2 \cdot b^{(\alpha)}), (1/2 \cdot a^{(\alpha)}], \\ \ldots &= \ldots, \\ \ldots &= \ldots, \\ A_n{}^{-1(\alpha)} &= [(1/n \cdot b^{(\alpha)}), (1/n \cdot a^{(\alpha)}]. \end{aligned} \tag{3.33}$$

From this it is obvious that the sequence

$$\lim_{n \to \infty} A_n{}^{-1(\alpha)} = [0, 0] = 0, \qquad \text{if } a^{(0)} > 0.$$

We observe from these equations that the sequence (3.33) is convergent.

For purposes of clarity we now use the numerical values given in Tables 3.16 and 3.17 on pages **214** and **215**.

EXAMPLE 3.15

Let us generalize the sequence $\sigma_n = (1/n!), \quad n \in N,$ to fuzzy numbers. Using (3.30), we have

$$(A^{(\alpha)}!)^{-1} = [1/(n!(b^{(\alpha)})^n), 1/(n!(a^{(\alpha)})^n], \qquad a^{(0)} > 0.$$

This sequence converges quickly to $[0, 0] = 0$ for increasing n.

Table 3.16 (Example 3.14)

α	$A_1(\alpha)$(given)	$A_2(\alpha)$	$A_3(\alpha)$	$A_4(\alpha)$
0	[0.1, 2]	[0.2, 4]	[0.3, 6]	[0.4, 8]
0.1	[0.2, 1.9]	[0.4, 3.8]	[0.6, 5.7]	[0.8, 7.6]
0.2	[0.3, 1.8]	[0.6, 3.6]	[0.9, 5.4]	[1.2, 7.2]
0.3	[0.4, 1.7]	[0.8, 3.4]	[1.2, 5.1]	[1.6, 6.8]
0.4	[0.5, 1.6]	[1, 3.2]	[1.5, 4.8]	[2, 6.4]
0.5	[0.6, 1.5]	[1.2, 3]	[1.8, 4.5]	[2.4, 6]
0.6	[0.7, 1.4]	[1.4, 2.8]	[2.1, 4.2]	[2.8, 5.6]
0.7	[0.8, 1.3]	[1.6, 2.6]	[2.4, 3.9]	[3.2, 5.2]
0.8	[0.9, 1.2]	[1.8, 2.4]	[2.7, 3.6]	[3.6, 4.8]
0.9	[1, 1.1]	[2, 2.2]	[3, 3.3]	[4, 4.4]
1	1	2	3	4

EXAMPLE 3.16

Let

$$\sigma_n = n/(n-1), \qquad n = 2, 3, 4, \ldots .$$

Now let us put

$$A_1^{(\alpha)} = [a^{(\alpha)}, b^{(\alpha)}] \subset R_0^+.$$

Using (3.29) we obtain

$$\underbrace{A_1^{(\alpha)}(+)A_1^{(\alpha)}\cdots(+)A_1^{(\alpha)}}_{n \text{ times}} = [na^{(\alpha)}, nb^{(\alpha)}],$$

$$\underbrace{A_1^{(\alpha)}(+)A_1^{(\alpha)}\cdots(+)A_1^{(\alpha)}}_{(n-1) \text{ times}} = [(n-1)a^{(\alpha)}, (n-1)b^{(\alpha)}],$$

$$\begin{aligned}\sigma^{(\alpha)}(n) &= [na^{(\alpha)}, nb^{(\alpha)}](:)[(n-1)a^{(\alpha)}, (n-1)b^{(\alpha)}]\\ &= [(n/(n-1))(a^{(\alpha)}/b^{(\alpha)}), (n/(n-1))(b^{(\alpha)}/a^{(\alpha)})],\\ \lim_{n\to\infty}\sigma^{(\alpha)}(n) &= [a^{(\alpha)}/b^{(\alpha)}, b^{(\alpha)}/a^{(\alpha)}],\\ &= A_1^{(\alpha)}(:)A_1^{(\alpha)},\end{aligned}$$

which is convergent.

Table 3.17 (Example 3.14)

α	$A_1{-1}(\alpha)$	$A_2{-1}(\alpha)$	$A_3{-1}(\alpha)$	$A_4{-1}(\alpha)$
0	[0.500, 10]	[0.250, 5]	[0.166, 3.333]	[0.125, 2.500]
0.1	[0.526, 5]	[0.263, 2.500]	[0.175, 1.666]	[0.131, 1.250]
0.2	[0.555, 3.333]	[0.277, 1.666]	[0.185, 1.111]	[0.138, 0.833]
0.3	[0.588, 2.500]	[0.294, 1.250]	[0.196, 0.833]	[0.147, 0.625]
0.4	[0.625, 2]	[0.312, 1]	[0.208, 0.666]	[0.156, 0.500]
0.5	[0.666, 1.666]	[0.335, 0.833]	[0.262, 0.555]	[0.166, 0.416]
0.6	[0.714, 1.428]	[0.357, 0.714]	[0.238, 0.476]	[0.178, 0.350]
0.7	[0.769, 1.250]	[0.384, 0.625]	[0.256, 0.416]	[0.192, 0.312]
0.3	[0.833, 1.111]	[0.416, 0.555]	[0.277, 0.370]	[0.208, 0.277]
0.4	[0.303, 1]	[0.454, 0.500]	[0.303, 0.333]	[0.227, 0.250]
1	1	0.500	0.333	0.250

We recall that

$$A_1^{(\alpha)} (\cdot) A_1^{(\alpha)} \neq A_1^{(\alpha)}$$

and

$$A_1^{(\alpha)} (:) A_1^{(\alpha)} \neq 1,$$

except for

$$A_1^{(\alpha)} = [1, 1] = 1.$$

Series with Fuzzy Numbers

We now have to recall that a series is a sum of terms in a given sequence. Thus

$$S(n) = \sum_{\nu = n_0}^{n} \sigma_\nu .$$

A series is infinite if $n \to \infty$ and is finite when n is finite.

A series is said to be convergent iff

$$\lim_{n \to \infty} S(n) = S^*.$$

Let us extend the concept of fuzziness to series by giving some examples.

EXAMPLE 3.17

Let

$$S(n) = 1 + 1/2 + 1/3 + \cdots + 1/n. \quad \textbf{(3.34)}$$

We can fuzzify the above sequence by defining

$$A_1^{(\alpha)} = [a^{(\alpha)}, b^{(\alpha)}] \subset R_0^+. \quad \textbf{(3.35)}$$

It is important at this point not to confuse

$$A_1^{-1(\alpha)} \qquad \text{with} \qquad A_1^{(\alpha)}. \quad \textbf{(3.36)}$$

We know that these are different segments of R_0^+,

$$A_1^{-1(\alpha)} = [1/b^{(\alpha)}, 1/a^{(\alpha)}]. \quad \textbf{(3.37)}$$

To generalize (3.34), we must decide whether to use (3.35) or (3.37) for the first term 1. For purposes of homogenity, we will select (3.37). We then have

$$A_1^{-1(\alpha)} = [1/b^{(\alpha)}, 1/a^{(\alpha)}],$$
$$A_2^{-1(\alpha)} = [1/2b^{(\alpha)}, 1/2a^{(\alpha)}],$$
$$A_3^{-1(\alpha)} = [1/3b^{(\alpha)}, 1/3a^{(\alpha)}],$$
$$\ldots = \ldots,$$
$$\ldots = \ldots,$$
$$A_n^{-1(\alpha)} = [1/nb^{(\alpha)}, 1/na^{(\alpha)}],$$

from which we have

$$\begin{aligned} S^{(\alpha)}(n) &= A_1^{-1(\alpha)} (+) A_2^{-1(\alpha)} (+) A_3^{-1(\alpha)} (+) \cdots (+) A_n^{-1(\alpha)} \\ &= [(1 + 1/2 + 1/3 + \cdots + 1/n)(1/b^{(\alpha)}), (1 + 1/2 + 1/3 + \cdots \\ &\quad + 1/n) 1/a^{(\alpha)}] \\ &= [S(n) \cdot 1/b^{(\alpha)}, S(n) \cdot 1/a^{(\alpha)}]. \end{aligned}$$

This series does not converge when $n \to \infty$ because $S(n)$ is not convergent.

EXAMPLE 3.18

Let

$$S(n) = 1 + 1/1! + 1/2! + 1/3! + \cdots + 1/n!. \qquad \textbf{(3.38)}$$

To make this series fuzzy we define

$$A_1^{(\alpha)} = [a^{(\alpha)}, b^{(\alpha)}] \subset R_0^+$$

and take $1 = [1, 1]$. We then have

$$A_1^{-1(\alpha)} = [1/b^{(\alpha)}, 1/a^{(\alpha)}]$$

and

$$A_2^{-1(\alpha)} = [1/(2!b^{(\alpha)2}), 1/(2!a^{(\alpha)2})],$$
$$A_3^{-1(\alpha)} = [1/(3!b^{(\alpha)3}), 1/(3!a^{(\alpha)3}],$$
$$\ldots = \ldots,$$
$$\ldots = \ldots,$$
$$A_n^{-1(\alpha)} = [1/(n!b^{(\alpha)n}), 1/(n!a^{(\alpha)n})].$$

Then

$$S^{(\alpha)}(n) = 1\,(+)\,A_1^{-1(\alpha)}\,(+)\,A_2^{-1(\alpha)}\,(+)\,A_3^{-1(\alpha)}\,(+)\cdots(+)\,A_n^{-1(\alpha)}$$
$$= [1 + 1/b^{(\alpha)} + 1/2!b^{(\alpha)2} + 1/3!b^{(\alpha)3} + \ldots + 1/n!b^{(\alpha)n},$$
$$1 + 1/a^{(\alpha)} + 1/2!b^{(\alpha)2} + 1/3!a^{(\alpha)3} + \ldots + 1/n!a^{(\alpha)n}].$$

It is well known that the infinite series

$$S(x) = \sum_{n=0}^{\infty} 1/(n!x^n)$$

converges for $x > 0$, which is the case here. This results in the infinite series

$$S^{(\alpha)} = [\sum_{n=0}^{\infty} 1/n!b^{(\alpha)n},\ \sum_{n=0}^{\infty} 1/n!a^{(\alpha)n}], \qquad \textbf{(3.39)}$$

which converges. Let us write

$$e(A_1^{(\alpha)}) = S^{(\alpha)}$$

to show that the fuzzy series is partly constructed from $A_1^{(\alpha)}$ if

$$A_1^{(\alpha)} = [1, 1].$$

Then

$$e(A_1^{(\alpha)}) = [e, e] = e,$$

where e is the base of the Naperian logarithm.

Let us return to (3.38) and note that

$$S(n, x) = 1 + 1/x + 1/2!x^2 + 1/3!x^3 + \cdots$$
$$+ 1/n!x^n, \qquad x > 0,$$

has the limit

$$\lim_{n \to \infty} S(n, x) = e^{1/x}.$$

Hence, (3.39) has the limit

$$e(A_1^{(\alpha)}) = [e^{(1/b^{(\alpha)})}, e^{(1/a^{(\alpha)})}]. \qquad \textbf{(3.40)}$$

A numerical example is shown in Table 3.18.

Table 3.18

α	$A_1(\alpha)$	$e(A_1(\alpha))$
0	[0.1, 2]	[1.6487, 22.026]
0.1	[0.2, 1.9]	[1.6926, 143.41]
0.2	[0.3, 1.8]	[1.7429, 28.031]
0.3	[0.4, 1.7]	[1.8008, 12.182]
0.4	[0.5, 1.6]	[1.8682, 7.3890]
0.5	[0.6, 1.5]	[1.9488, 5.2944]
0.6	[0.7, 1.4]	[2.0427, 4.1727]
0.7	[0.8, 1.3]	[2.1581, 3.4903]
0.8	[0.9, 1.2]	[2.3009, 3.0377]
0.9	[1, 1.1]	[2.4820, 2.7182]
1	1	2.7182

Some interesting properties can be established from (3.40). Thus

$$e(A_1^{(\alpha)})\,(\cdot)\,e(A_1^{(\alpha)}) = [e^{1/b^{(\alpha)}}, e^{1/a^{(\alpha)}}](\cdot)[e^{1/b^{(\alpha)}}, e^{1/a^{(\alpha)}}]$$
$$= [e^{2/b^{(\alpha)}}, e^{2/a^{(\alpha)}}],$$

$$e^{n}(A^{(\alpha)}) = [e^{n/b^{(\alpha)}}, e^{n/a^{(\alpha)}}],$$

$$e^{n_1}(A_1^{(\alpha)})\,(\cdot)\,e^{n_2}(A_1^{(\alpha)}) = e^{n_2}(A_1^{(\alpha)})\,(\cdot)\,e^{n_2}(A_1^{(\alpha)}), \tag{3.41}$$

$$(e^{n_1}(A_1^{(\alpha)}))^{n_2} = (e^{n_2}(A_1^{(\alpha)}))^{n_1} = (e^{n_1\cdot n_2}(A_1^{(\alpha)}). \tag{3.42}$$

We can now extend equations (3.41) to (3.42) to

$\nu \in R^+$:

$$e^{\nu}(A_1^{(\alpha)}) = [e^{\nu/b^{(\alpha)}}, e^{\nu/a^{(\alpha)}}],$$

and we are able to define the exponential function generated by $A_1^{(\alpha)}$ as

$$e^{x}(A_1^{(\alpha)}), \qquad x \in R^+.$$

An extension of the results from $x \in R^+$ to $x \in R$ is given in Appendix A.

Remark

We must be careful not to confuse $e^x(A_1{}^{(\alpha)})$ with $e^{X(\alpha)}$, which is another type of fuzzy function.

$$e^x(A_1{}^{(\alpha)}) = [e^{x/b_1{}^{(\alpha)}}, e^{x/a_1{}^{(\alpha)}}], \tag{3.43}$$

and if

$$X^{(\alpha)} = [x_1{}^{(\alpha)}, x_2{}^{(\alpha)}] \subset R^+,$$

then

$$e^{X^{(\alpha)}} = [e^{x_1{}^{(\alpha)}}, e^{x_2{}^{(\alpha)}}]. \tag{3.44}$$

We call (3.43) the *fuzzy exponential of the first kind* and (3.44) the *fuzzy exponential function of the second kind*.

Let us now examine this second kind of exponential function in more detail.

$$e^{X^{(\alpha)}} = 1\,(+)\,X^{(\alpha)}\,(+)\,(1/2!)X^{(\alpha)2}\,(+)\,(1/3!)X^{(\alpha)3}\,(+)\cdots$$
$$(+)\,(1/n!)X^{(\alpha)n}\,(+)\cdots,$$

where

$$X^{(\alpha)^n} = \underbrace{X^{(\alpha)}\,(\cdot)\,X^{(\alpha)}\cdots(\cdot)\,X^{(\alpha)}}_{n \text{ times}},\ X^{(\alpha)} \subset R^+,\ \alpha \in [0,1].$$

We then have

$$e^{X^{(\alpha)}}\,(\cdot)\,e^{Y^{(\alpha)}} = e^{Y^{(\alpha)}}\,(\cdot)\,e^{X^{(\alpha)}} = e^{X^{(\alpha)}\,(+)\,Y^{(\alpha)}},$$

$$(e^{X^{(\alpha)}})^{Y^{(\alpha)}} = (e^{Y^{(\alpha)}})^{X^{(\alpha)}} = e^{X^{(\alpha)}\,(\cdot)\,Y^{(\alpha)}}.$$

By means of Example 3.19 we show the difference that exists between

$$e^2(A_1{}^{(\alpha)}) \quad \text{and} \quad e^{A_1(\alpha)^2}.$$

EXAMPLE 3.19

Consider the following examples:

$$e^2(A_1{}^{(\alpha)}) = [e^{2/b^{(\alpha)}}, e^{2/a^{(\alpha)}}]$$

and

$$e^{A_1(\alpha)^2} = [e^{2a^{(\alpha)}}, e^{2b^{(\alpha)}}].$$

Then we have the results shown in Table 3.19.

Table 3.19

α	$A_1(\alpha)$	$e^2(A_1(\alpha))$	$e^{A_1(\alpha)^2}$
0	[0.1, 2]	[2.7182, 4.8516 × 10⁹]	[1.2214, 54.598]
0.1	[0.2, 1.9]	[2.8651, 2.2026 × 10⁵]	[1.4918, 44.701]
0.2	[0.3, 1.8]	[3.0377, 785.77]	[1.9221, 36.598]
0.3	[0.4, 1.7]	[3.2429, 148.41]	[2.2255, 29,964]
0.4	[0.5, 1.6]	[3.4903, 54.598]	[2.7182, 24.532]
0.5	[0.6, 1.5]	[3.7936, 28.031]	[3.3201, 20.095]
0.6	[0.7, 1.4]	[4.1727, 17.411]	[4.0551, 16,444]
0.7	[0.8, 1.3]	[4.6574, 12.182]	[4.9530, 13.463]
0.8	[0.9, 1.2]	[5.2944, 9.2278]	[6.0496, 11.023]
0.9	[1, 1.1]	[6.1606, 7.3890]	[7.3890, 9.025]
1	1	7.3890	7.3890

3.6 FUNCTIONS OF FUZZY NUMBERS

Because the term fuzzy variable has been used in different ways by different authors, we wish to stress that in this book we use the term fuzzy number to mean fuzzy variable.[3] To begin, let us examine the first *kind* of fuzzy function.

Integer Powers of a Fuzzy Number

Let **X** be a fuzzy number in R:

$$\mathbf{X}^n = \underbrace{\mathbf{X}(\cdot)\,\mathbf{X}(\cdot)\cdots(\cdot)\,\mathbf{X}}_{n\text{ times}}.$$

1). First case: $\mathbf{X} \subset R^+$; here $\forall\ \alpha \in [0, 1]$; therefore from $X^{(\alpha)} = [a^{(\alpha)}, b^{(\alpha)}]$, we have $X^{(\alpha)n} = [a^{(\alpha)n}, b^{(\alpha)n}]$.

2). Second and general case: $\mathbf{X} \subset R$. This is a much more complicated case and requires the examination of the parity of n (the *evenness* or *oddness*). See Table 3.20.

Table 3.20

	$[a, b]^1$	$[a, b]^2$	$[a, b]^3$	$[a, b]^4$	$[a, b]^5$
$a \le b \le 0$	$[a, b]$	$[b^2, a^2]$	$[a^3, b^3]$	$[b^4, a^4]$	$[a^5, b^5]$
$a \le 0 \le b$ $\lvert a\rvert \le \lvert b\rvert$	$[a, b]$	$[ab, b^2]$	$[ab^2, b^3]$	$[ab^3, b^4]$	$[ab^4, b^5]$
$a \le 0 \le b$ $\lvert a\rvert \ge \lvert b\rvert$	$[a, b]$	$[ab, a^2]$	$[a^3, a^2b]$	$[a^3b, a^4]$	$[a^5, a^4b]$
$0 \le a \le b$	$[a, b]$	$[a^2, b^2]$	$[a^3, b^3]$	$[a^4, b^4]$	$[a^5, b^5]$

For $n \in N_0$:

1. $a \le b \le 0$:

$$[a, b]^n = [a^n, b^n], \qquad \text{if } n \text{ is odd,}$$
$$= [b^n, a^n], \qquad \text{if } n \text{ is even.}$$

2. $a \le 0 \le b, |a| \le |b|$:

$$[a, b]^n = [ab^{n-1}, b^n].$$

3. $a \le 0 \le b, |a| \ge |b|$:

$$[a, b]^n = [a^n, a^{n-1} b], \qquad n \text{ odd,}$$
$$= [a^{n-1} b, a^n], \qquad n \text{ even.}$$

4. $0 \le a \le b$:

$$[a, b]^n = [a^n, b^n].$$

Domain of Regularity

If, for every fuzzy number $\mathbf{X} \subset R$, there is a one-to-one correspondence with $\mathbf{Y} \subset R$, we can define a mapping function from R in R denoted as $f(\mathbf{X})$ by

$\forall \alpha \in [0, 1]$:

$$\mathbf{Y}^{(\alpha)} = f_\alpha(\mathbf{X}^{(\alpha)}),$$

which is shown in Figure 3.3.

It is easy to see, however, that every function f_α at the level α does not necessarily give a segment of R that we can call the interval of confidence. Let us consider, therefore,

$$x^{(\alpha)} \in [a^{(\alpha)}, b^{(\alpha)}].$$

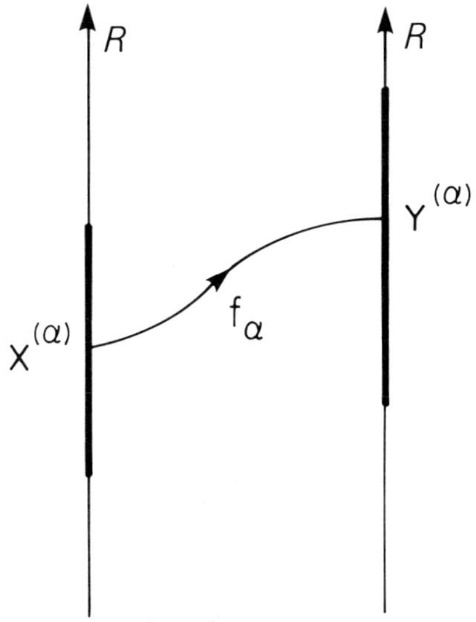

Figure 3.3 Mapping $\mathbf{y}^{(\alpha)} = f_\alpha\,(\mathbf{x}^{(\alpha)})$.

In Figure 3.4, we see that for every $x^{(\alpha)}$ there corresponds one and only one $f(x^{(\alpha)})$ such that

$$f(x^{(\alpha)}) \in [f(a^{(\alpha)}), f(b^{(\alpha)})]. \qquad \textbf{(3.45)}$$

If $\forall\ \alpha \in [0, 1]$ and equation (3.45) is valid, call f a *regular function* for the fuzzy number $\mathbf{X} \subset R$. From this definition the interval of confidence of level α is $[a^{(\alpha)}, b^{(\alpha)}]$, which implies that $f(x)$ is monotonically increasing in the segment $[a^{(\alpha)}, b^{(\alpha)}]$.

Now let us consider Figure 3.5. We see that for every $x^{(\alpha)}$ there is one and only one $f(x^{(\alpha)})$ such that

$$f(x^{(\alpha)}) \in [f(b^{(\alpha)}), f(a^{(\alpha)})]. \qquad \textbf{(3.46)}$$

If $\forall\ \alpha \in [0, 1]$ and equation (3.46) is valid we call f a *nonregular function* for the fuzzy number $\mathbf{X} \subset R$. From this definition we know that the interval of confidence of level α decreases in the segment $[a^{(\alpha)}, b^{(\alpha)}]$.

For a number of reasons we may not be able to define a function as regular or nonregular. In this case, we use the qualification *irregular*. For instance, let us examine Figure 3.6. We see that for

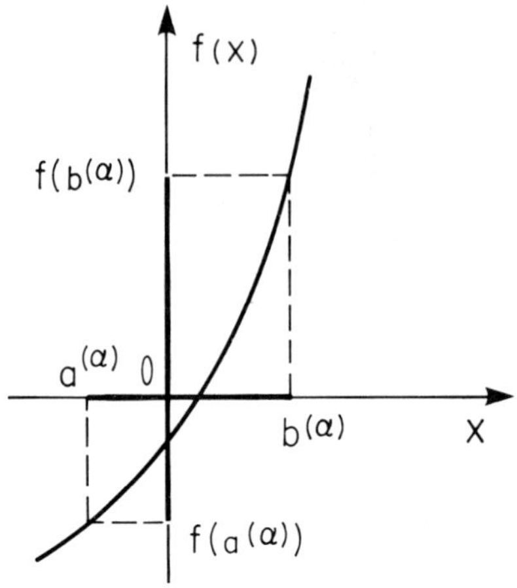

Figure 3.4 A regular function: $f(x^{(\alpha)}) \in [f(a^{(\alpha)}), f(b^{(\alpha)})]$.

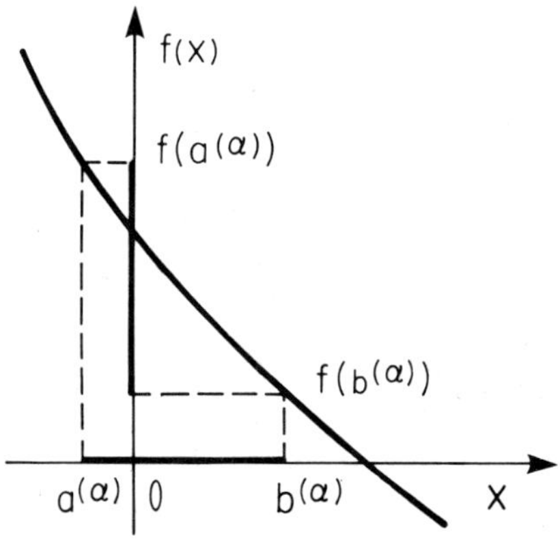

Figure 3.5 A nonregular function.

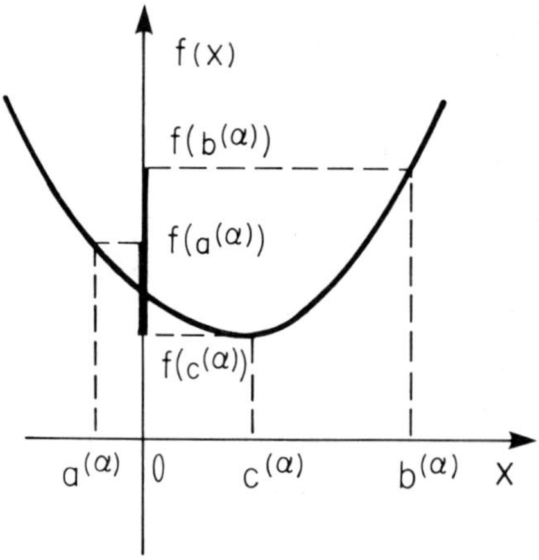

Figure 3.6 An irregular function.

$$x^{(\alpha)} \in [a^{(\alpha)}, c^{(\alpha)}]$$

the function f is irregular, and for

$$x^{(\alpha)} \in [c^{(\alpha)}, b^{(\alpha)}],$$

is regular. Hence, if f is irregular, the corresponding fuzzy subset is irregular.

Now, let us consider a domain $D \subset R$, where the function f of a fuzzy number is regular[4] in D iff

$\forall\ \alpha \in [0, 1], \forall [a^{(\alpha)}, b^{(\alpha)}] \subset D$:

$$f(X^{(\alpha)}) = f([a^{(\alpha)}, b^{(\alpha)}])$$

$$= [f(a^{(\alpha)}), f(b^{(\alpha)})].$$

We now show that f is nonregular in D' iff

$\forall\ \alpha \in [0, 1], \forall [a^{(\alpha)}, b^{(\alpha)}] \subset D'$:

$$f(X^{(\alpha)}) = f([a^{(\alpha)}, b^{(\alpha)}])$$

$$= [f(b^{(\alpha)}), f(a^{(\alpha)})].$$

The domain D is called the *domain of regularity* and the domain D', the *domain of nonregularity.*

EXAMPLE 3.20

Consider the following function of a fuzzy number:

$$f(\mathbf{x}) = \mathbf{X}\,(+)\,\mathbf{X}^2.$$

This function is regular in the domain $D = R^+$. Indeed, let

$$X^{(\alpha)} = [a^{(\alpha)}, b^{(\alpha)}] \subset R^+.$$

Then

$$X^{(\alpha)^2} = [a^{(\alpha)^2}, b^{(\alpha)^2}] \subset R^+,$$

$$f(X^{(\alpha)}) = X^{(\alpha)}\,(+)\,X^{(\alpha)2}$$

$$= [a^{(\alpha)} + a^{(\alpha)2}, b^{(\alpha)} + b^{(\alpha)2}]$$

$$= [f(a^{(\alpha)}), f(b^{(\alpha)})].$$

This function is irregular in $D = R$. Let $X^{(\alpha)} = [a'^{(\alpha)}, b'^{(\alpha)}]$ with $a'^{(\alpha)} < 0$ and $b'^{(\alpha)} > 0$:

$$|a'^{(\alpha)}| > |b'^{(\alpha)}|,$$

$$X^{(\alpha)2} = [a'^{(\alpha)}\,(\cdot)\,b'^{(\alpha)}, b'^{(\alpha)2}],$$

$$f(X^{(\alpha)}) = X^{(\alpha)} (+) X^{(\alpha)2}$$
$$= [a'^{(\alpha)}, b'^{(\alpha)}] (+) [a'^{(\alpha)} b'^{(\alpha)}, b'^{(\alpha)2}]$$
$$= [a'^{(\alpha)} + a'^{(\alpha)} b'^{(\alpha)}, b'^{(\alpha)} + b'^{(\alpha)2}].$$

EXAMPLE 3.21

Let $f(X) = X^n$. Then $D = R^-$, $X^{(\alpha)n}$ is regular if $n = 1, 3, 5, \ldots$, and irregular if $n = 2, 4, 6, \ldots$. In $D = R$, $X^{(\alpha)n}$ is irregular. In $D = R^+$, $X^{(\alpha)n}$ is regular for $n = 1, 2, 3, 4, \ldots$.

THEOREM 3.4

If $f_1(\mathbf{X})$ and $f_2(\mathbf{X})$ are regular in $D \subset R$, then $f(\mathbf{X}) = f_1(\mathbf{X}) (+) f_2(\mathbf{X})$ is also regular in D.

Proof:

By hypothesis we have

$\forall\, \alpha \in [0, 1]$:

$$f_1(X^{(\alpha)}) = f_1([a^{(\alpha)}, b^{(\alpha)}])$$
$$= [f_1(a^{(\alpha)}), f_1(b^{(\alpha)})],$$
$$f_2(X^{(\alpha)}) = f_2([a^{(\alpha)}, b^{(\alpha)}])$$
$$= [f_2(a^{(\alpha)}), f_2(b^{(\alpha)})],$$
$$f(x^{(\alpha)}) = f_1(X^{(\alpha)}) (+) f_2(X^{(\alpha)})$$
$$= [f_1(a^{(\alpha)}) + f_2(a^{(\alpha)}), f_1(b^{(\alpha)}) + f_2(b^{(\alpha)})]$$
$$= [f(a^{(\alpha)}), f(b^{(\alpha)})].$$

THEOREM 3.5

If $f_1(\mathbf{X})$ and $f_2(\mathbf{X})$ are regular in $D \subset R^+$, then $f(\mathbf{X}) = f_1(\mathbf{X}) (\cdot) f_2(\mathbf{X})$ is also regular in D.

The proof of this theorem follows the same procedure as used in the case of $(+)$.

THEOREM 3.6

If $f(\mathbf{X})$ is regular in $D \subset R$, then, for $k\mathbf{X} \subset D$, $k \geq 0$, $f(k\mathbf{X})$ is also regular in D because $k\mathbf{X}$ remains in D.

THEOREM 3.7

If $f(\mathbf{X})$ is regular in $D \subset R$, then $F(\mathbf{X}) = kf(\mathbf{X})$, $k \geq 0$, is also regular in D.

Proof:

$\forall\ \alpha \in [0, 1]$:

$$f(X^{(\alpha)}) = f([a^{(\alpha)}, b^{(\alpha)}])$$

$$= [f(a^{(\alpha)}), f(b^{(\alpha)})]$$

and

$$F(X^{(\alpha)}) = kf(X^{(\alpha)})$$

$$= k[f(a^{(\alpha)}), f(b^{(\alpha)})]$$

$$= [kf(a^{(\alpha)}), kf(b^{(\alpha)})]$$

$$= [F(a^{(\alpha)}), F(b^{(\alpha)})].$$

THEOREM 3.8

If for every $n \in$ N a sequence $\sigma_m(\mathbf{X})$ is regular in $D \subset R$, then the series

$$S_n(\mathbf{X}) = (\sum_{k=0}^{\infty})\, \sigma_k(\mathbf{X}),$$

where (Σ) is $(+)$, is regular in D. This is a direct consequence of Theorem 3.4.

THEOREM 3.9

Let a series $S_n(\mathbf{X})$ be regular in $D \subset R$. If

1. $\forall\ \alpha \in [0, 1]$:

$$S_n(X^{(\alpha)}) = [S_n(a^{(\alpha)}, S(b^{(\alpha)})],$$

where

$$X^{(\alpha)} = [a^{(\alpha)}, b^{(\alpha)}].$$

2. $\lim_{n\to\infty} S_n(a^{(\alpha)}) = S^*(a^{(\alpha)})$,

$\lim_{n\to\infty} S_n(b^{(\alpha)}) = S^*(b^{(\alpha)})$.

Then

$$\begin{aligned}\lim_{n\to\infty} S_n(X^{(\alpha)}) &= [\lim_{n\to\infty} S_n(a^{(\alpha)}), \lim_{n\to\infty} S_n(b^{(\alpha)})] \\ &= [S^*(a^{(\alpha)}), S^*(b^{(\alpha)})] \\ &= S^*(X^{(\alpha)}).\end{aligned}$$

This is a direct consequence of the convergence and regularity of the function.

Note that a function can be regular and not convergent and vice versa.

EXAMPLE 3.22

Let us consider a series or exponential function of the second kind.

$$e^{\mathbf{X}} = 1\,(+)\,\mathbf{X}\,(+)\,(1/2!)\mathbf{X}^2\,(+)\,(1/3!)\mathbf{X}^3\,(+)\cdots (+)\,(1/n!)\mathbf{X}^n\,(+)\cdots,$$

where $X^{(\alpha)} = [a^{(\alpha)}, b^{(\alpha)}]$.

We study the domain of regularity and its convergence. Consider

$$\begin{aligned}e^{X^{(\alpha)}} = {} & [1,1]\,(+)\,[a^{(\alpha)}, b^{(\alpha)}]\,(+)\,(1/2!)\,[a^{(\alpha)}, b^{(\alpha)}]^2 \\ & (+)\,(1/3!)\,[a^{(\alpha)}, b^{(\alpha)}]^3\,(+)\,(1/4!)\,[a^{(\alpha)}, b^{(\alpha)}]^4 \\ & (+)\,(1/5!)\,[a^{(\alpha)}, b^{(\alpha)}]^5\,(+)\cdots \\ & (+)\,(1/n!)\,[a^{(\alpha)}, b^{(\alpha)}]^n\,(+)\cdots.\end{aligned}$$

We examine the following four cases, where $\alpha \in [0, 1]$:

1. $a^{(\alpha)} \le b^{(\alpha)} \le 0$,

2. $a^{(\alpha)} \le 0 \le b^{(\alpha)}, |a^{(\alpha)}| \le |b^{(\alpha)}|$,

3. $a^{(\alpha)} \le 0 \le b^{(\alpha)}, |a^{(\alpha)}| \ge |b^{(\alpha)}|,$

4. $0 \le a^{(\alpha)} \le b^{(\alpha)}.$

Using Table 3.20 we find that

1. $a^{(\alpha)} \le b^{(\alpha)} \le 0,$

$$e^{X^{(\alpha)}} = [1/2(e^{a^{(\alpha)}} - e^{-a^{(\alpha)}} + e^{b^{(\alpha)}} + e^{-b^{(\alpha)}}), 1/2\,(e^{b^{(\alpha)}} - e^{-b^{(\alpha)}} + e^{a^{(\alpha)}} + e^{-a^{(\alpha)}})].$$

2. $a^{(\alpha)} \le 0 \le b^{(\alpha)}, \qquad |a^{(\alpha)}| \le |b^{(\alpha)}|,$

$$e^{X^{(\alpha)}} = 1 + (a^{(\alpha)}/b^{(\alpha)})\,(e^{b^{(\alpha)}} - 1), e^{b^{(\alpha)}}].$$

3. $a^{(\alpha)} \le 0 \le b^{(\alpha)}, \qquad |a^{(\alpha)}| \ge |b^{(\alpha)}|,$

$$e^{X^{(\alpha)}} = [\,\{2(a^{(\alpha)} - b^{(\alpha)}) + b^{(\alpha)}\,(e^{a^{(\alpha)}} + e^{-a^{(\alpha)}}) + a^{(\alpha)}\,(e^{a^{(\alpha)}} - e^{-a^{(\alpha)}})\}/2\alpha^{(\alpha)}, \{(a^{(\alpha)} + b^{(\alpha)})e^{a^{(\alpha)}} + (a^{(\alpha)} - b^{(\alpha)})e^{-a^{(a)}}\}/2a^{(\alpha)}].$$

4. $0 \ge a^{(\alpha)} \ge b^{(\alpha)},$

$$e^{X^{(\alpha)}} = [e^{a^{(\alpha)}}, e^{b^{(\alpha)}}].$$

We thus see that the domain of regularity of $e^{\mathbf{X}}$ is R^+. Since this series is convergent in R, it is regular and convergent in R^+ as well.

One point of concern pertains to Figure 3.7, which shows the function $y = e^x, \quad x \in R$. This function is not defined by the series

$$y = 1 + x + x^2/2! + x^3/3! + \ldots + x^n/n! + \ldots$$

but like a special case of $y = a^x, \quad a > 0$.

The function in Figure 3.7 is continuous and increases monotonically in R. Consequently, $e^{\mathbf{X}}$ should be regular in R, but is not, because a series of a fuzzy function is regular only if the domain in which it is developed is regular.

EXAMPLE 3.23

Let us now consider the hyperbolic cosine function and hyperbolic sine function. If we consider their respective definitions by a series,

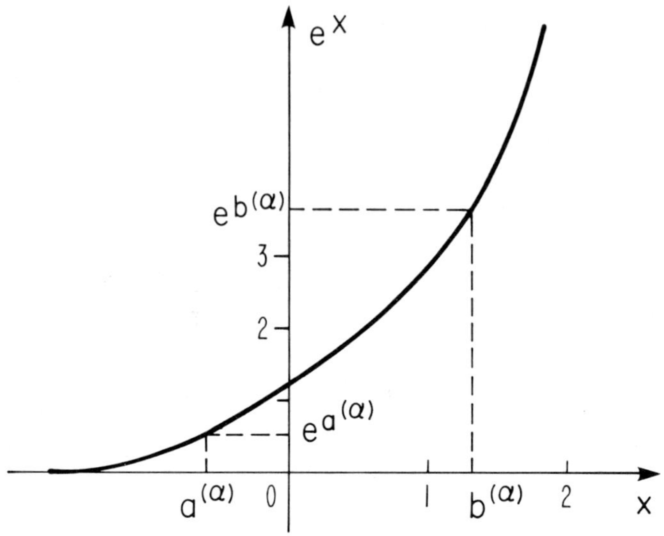

Figure 3.7 An example of a fuzzy function that is not regular.

we find that

$$X^{(\alpha)} = [a^{(\alpha)}, b^{(\alpha)}], \qquad \alpha \in [0,1],$$

$$\begin{aligned}
\cosh X^{(\alpha)} &= \cosh [a^{(\alpha)}, b^{(\alpha)}] \\
&= [\cosh b^{(\alpha)}, \cosh a^{(\alpha)}], \qquad |a^{(\alpha)}| \le |b^{(\alpha)}| \le 0 \\
&= [1 + (a^{(\alpha)}/b^{(\alpha)})(\cosh b^{(\alpha)} - 1), \cosh b^{(\alpha)}], \\
&\quad a^{(\alpha)} \le 0 \le b^{(\alpha)}, |a^{(\alpha)}| \le |b^{(\alpha)}| \\
&= [1 + (b^{(\alpha)}/a^{(\alpha)})(\cosh a^{(\alpha)} - 1), \cosh a^{(\alpha)}], \\
&\quad a^{(\alpha)} \le 0 \le b^{(\alpha)}, |a^{(\alpha)}| \ge |b^{(\alpha)}| \\
&= [\cosh a^{(\alpha)}, \cosh b^{(\alpha)}], \qquad 0 \le a^{(\alpha)} \le b^{(\alpha)}.
\end{aligned} \tag{3.47}$$

We see, therefore, that using the definition of a series, the domain of regularity is R^+.

Let us now define cosh **X** using the exponential function cosh x $= [e^x + e^{-x}]/2$. Since this function decreases monotonically in R^- and increases monotonically in R^+, cosh **X** is nonregular in R^- and regular in R^+. This is what we found in (3.47).

Let us examine now what appears to be a case of a hyperbolic sine function. From the series we have

$$\sinh X^{(\alpha)} = \sinh [a^{(\alpha)}, b^{(\alpha)}]$$

$$= [\sinh a^{(\alpha)}, \sinh b^{(\alpha)}], \qquad a^{(\alpha)} \leq b^{(\alpha)} \leq 0$$

$$= [(a^{(\alpha)}/b^{(\alpha)}) \sinh b^{(\alpha)}, \sinh b^{(\alpha)}],$$

$$a^{(\alpha)} \leq 0 \leq b^{(\alpha)}, |a^{(\alpha)}| \leq |b^{(\alpha)}|$$

$$= [\sinh a^{(\alpha)}, (b^{(\alpha)}/a^{(\alpha)}) \sinh a^{(\alpha)}],$$

$$a^{(\alpha)} \leq 0 \leq b^{(\alpha)}, |a^{(\alpha)}| \geq |b^{(\alpha)}|$$

$$= [\sinh a^{(\alpha)}, \sinh b^{(\alpha)}], \qquad 0 \leq a^{(\alpha)} \leq b^{(\alpha)}.$$

This shows that the sinh **X** is regular both in R^- and in R^+, but not in $R = R^- \cup R^+$. It is irregular for every segment for which $x = 0$. On the other hand, if we define sinh **X** by the exponential function $\sinh X = [e^X - e^{-X}]/2$, we find that this function increases monotonically in R, and for this reason sinh **X** is regular in R.

EXAMPLE 3.24

Let us examine the case of a function that is everywhere irregular: $f(\mathbf{X}) = \mathbf{X}\,(-)\,\mathbf{X}^3$. We then have

$\forall\, \alpha \in [0, 1]$:

$$f(X^{(\alpha)}) = [a^{(\alpha)}, b^{(\alpha)}]\,(-)\,[a^{(\alpha)}, b^{(\alpha)}]^3.$$

1. $a^{(\alpha)} \leq b^{(\alpha)} \leq 0$

$$f(X^{(\alpha)}) = [a^{(\alpha)}, b^{(\alpha)}]\,(-)\,[a^{(\alpha)3}, b^{(\alpha)3}]$$

$$= [a^{(\alpha)} - b^{(\alpha)3}, b^{(\alpha)} - a^{(\alpha)3}].$$

2. $a^{(\alpha)} \leq 0 \leq b^{(\alpha)}, \qquad |a^{(\alpha)}| \leq |b^{(\alpha)}|$

$$f(X^{(\alpha)}) = [a^{(\alpha)}, b^{(\alpha)}]\,(-)\,[a^{(\alpha)}\, b^{(\alpha)2}, b^{(\alpha)3}]$$

$$= [a^{(\alpha)} - b^{(\alpha)3}, b^{(\alpha)} - a^{(\alpha)} b^{(\alpha)2}].$$

3. $a^{(\alpha)} \leq 0 \leq b^{(\alpha)}, \qquad |a^{(\alpha)}| \geq |b^{(\alpha)}|$

$$f(X^{(\alpha)}) = [a^{(\alpha)}, b^{(\alpha)}]\,(-)\,[a^{(\alpha)3}, a^{(\alpha)2}\, b^{(\alpha)}]$$

$$= [a^{(\alpha)} - a^{(\alpha)2} b^{(\alpha)}, b^{(\alpha)}\, a^{(\alpha)3}].$$

4. $0 \le a^{(\alpha)} \le b^{(\alpha)}$

$$f(X^{(\alpha)}) = [a^{(\alpha)}, b^{(\alpha)}](-)[a^{(\alpha)3}, b^{(\alpha)3}]$$

$$= [a^{(\alpha)} - b^{(\alpha)3}, b^{(\alpha)} - a^{(\alpha)3}].$$

THEOREM 3.10

Every function $f(\mathbf{X}) = \mathbf{X}n$, $n = 1, 2, 3, 4, \ldots$ is regular in R^+ only, while every function $f(\mathbf{X}) = -\mathbf{X}^n$, $n = 2, 4, 6, \ldots$ is regular in R^- only.

The proof of this theorem follows from that given for Table 3.20.

3.7 DERIVATIVE OF A FUNCTION OF FUZZY NUMBERS

We now examine the derivative of the function $f(\mathbf{X})$, where $f(\mathbf{X})$ is a nonnegative, continuous function that increases monotonically in the segment $[A, B] \subset R^+$. The derivative is therefore nonnegative in this segment.

Let us consider the intervals of confidence of level α for $f(\mathbf{X})$.

$\forall\, \alpha \in [0, 1]$:

$$f(X^{(\alpha)}) = f[x_1^{(\alpha)}, x_2^{(\alpha)}]$$

$$= [f(x_1^{(\alpha)}), f(x_2^{(\alpha)})],$$

where $[x_1^{(\alpha)}, x_2^{(\alpha)}] \subset [A, B] \subset R^+$, which is shown in Figure 3.8.

$$\forall\, x \in [x_1^{(\alpha)}, x_2^{(\alpha)}].$$

We can define the derivative of $f(x)$ as $f'(x)$. From $f'(x_1^{(\alpha)})$ to $f'(x_2^{(\alpha)})$, we can construct segments that are continuous and increase monotonically. The derivative of $f(\mathbf{X})$ is a function of $\mathbf{X}$, where the interval of confidence of level α is

$\forall\, \alpha \in [0, 1]$:

$$f'(X)^{(\alpha)} = [f'(x_1^{(\alpha)}), f'(x_2^{(\alpha)})].$$

It is therefore clear that $f(\mathbf{X})$ and $f'(\mathbf{X})$ have membership functions that are convex. That is, because $\mathbf{X}$ is a fuzzy number, it exists as

$$x^{(\alpha=1)} \text{ for which } \mu_{\mathbf{X}}(x^{(\alpha=1)}) = 1 \text{ and } \mu_{f(\mathbf{X})}[f(x^{(\alpha=1)})] = 1.$$

The function $f(\mathbf{X})$ is therefore normal if $\mathbf{X}$ is normal, and the same is true of $f'(\mathbf{X})$. This can be seen more clearly from the following example.

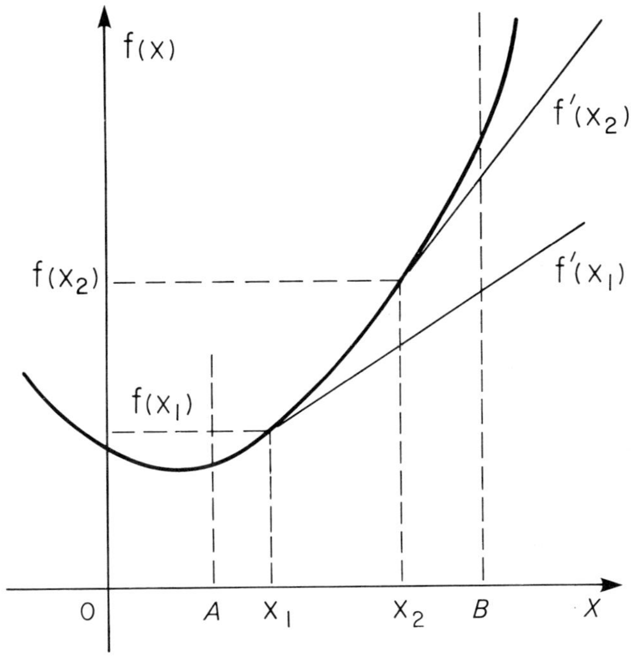

Figure 3.8 Derivative of a function $f(\mathbf{x}) \subset R^+$.

EXAMPLE 3.25

Let us consider the fuzzy number **X** such that

$\forall\, x \in R^+$:

$$
\begin{aligned}
\mu_{\mathbf{X}}(x) &= 0, & x \le 1,\\
&= x/2 - 1/2, & 1 \le x \le 3,\\
&= -x + 4, & 3 \le x \le 4,\\
&= 0, & x \ge 4.
\end{aligned}
$$

This TFN is shown in Figure 3.9.

Consider now the function $f(\mathbf{X}) = \mathbf{X}^2\ (+)\ \mathbf{X}$, which is regular in R^+. Further, $f(x) = x^2 + x$ is nonnegative, continuous, and increasing monotonically in R^+. We can therefore write, for every interval of confidence of level α,

$\forall\, \alpha \in [0, 1]$:

$$
\begin{aligned}
f(X^{(\alpha)}) &= X^{(\alpha)2}\ (+)\ X^{(\alpha)}\\
&= [x_1{}^{(\alpha)}, x_2{}^{(\alpha)}]^2\ (+)\ [x_1{}^{(\alpha)}, x_2{}^{(\alpha)}]\\
&= [x_1{}^{(\alpha)2} + x_1{}^{(\alpha)}, x_2{}^{(\alpha)2} + x_2{}^{(\alpha)}].
\end{aligned}
$$

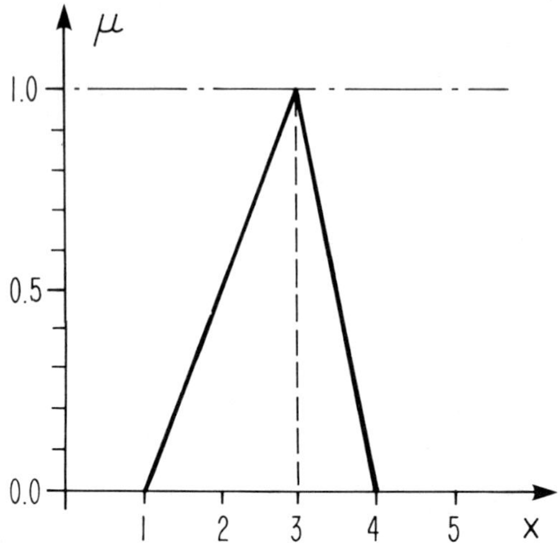

Figure 3.9 Membership of a fuzzy number (Example 3.25).

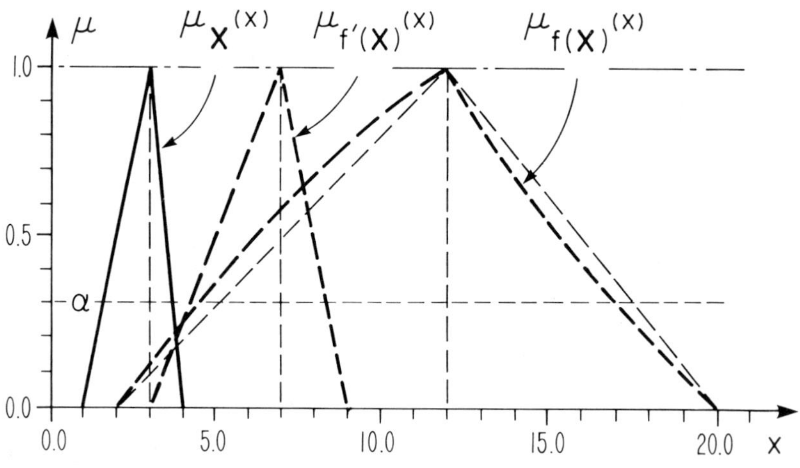

Figure 3.10 Membership of a fuzzy function and its derivative (Example 3.25).

The derivative of $f(x) = x^2 + x$ is $f'(x) = 2x + 1$. Hence $f'(\mathbf{X}) = 2\mathbf{X}\ (+)\ 1$. Let us now compute the intervals of confidence for levels $\alpha = 0, 0.1, 0.2, 0.3, \ldots, 0.9, 1$ for $\mathbf{X}$, $f(\mathbf{X})$, and $f'(\mathbf{X})$ (see Table 3.21).

These results are also shown in Figure 3.10. At the point $x = 1.4$, for example at the level $\alpha = 0.3$, we have $f(X^{(\alpha=0.3)}) = [4.16, 17.39]$ and $f'(X^{(\alpha=0.3)}) = [4.2, 8.4]$.

Table 3.21 (Example 3.25)

α	$[x_1(\alpha), x_2(\alpha)]$	$[x_1(\alpha)2 + x_1, x_2(\alpha)2 + x_2(\alpha)]$	$[2x_1(\alpha) + 1, 2x_2(\alpha) + 1]$
0	[1, 4]	[2, 20]	[3, 9]
0.1	[1.2, 3.9]	[2.64, 19.11]	[3.4, 8.8]
0.2	[1.4, 3.8]	[3.36, 18.24]	[3.8, 8.6]
0.3	[1.6, 3.7]	[4.16, 17.39]	[4.2, 8.4]
0.4	[1.8, 3.6]	[5.04, 16.56]	[4.6, 8.2]
0.5	[2, 3.5]	[6, 15.75]	[5, 8]
0.6	[2.2, 3.4]	[7.04, 14.96]	[5.4, 7.8]
0.7	[2.4, 3.3]	[8.16, 14.19]	[5.8, 7.6]
0.8	[2.6, 3.2]	[9.36, 13.44]	[6.2, 7.4]
0.9	[2.8, 3.1]	[10.64, 12.71]	[6.6, 7.2]
1	3	12	7

Points of Correction

For a given function $f(x)$, $x \in R$, every point where x, $f(x)$, $f'(x)$, $f''(x)$, . . . change in sign will be called a *point of correction* in the domain of interest. These points of correction are very important because they affect the membership function $\mu_{f(\mathbf{X})}$ for a given $\mu_{\mathbf{X}}$, as shown in Figure 3.11.

EXAMPLE 3.26

Consider again $f(x) = x^2 + x$, which is shown in Figure 3.11. Its derivative is $f'(x) = 2x + 1$, which is also shown in Figure 3.11.

Now consider the TFN that is given in Figure 3.12, where

$$\begin{aligned}\mu_{\mathbf{X}}(x) &= 0, & 0 \le -5/4,\\ &= 4x/3 + 5/3, & -5/4 \le x \le -1/2,\\ &= -4x/5 + 3/5, & -1/2 \le x \le 3/4,\\ &= 0, & x \ge 3/4.\end{aligned} \tag{3.48}$$

Now let us consider the function of the fuzzy numbers $f(\mathbf{X}) = \mathbf{X}^2\,(+)\,\mathbf{X}$. This function is regular in R^+ and so is its derivative, $f'(\mathbf{X}) = 2 \cdot \mathbf{X}\,(+)\,1$.

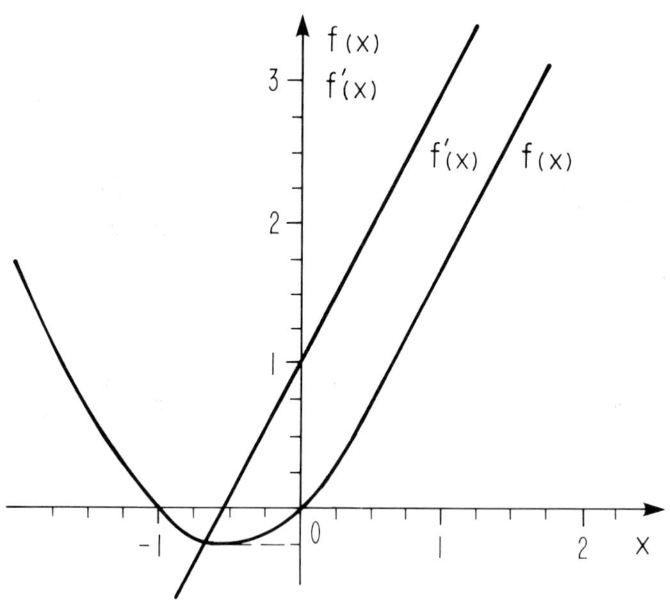

Figure 3.11 Derivative of a fuzzy function (Example 3.26).

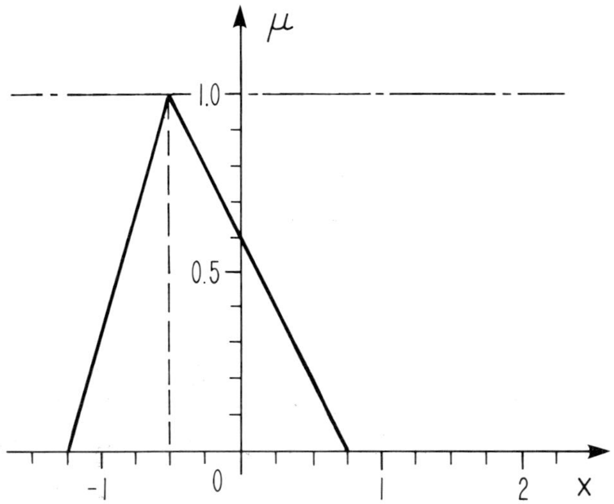

Figure 3.12 A TFN (Example 3.26).

This function $f(\mathbf{X})$ has the following points of correction from $-\infty$ to $+\infty$ on $0-x$ axis:

$$x=-1, \qquad x=-0.5, \qquad x=0 \tag{3.49}$$

As defined by (3.48), **X** has a membership function that is null for $x \leq -1.25$ and for $x \geq 0.75$. For this reason we must first examine the correction points that belong to $[-1.25, 0.75]$. The three points given in (3.49) belong to this segment. Let us therefore decompose $[-1.25, 0.75]$ into four segments, namely, $[-1.25, -1]$, $[-1, -0.5]$, $[-0.5, 0]$, $[0, 0.75]$. Consequently, for each $\alpha \in [0, 1]$ we can write (3.48) as follows: $X^{(\alpha)} = [3\alpha/4 - 5/4, 3/4 - 5\alpha/4]$. Now, for each $\alpha = 0, 0.1, 0.2, \ldots, 0.9, 1$, we prepare a table (Table 3.22) that gives $X^{(\alpha)}$, $f(X^{(\alpha)})$, and $f'(X^{(\alpha)})$. We can now compute the membership functions when considering the reunion of the segments appearing in Table 3.22. The results are given in Table 3.23.

$$\begin{aligned} f(X^{(0.3)}) &= X^{(0.3)2}\,(+)\,X^{(0.3)} \\ &= (1) \cup (2) \cup (3) \cup (4) \qquad \text{for the row } \alpha=0.3 \text{ of } f(X^{(\alpha)}) \\ &= [-0.025, 0.0506] \cup [-0.750, 0.500] \cup [-0.500, 0.250] \cup \\ &\quad\ [0, 0.5156] \\ &= [-0.7500, 0.5156]. \end{aligned}$$

The results of Table 3.23 are shown in Figure 3.13, which also gives the second derivative of this function. Its membership function defines the nonfuzzy number 2 for every **X**.

Table 3.22

		(1)	(2)	(3)	(4)
	α	$x \in [-1.25, -1]$	$x \in [-1, -0.5]$	$x \in [-0.5, -0]$	$x \in [0, 0.75]$
	0	[−1.250, −1]	[−1, −0.500]	[−0.500, 0]	[0, 0.750]
	0.1	[−1.175, −1]	[−1, −0.500]	[−0.500, 0]	[0, 0.625]
	0.2	[−1.100, −1]	[−1, −0.500]	[−0.500, 0]	[0, 0.500]
	0.3	[−1.025, −1]	[−1, −0.500]	[−0.500, 0]	[0, 0.375]
	0.4	∅	[−0.950, −0.500]	[−0.500, 0]	[0, 0.250]
$X^{(\alpha)}$	0.5	∅	[−0.875, −0.500]	[−0.500, 0]	[0, 0.125]
	0.6	∅	[−0.800, −0.500]	[−0.500, 0]	0
	0.7	∅	[−0.725, −0.500]	[−0.500, −0.125]	∅
	0.8	∅	[−0.650, −0.500]	[−0.500, −0.250]	∅
	0.9	∅	[−0.575, −0.500]	[−0.500, −0.375]	∅
	1	∅	−0.500	−0.500	∅

	0	[−0.250, 0.5625]	[−0.750, 0.5000]	[−0.5000, 0.250]	[0, 1.3125]
	0.1	[−0.175, 0.3806]	[−0.750, 0.5000]	[−0.5000, 0.250]	[0, 1.0156]
	0.2	[−0.100, 0.2100]	[−0.750, 0.5000]	[−0.5000, 0.250]	[0, 0.7500]
	0.3	[−0.025, 0.0506]	[−0.750, 0.5000]	[−0.5000, 0.250]	[0, 0.5156]
	0.4	∅	[−0.700, 0.4025]	[−0.5000, 0.250]	[0, 0.3125]
$f(X^{(\alpha)})$	0.5	∅	[−0.625, 0.2656]	[−0.5000, 0.250]	[0, 0.1406]
	0.6	∅	[−0.550, 0.1400]	[−0.5000, 0.250]	0
	0.7	∅	[−0.475, 0.0256]	[−0.4844, 0.125]	∅
	0.8	∅	[−0.400, −0.0775]	[−0.4375, 0]	∅
	0.9	∅	[−0.325, −0.1634]	[−0.3594, −0.125]	∅
	1	∅	−0.25	−0.25	∅

Table 3.22 (*continued*)

		(1)	(2)	(3)	(4)
	α	$x \in [-1.25, -1]$	$x \in [-1, -0.5]$	$x \in [-0.5, -0]$	$x \in [0, 0.75]$
	0	[−1.50, −1]	[−1, 0]	[0, 1]	[1, 2.50]
	0.1	[−1.35, −1]	[−1, 0]	[0, 1]	[1, 2.25]
	0.2	[−1.20, −1]	[−1, 0]	[0, 1]	[1, 2]
	0.3	[−1.05, −1]	[−1, 0]	[0, 1]	[1, 1.75]
	0.4	$\varnothing$	[−0.90, 0]	[0, 1]	[1, 1.50]
$f'(X^{(\alpha)})$	0.5	$\varnothing$	[0.75, 0]	[0, 1]	[1, 1.25]
	0.6	$\varnothing$	[0.60, 0]	[0, 1]	1
	0.7	$\varnothing$	[0.45, 0]	[0, 0.75]	$\varnothing$
	0.8	$\varnothing$	[0.30, 0]	[0, 0.50]	$\varnothing$
	0.9	$\varnothing$	[0.15, 0]	[0, 0.25]	$\varnothing$
	1	$\varnothing$	0	0	$\varnothing$

Table 3.23

α	$x^{(\alpha)}$	$f(X^{(\alpha)}) = X^{(\alpha)2} (+) X^{(\alpha)}$	$f'(X^{(\alpha)}) = 2 \cdot X^{(\alpha)} (+) 1$
0	[−1.250, 0.750]	[−0.7500, 1.3125]	[−1.50, 2.50]
0.1	[−1.175, 0.625]	[−0.7500, 1.0156]	[−1.35, 2.25]
0.2	[−1.100, 0.500]	[−0.7500, 0.7500]	[−1.20, 2]
0.3	[−1.025, 0.375]	[−0.7500, 0.5156]	[−1.05, 1.75]
0.4	[−0.950, 0.250]	[−0.7000, 0.4025]	[−0.90, 1.50]
0.5	[−0.875, 0.125]	[−0.6250, 0.2656]	[−0.75, 1.25]
0.6	[−0.800, 0]	[−0.5500, 0.2500]	[−0.60, 1]
0.7	[−0.725, −0.125]	[−0.4844, 0.1250]	[−0.45, 0.75]
0.8	[−0.650, −0.250]	[−0.4375, 0]	[−0.30, 0.50]
0.9	[−0.575, −0.375]	[−0.3594, −0.1250]	[−0.15, 0.25]
1	−0.50	−0.25	0

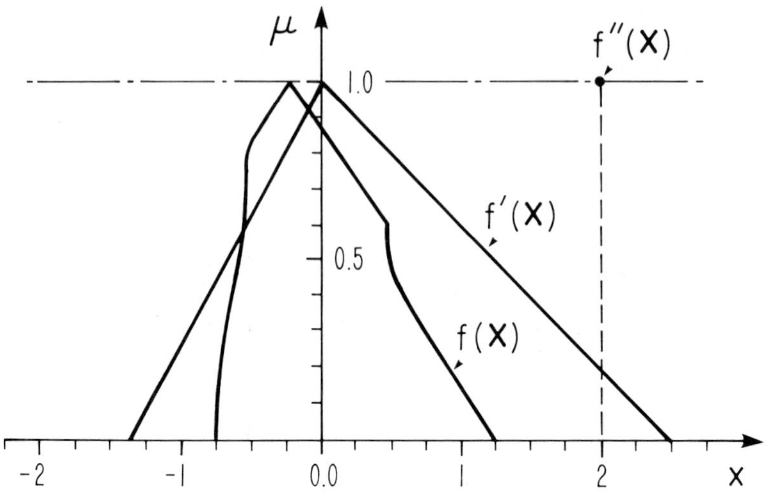

Figure 3.13 $f(\mathbf{x})$, $f'(\mathbf{x})$, and $f''(\mathbf{x})$ (Example 3.26).

3.8 FUZZY TRIGONOMETRIC FUNCTIONS, FUZZY HYPERBOLIC FUNCTIONS, AND FUZZY COMPLEX NUMBERS

Let us recall the following classical trigonometric hyperbolic and exponential equations:

$$e^x = 1 + x + x^2/2! + x^3/3! + \cdots + x^n/n! + \cdots,$$

$$a^x = 1 + x \ln a + [x^2 \ln^2 a]/2! + [x^3 \ln^3 a]/3! + \cdots + [x^n \ln^n a]/n! + \cdots,$$

$$\cos x = 1 - x^2/2! + x^4/4! - \cdots + [x^{4n}]/(4n)! - [x^{4n+2}]/(4n+2)! + \cdots,$$

$$\sin x = x - x^3/3! + x^5/5! - \cdots + [x^{4n+1}]/(4n+1)! - [x^{4n+2}]/(4n+2)! + \cdots,$$

$$\cosh x = 1 + x^2/2! + x^4/4! + \cdots + [x^{2n}]/(2n)! + \cdots,$$

$$\sinh x = x + x^3/3! + x^5/5! + \cdots + [x^{2n+1}]/(2n+1)! + \cdots,$$

$$\cos x = [e^{jx} + e^{-jx}]/2, \qquad j = \sqrt{-1}. \tag{3.50}$$

$$\sin x = [e^{jx} - e^{-jx}]/2j, \qquad j = \sqrt{-1}.$$

$$\cosh x = [e^{x} + e^{-x}]/2.$$

$$\sinh x = [e^{x} - e^{-x}]/2.$$

Every function derived from these exponential functions will be of two types with fuzzy numbers: The *first* and the *second* kind.

Trigonometric Functions and Hyperbolic Functions of the First Kind

Consider the following fuzzy number:

$$A_1^{(\alpha)} = [a^{(\alpha)}, b^{(\alpha)}] \subset R_0^+,$$

with

$$A_1^{(\alpha=1)} = 1.$$

Here $\mathbf{A}_1 = 1$ is a fuzzy number. With ordinary numbers, such a 1 is unique, but this is not the case with fuzzy numbers.

We have shown in section 3.5 how to derive an exponential fuzzy function from $\mathbf{A}_1$, as follows:

$$e^{x}(A_1^{(\alpha)}) = [e^{x/b^{(\alpha)}}, e^{x/a^{(\alpha)}}]. \tag{3.51}$$

Now let us examine how to derive trigonometric and hyperbolic functions of the first kind from (3.51) in R_0^+.

We have

$$\cos A_1^{(\alpha)}(x) = [e^{jx}(A_1^{(\alpha)}) + e^{-jx}(A_1^{(\alpha)})]/2, \qquad j = \sqrt{-1}. \tag{3.52}$$

Now in passing from (3.50) to (3.52), we use the fact that we remain in R_0^+ and

$$e^{-jx}(A_1^{(\alpha)}) = [1, 1](:)[e^{jx/b^{(\alpha)}}, e^{jx/a^{(\alpha)}}]$$

$$= [e^{-jx/a^{(\alpha)}}, e^{-jx/b^{(\alpha)}}].$$

From

$$\cos A_1^{(\alpha)}(x) = 1/2([e^{jx/b^{(\alpha)}}, e^{jx/a^{(\alpha)}}](+)[e^{-jx/a^{(\alpha)}}, e^{-jx/b^{(\alpha)}}]$$

$$= [(e^{jx/b^{(\alpha)}} + e^{-jx/a^{(\alpha)}})/2, (e^{jx/a^{(\alpha)}} - e^{-jx/b^{(\alpha)}})/2]$$

$$x/a^{(\alpha)}, x/b^{(\alpha)} \in [0, \pi/2]. \tag{3.53}$$

We do not find any bounds of the cosine function except when $b^{(\alpha)} = a^{(\alpha)}$. Similarly,

$$\sin A_1^{(\alpha)}(x) = [(e^{jx/b^{(\alpha)}} - e^{-jx/a^{(\alpha)}})/2j, (e^{jx/a^{(\alpha)}} - e^{-jx/b^{(\alpha)}})/2j],$$

$$x/a^{(\alpha)}, x/b^{(\alpha)} \in [0, \pi/2]. \qquad \textbf{(3.54)}$$

$$\cosh A_1^{(\alpha)}(x) = [(e^{x/b^{(\alpha)}} + e^{-x/a^{(\alpha)}})/2, (e^{x/a^{(\alpha)}} + e^{x/b^{(\alpha)}})/2]$$

$$x/a^{(\alpha)}, x/b^{(\alpha)} \in R^+. \qquad \textbf{(3.55)}$$

$$\sinh A_1^{(\alpha)}(x) = [(e^{x/b^{(\alpha)}} - e^{-x/a^{(\alpha)}})/2, (e^{x/a^{(\alpha)}} - e^{-x/b^{(\alpha)}})/2],$$

$$x/a^{(\alpha)}, x/b^{(\alpha)} \in R. \qquad \textbf{(3.56)}$$

The remarks that were made about (3.53) are also relevant for (3.54), (3.55), and (3.56).

Trigonometric and Hyperbolic Function of the Second Kind

A fuzzy number **X** is given by

$\forall\, \alpha \in [0, 1]$:

$$X^{(\alpha)} = [x_1^{(\alpha)}, x_2^{(\alpha)}].$$

From this we can construct $e^{\mathbf{X}}$ as follows:

$$\begin{aligned} e^{X^{(\alpha)}} &= 1\,(+)\,X^{(\alpha)}\,(+)\,(1/2!)X^{(\alpha)2}\,(+)\,(1/3!)X^{(\alpha)3}\,(+)\cdots \\ &\quad (+)\,(1/n!)X^{(\alpha)n}\,(+)\cdots \\ &= [e^{x_1^{(\alpha)}}, e^{x_2^{(\alpha)}}]. \end{aligned}$$

Trigonometric and hyperbolic functions can be constructed in the same way. For example,

$$\cos X^{(\alpha)} = [(e^{jx_1^{(\alpha)}} + e^{-jx_1^{(\alpha)}})/2, (e^{jx_2^{(\alpha)}} + e^{-jx_2^{(\alpha)}})/2],$$

$$x_1^{(\alpha)}, x_2^{(\alpha)} \in [0, \pi/2]. \qquad \textbf{(3.57)}$$

$$\sin X^{(\alpha)} = [(e^{jx_1^{(\alpha)}} - e^{-jx_1^{(\alpha)}})/2j, (e^{jx_2^{(\alpha)}} - e^{-jx_2^{(\alpha)}})/2j],$$

$$x_1^{(\alpha)}, x_2^{(\alpha)} \in [0, \pi/2]. \qquad \textbf{(3.58)}$$

$$\cosh X^{(\alpha)} = [(e^{x_1^{(\alpha)}} + e^{-x_1^{(\alpha)}})/2, (e^{x_2^{(\alpha)}} - e^{-x_2^{(\alpha)}})/2],$$

$$x_1^{(\alpha)}, x_2^{(\alpha)} \in R^+.$$

$$\sinh X^{(\alpha)} = [(e^{x_1^{(\alpha)}} - e^{-x_1^{(\alpha)}})/2, (e^{x_2^{(\alpha)}} - e^{-x_2^{(\alpha)}})/2], \qquad x_1^{(\alpha)}, x_2^{(\alpha)} \in R.$$

Remark

Equations (3.53), (3.54) and (3.57) and (3.58) are limited to the first quadrant for trigonmetric functions, to R^+ for hyperbolic cosine functions, and to R for hyperbolic sine functions. If other quadrants are to be used, it is necessary to develop the relevant segments, taking into account the signs and locations of the bounds.

An Alternative Way of Presenting Trigonometric Functions

We now present the concept of the fuzzy angle θ, which we represent by its intervals of confidence of level α:

$$\theta^{(\alpha)} = [\theta_1^{(\alpha)}, \theta_2^{(\alpha)}].$$

We simplify this expression by limiting θ in the domain

$\forall\ \alpha \in [0, 1]$:

$$0 \leq (\theta_2^{(\alpha)} - \theta_1^{(\alpha)}) \leq \pi/2,$$

from which

$0 \leq \theta_1^{(\alpha)} \leq \theta_2^{(\alpha)} \leq \pi/2$:

$$\sin\theta^{(\alpha)} = [\sin\theta_1^{(\alpha)}, \sin\theta_2^{(\alpha)}],$$
$$\cos\theta^{(\alpha)} = [\cos\theta_2^{(\alpha)}, \cos\theta_1^{(\alpha)}],$$
$$\tan\theta^{(\alpha)} = [\tan\theta_1^{(\alpha)}, \tan\theta_2^{(\alpha)}].$$

$0 \leq \theta_1^{(\alpha)} \leq \pi/2 \leq \theta_2^{(\alpha)} \leq \pi$:

$$\sin\theta^{(\alpha)} = [\sin\theta_1^{(\alpha)} \wedge \sin\theta_2^{(\alpha)}, 1],$$
$$\cos\theta^{(\alpha)} = [\cos\theta_2^{(\alpha)}, \cos\theta_1^{(\alpha)}],$$
$$\tan\theta^{(\alpha)} = (-\infty, \tan\theta_2^{(\alpha)}] \cup [\tan\theta_1^{(\alpha)}, \infty)^*$$

$\pi/2 \leq \theta_1^{(\alpha)} \leq \theta_2^{(\alpha)} \leq \pi$:

$$\sin\theta^{(\alpha)} = [\sin\theta_2^{(\alpha)}, \sin\theta_1^{(\alpha)}],$$
$$\cos\theta^{(\alpha)} = [\cos\theta_2^{(\alpha)}, \cos\theta_1^{(\alpha)}],$$
$$\tan\theta^{(\alpha)} = [\tan\theta_1^{(\alpha)}, \tan\theta_2^{(\alpha)}].$$

$\pi/2 \leq \theta_1^{(\alpha)} \leq \pi \leq \theta_2^{(\alpha)} \leq 3\pi/2$:

$$\sin\theta^{(\alpha)} = [\sin\theta_2^{(\alpha)}, \sin\theta_1^{(\alpha)}],$$

$$\cos\theta^{(\alpha)} = [-1, \cos\theta_1^{(\alpha)} \vee \cos\theta_2^{(\alpha)}],$$
$$\tan\theta^{(\alpha)} = [\tan\theta_1^{(\alpha)}, \tan\theta_2^{(\alpha)}].$$

$\pi \le \theta_1^{(\alpha)} \le \theta_2^{(\alpha)} \le 3\pi/2$:

$$\sin\theta^{(\alpha)} = [\sin\theta_2^{(\alpha)}, \sin\theta_1^{(\alpha)}],$$
$$\cos\theta^{(\alpha)} = [\cos\theta_1^{(\alpha)}, \cos\theta_2^{(\alpha)}],$$
$$\tan\theta^{(\alpha)} = [\tan\theta_1^{(\alpha)}, \tan\theta_2^{(\alpha)}].$$

$\pi \le \theta_1^{(\alpha)} \le 3\pi/2 \le \theta_2^{(\alpha)} \le 2\pi$:

$$\sin\theta^{(\alpha)} = [-1, \sin\theta_1^{(\alpha)} \vee \sin\theta_2^{(\alpha)}],$$
$$\cos\theta^{(\alpha)} = [\cos\theta_1^{(\alpha)}, \cos\theta_2^{(\alpha)}],$$
$$\tan\theta^{(\alpha)} = (-\infty, \tan\theta_2^{(\alpha)}] \cup [\tan\theta_1^{(\alpha)}, \infty).*$$

$3\pi/2 \le \theta_1^{(\alpha)} \le \theta_2^{(\alpha)} \le 2\pi$:

$$\sin\theta^{(\alpha)} = [\sin\theta_1^{(\alpha)}, \sin\theta_2^{(\alpha)}],$$
$$\cos\theta^{(\alpha)} = [\cos\theta_1^{(\alpha)}, \cos\theta_2^{(\alpha)}],$$
$$\tan\theta^{(\alpha)} = [\tan\theta_1^{(\alpha)}, \tan\theta_2^{(\alpha)}].$$

$3\pi/2 \le \theta_1^{(\alpha)} \le 2\pi \le \theta_2^{(\alpha)} \le 5\pi/2$:

$$\sin\theta^{(\alpha)} = [\sin\theta_1^{(\alpha)}, \sin\theta_2^{(\alpha)}],$$
$$\cos\theta^{(\alpha)} = [\cos\theta_1^{(\alpha)} \wedge \cos\theta_2^{(\alpha)}, 1],$$
$$\tan\theta^{(\alpha)} = [\tan\theta_1^{(\alpha)}, \tan\theta_2^{(\alpha)}].$$

In the above expressions, equations marked by * do not give a fuzzy number.

At this point an important question arises. What happens whan α increases from 0 to 1? It is fundamental to the two functions sine and cosine but not to the function tangent, which does not produce a fuzzy number. There is no problem when the segments $[\theta_1^{(\alpha)}, \theta_2^{(\alpha)}]$ are completely localized in R and the function being considered is monotonic. Now let us examine the case where the function is not monotonic, for instance, for a cosine for $\theta = \pi$. When we pass from α_1 to α_2, where $\alpha_2 > \alpha_1$, the segment $\cos[\theta_1^{(\alpha_2)}, \theta_2^{(\alpha_2)}]$ must be included in the segment $\cos[\theta_1^{(\alpha_1)}, \theta_1^{(\alpha_1)}]$. We then have

$$\cos[\theta_1^{(\alpha_2)}, \theta_2^{(\alpha_2)}] = [-1, \cos\theta_1^{(\alpha_2)} \vee \cos\theta_2^{(\alpha_2)}]$$

and

$$\cos[\theta_1^{(\alpha_1)}, \theta_1^{(\alpha_1)}] = [-1, \cos\theta_1^{(\alpha_1)} \vee \cos\theta_1^{(\alpha_1)}].$$

These results are shown in Figure 3.14 and are seen to be valid.

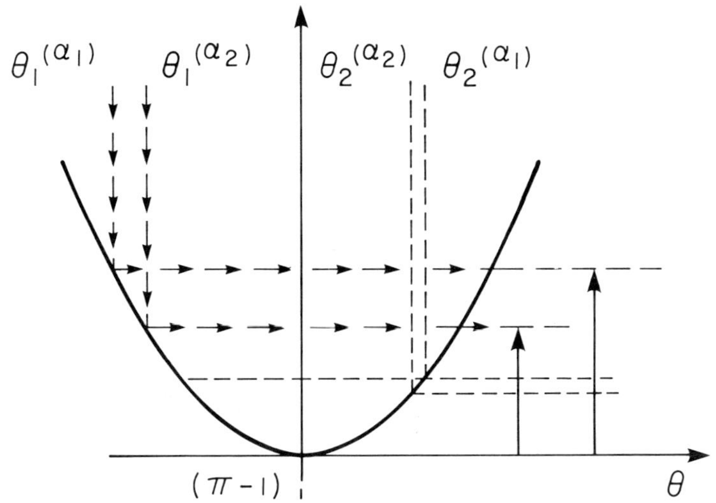

Figure 3.14 A valid inclusion for segment $[\theta_1(\alpha), \theta_2(\alpha)]$.

The results for $\theta = 2\pi$ are valid for the sine function for $\theta = \pi/2$ and $\theta = 3\pi/2$. Figures 3.15 and 3.16 show these results (pages **248** and **249**).

Fuzzy Complex Numbers

Let us recall that for an ordinary complex number we can consider the couplet (ρ, θ), $\rho \in R^+$, $\theta \in R$, where ρ is the *modulus* and θ the *argument*

A product is defined as

$$(\rho_1, \theta_1) * (\rho_2, \theta_2) = (\rho_1 \cdot \rho_2, \theta_1 + \theta_2).$$

It is usual to represent a complex number by z and the set of complex numbers by C, where C has the structure of a commutative group. Let

$$z_1 = (\rho_1, \theta_1), z_2 = (\rho_2, \theta_2), z_3 = (\rho_3, \theta_3).$$

We then have

$$\forall\ z_1, z_2, z_3 \in C:$$

$(z_1 * z_2) * z_3 = z_1 * (z_2 * z_3)$,	associative.
$z_1 * z_2 = z_2 * z_1$,	commutative.
$z_1 * 1 = 1 * z_1 = z_1$,	neutral is $z = (1, 0) = 1$.

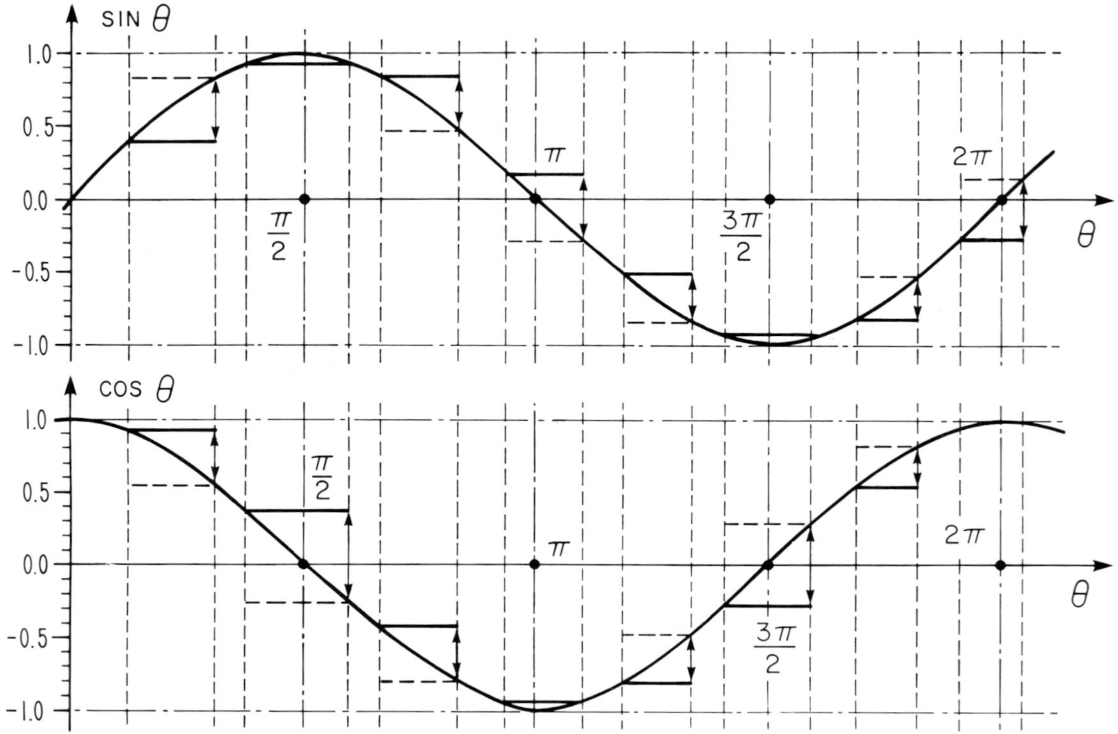

Figure 3.15 Segments on sine and cosine curves.

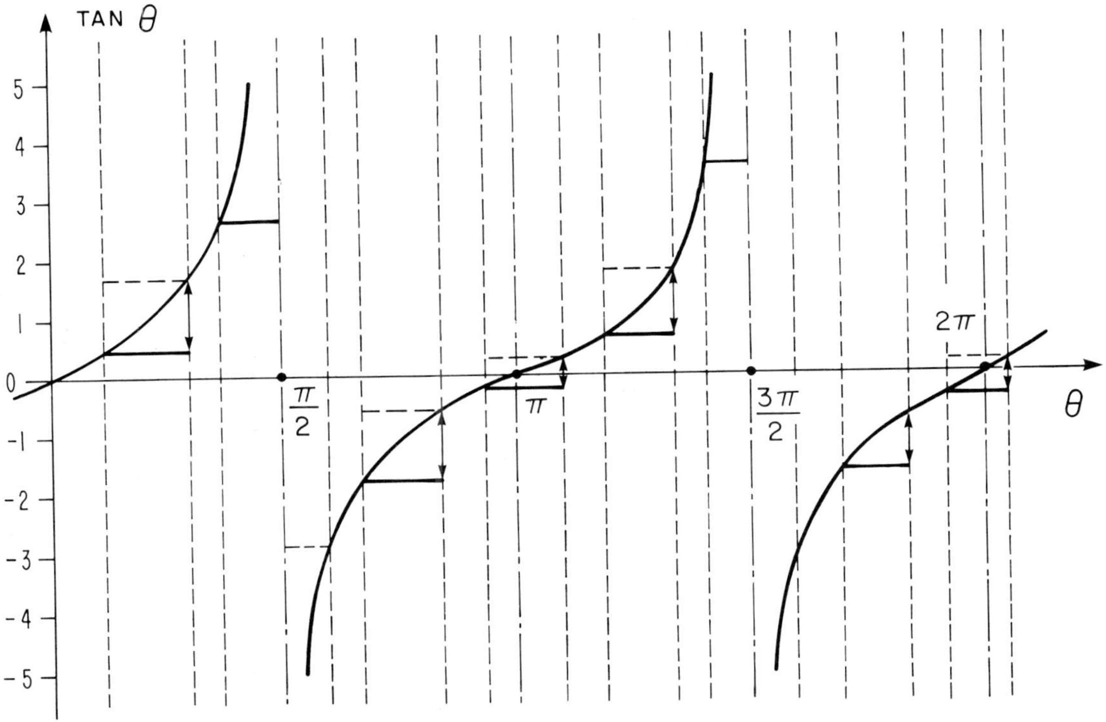

Figure 3.16 Segments on a tangent curve.

The inverse of $z=(\rho,\theta)$ is

$$z^{-1}=(1/\rho,-\theta),$$

and we then have

$$z*z^{-1}=z^{-1}*z=1.$$

Using the symbol $*^{-1}$ for division, we have

$$z_1*^{-1}z_2=z_1*z_2^{-1}=(\rho_1/\rho_2,\theta_1-\theta_2).$$

To make a fuzzy number complex, the following procedure is used where two intervals of confidence are considered:

$\forall\ \alpha\in[0,1]$:

$$\rho^{(\alpha)}=[\rho_1^{(\alpha)},\rho_2^{(\alpha)}]\subset R^+,$$
$$\theta^{(\alpha)}=[\theta_1^{(\alpha)},\theta_2^{(\alpha)}]\subset R.$$

This gives the couplet $Z^{(\alpha)}=(\rho^{(\alpha)},\theta^{(\alpha)})$, $\alpha\in[0,1]$ or $\mathbf{Z}=(\rho,\theta)$. Numbers such as $\mathbf{Z}$ are called *fuzzy complex numbers*. For such numbers an operation *product* is defined that is also called *multiplication*:

$$\begin{aligned}\mathbf{Z}_a^{(\alpha)}(*)\mathbf{Z}_b^{(\alpha)}&=(\rho_a^{(\alpha)},\theta_a^{(\alpha)})(*)(\rho_b^{(\alpha)},\theta_b^{(\alpha)})\\&=(\rho_a^{(\alpha)}(\cdot)\rho_b^{(\alpha)},\theta_a^{(\alpha)}(+)\theta_b^{(\alpha)}).\end{aligned}\qquad\textbf{(3.57)}$$

We may easily check that

$$\begin{aligned}&(\mathbf{Z}_a(*)\mathbf{Z}_b)(*)\mathbf{Z}_c=\mathbf{Z}_a(*)(\mathbf{Z}_b(*)\mathbf{Z}_c), &&\text{associative,}\\&\mathbf{Z}_a(*)\mathbf{Z}_b=\mathbf{Z}_b(*)\mathbf{Z}_a, &&\text{commutative,}\\&\mathbf{Z}(*)1=1(*)\mathbf{Z}=\mathbf{Z}, &&\text{neutral is } 1=(1,0).\end{aligned}$$

Fuzzy complex numbers have only a pseudoinverse, defined as

$$\begin{aligned}Z^{-1(\alpha)}&=(\rho^{-1(\alpha)},-\theta^{(\alpha)})\\&=([1/\rho_2^{(\alpha)},1/\rho_1^{(\alpha)}],[-\theta_2^{(\alpha)},-\theta_1^{(\alpha)}])\end{aligned}$$

and

$$\begin{aligned}Z^{(\alpha)}(*)Z^{-1(\alpha)}&=([\rho_1^{(\alpha)},\rho_2^{(\alpha)}],[\theta_1^{(\alpha)},\theta_2^{(\alpha)}])\\&\quad(*)([1/\rho_2^{(\alpha)},1/\rho_1^{(\alpha)}],[-\theta_2^{(\alpha)},-\theta_1^{(\alpha)}])\\&=([\rho_1^{(\alpha)}/\rho_2^{(\alpha)},\rho_2^{(\alpha)}/\rho_1^{(\alpha)}],\\&\quad[\theta_1^{(\alpha)}-\theta_2^{(\alpha)},\theta_2^{(\alpha)}-\theta_1^{(\alpha)}])\\&\neq(1,0).\end{aligned}$$

In a fuzzy complex number $\mathbf{Z}$, $\rho^{(\alpha)}$ is called *modular interval of confidence of level* α, and $\theta^{(\alpha)}$ *the angular interval of confidence of level* α.

To divide $\mathbf{Z}_a$ by $\mathbf{Z}_b$ we multiply $\mathbf{Z}_a$ by $\mathbf{Z}_b^{-1}$:

$$Z_a^{(\alpha)}(*)^{-1}Z_b^{(\alpha)} = Z_a^{(\alpha)}(*)Z_b^{-1(\alpha)}$$

$$= ([\rho_{a_1}{}^{(\alpha)}/\rho_{b_2}{}^{(\alpha)}, \rho_{a_2}{}^{(\alpha)}/\rho_{b_1}{}^{(\alpha)}],\ [\theta_{a_1}{}^{(\alpha)} - \theta_{b_2}{}^{(\alpha)}, \theta_{a_2}{}^{(\alpha)} - \theta_{b_1}{}^{(\alpha)}]).$$

A fuzzy complex number can be represented in polar coordinates by the surface defined using $[\rho_1{}^{(\alpha)}, \rho_2{}^{(\alpha)}]$ and $[\theta_1{}^{(\alpha)}, \theta_2{}^{(\alpha)}]$. The argument is then an angle as shown in Figure 3.17, where we use the relations $x = \rho \cos\theta$, and $y = \rho \sin\theta$, where

$$\rho^2 = x^2 + y^2$$

and $\theta = \arctan y/x$ to pass from polar representation to Cartesian representation.

In the case of fuzzy complex numbers, surface *ABCD* in Figure 3.17 is not convex in the region between C and *D*. It is necessary, therefore, to define another type of fuzzy complex number, and we can do this by putting $\forall\ \alpha \in [0, 1]$:

$$\omega^{(\alpha)} = [z_1{}^{(\alpha)}, z_2{}^{(\alpha)}]$$

$$= [x_1{}^{(\alpha)}, x_2{}^{(\alpha)}] + j[y_1{}^{(\alpha)}, y_2{}^{(\alpha)}].$$

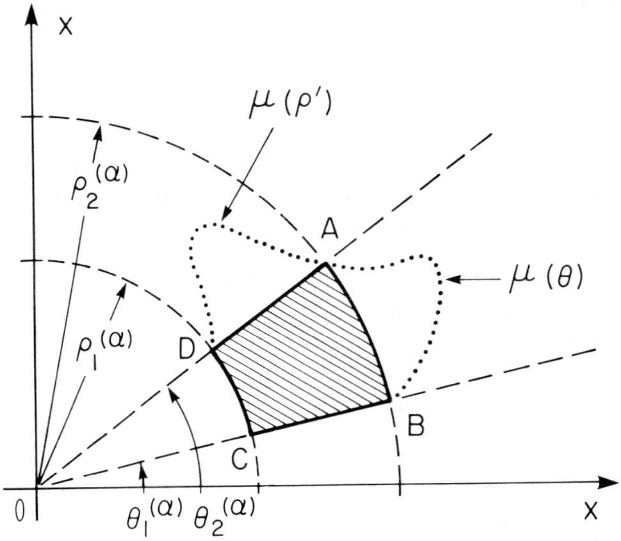

Figure 3.17 Representative of a fuzzy complex number using polar coordinates.

Of course, the concept that has been employed here is different from the one used in developing (3.57).

The imaginary number

$$j = \sqrt{-1}$$

comes from the relation $x^2 + 1 = 0$. Let us now consider this equation.

$$([x_1^{(\alpha)}, x_2^{(\alpha)}] (\cdot) [x_1^{(\alpha)}, x_2^{(\alpha)}]) (+) [1, 1] = [0, 0].$$

We will now present some interesting properties. Let us now return to (3.57), from which we see that

$$Z^{(\alpha)} (*) Z^{(\alpha)} = Z^{(\alpha)2}$$

$$= (\rho^{(\alpha)}, \theta^{(\alpha)}) (*) (\rho^{(\alpha)}, \theta^{(\alpha)})$$

$$= (\rho^{(\alpha)2}, 2 \cdot \theta^{(\alpha)},$$

$$Z^{(\alpha)n} = (\rho^{(\alpha)n}, n\theta^{(\alpha)}),$$

where n is an integer.

$$Z^{(\alpha)n_1} (*) Z^{(\alpha)n_2} = Z^{(\alpha)n_2} (*) Z^{(\alpha)n_1} = Z^{(\alpha)n_1n_2},$$

where n_1 and n_2 are integers.

$$(Z^{(\alpha)n_1})^{n_2} = (Z^{(\alpha)n_2})^{n_1} = Z^{(\alpha)n_1n_2}$$

where n_1 and n_2 are integers.

Many interesting properties arise from these considerations and will be treated elsewhere. Many powerful extensions of these concepts are also possible.

NOTES

1. This multiplication table modulo n is not symmetric because **A** starts with 1 and **B** starts with 1. A convenient shifting can give again symmetry.
2. The use of the word *interval* can be critized but we use this word for convenience.
3. See A. Kaufmann, *Introduction to the Theory of Fuzzy Subsets*, vol. 1, sect. 32, Academic Press, New York.
4. We can also call *isotonic* instead of *regular* and *antitionic* instead of *antiregular*.

CHAPTER 4

SOME COMPLEMENTS

4.1 INTRODUCTION

This last chapter is concerned with *semantical vagueness* and *degree of uncertainty*. When using fuzzy numbers it is sometimes useful, even necessary, to evaluate the distance between data and thus evaluate the data that have the highest level of presumption. Various methods may be employed, but the question of which method to use remains a problem, and realism and adaptability must determine the choice. Fuzzy numbers possess discriminating characteristics. Indices of fuzziness and nonprobabilistic entropies may not be particularly significant in describing the fuzziness of fuzzy numbers, but convexity and ordered configurations *are* important and indices based upon them must be developed. Uncertainty is a subjective and relative quantity that may or may not be important, especially when we are dealing with errors, which themselves may or may not be important, depending on the size of the error in relation to the data. In the following pages we try to bridge two novel theories: fuzzy theory and catastrophic theory. We present only a rough sketch that in itself has many extensions, and we hope that this will lead to some other interesting developments.

4.2 DEVIATION OR DIVERGENCE FOR A FUZZY NUMBER

A unimodal fuzzy number is one in which $\mu_{\mathbf{A}}(x) = 1$ for only one value of x, indicated by x_m.

The membership function of concern is divided into two parts, one at the left of x_m, $x_\ell(\alpha)$, and the other at the right of x_m, $x_r(\alpha)$. The quantity

$$\delta_\ell(\mathbf{A}) = \int_{\alpha=0}^{1} [x_m - x_\ell(\alpha)]\, d\alpha \tag{4.1}$$

is called the *left mean deviation* and the quantity

$$\delta_r(\mathbf{A}) = \int_{\alpha=0}^{1} [x_r(\alpha) - x_m]\, d\alpha \tag{4.2}$$

is called the *right mean deviation*.

And,

$$\begin{aligned}\delta(\mathbf{A}) &= \delta_\ell(\mathbf{A}) + \delta_r(\mathbf{A}) \\ &= \int_{\alpha=0}^{1} [x_r(\alpha) - x_\ell(\alpha)]\, d\alpha,\end{aligned} \tag{4.3}$$

is called the *mean deviation* of the fuzzy number **A**.

By construction, we have

$\forall\ \alpha \in [0, 1]$:

$$x_\ell(\alpha) \le x_m \le x_r(\alpha).$$

Hence

$$\delta_\ell(\mathbf{A}) \ge 0, \qquad \delta_r(\mathbf{A}) \ge 0, \quad \delta(\mathbf{A}) \ge 0.$$

Also, by construction, $\delta_\ell(\mathbf{A})$ represents the area to the left of x_m and $\delta_r(\mathbf{A})$ the area to the right, as shown in Figure 4.1, where $\delta(\mathbf{A})$ is the sum of the two areas.

The deviations defined above are of order 1. In the same way, deviations of order 2 can be defined by

$${}_2\delta_\ell(\mathbf{A}) = \int_{\alpha=0}^{1} [x_m - x_\ell(\alpha)]^2\, d\alpha,$$

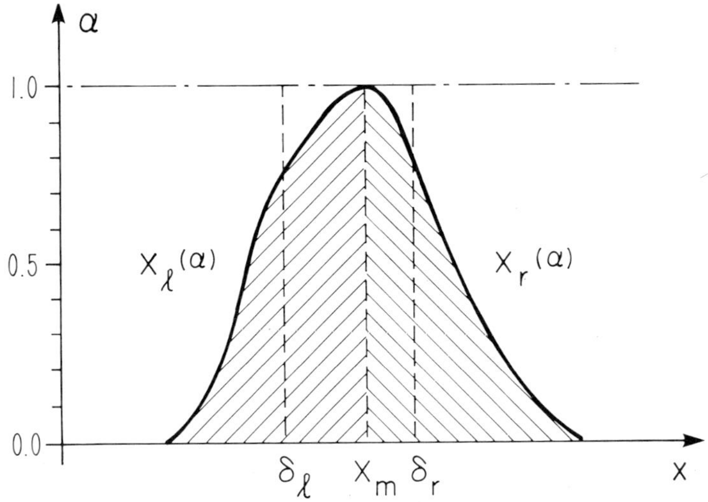

Figure 4.1 Definition of *first-order deviation* in a fuzzy number.

$$_2\delta_r(\mathbf{A}) = \int_{\alpha=0}^{1} [x_r(\alpha) - x_m]^2 \, d\alpha,$$

and

$$\begin{aligned} _2\delta(\mathbf{A}) &= {_2\delta_\ell}(\mathbf{A}) + {_2\delta_r}(\mathbf{A}) \\ &= \int_{\alpha=0}^{1} [x_m - x_\ell(\alpha)]^2 \, d\alpha + \int_{\alpha=0}^{1} [x_r(\alpha) - x_m]^2 \, d\alpha. \end{aligned} \quad \textbf{(4.4)}$$

More generally, the deviation of order n is defined as:

$$_n\delta_\ell(\mathbf{A}) = \int_{\alpha=0}^{1} [x_m - x_\ell(\alpha)]^n \, d\alpha,$$

$$_n\delta_r(\mathbf{A}) = \int_{\alpha=0}^{1} [x_r(\alpha) - x_m]^n \, d\alpha,$$

and

$$_n\delta(\mathbf{A}) = {_n\delta_\ell}(\mathbf{A}) + {_n\delta_r}(\mathbf{A}). \quad \textbf{(4.5)}$$

Of course, for $n \geq 2$, expressions of the type (4.4) and (4.5) cannot be simplified in the same way as was in (4.3).

Let us now examine some properties of deviations of order 1. We have, for example,

$$A_\alpha = [x_{a\ell}{}^{(\alpha)}, x_{ar}{}^{(\alpha)}],$$
$$B_\alpha = [x_{b\ell}{}^{(\alpha)}, x_{br}{}^{(\alpha)}],$$
$$A_\alpha\,(+)\,B_\alpha = [x_{a\ell}{}^{(\alpha)} + x_{b\ell}{}^{(\alpha)}, x_{ar}{}^{(\alpha)} + x_{br}{}^{(\alpha)}],$$
$$x_m(\mathbf{A}\,(+)\,\mathbf{B}) = x_m(\mathbf{A}) + x_m(\mathbf{B}),$$
$$\delta_\ell\,(\mathbf{A}\,(+)\,\mathbf{B}) = \delta_\ell\,(\mathbf{A}) + \delta_\ell\,(\mathbf{B}), \qquad \mathbf{(4.6)}$$
$$\delta_r(\mathbf{A}\,(+)\,\mathbf{B}) = \delta_r(\mathbf{A}) + \delta_r(\mathbf{B}), \qquad \mathbf{(4.7)}$$
$$\delta(\mathbf{A}\,(+)\,\mathbf{B}) = \delta(\mathbf{A}) + \delta(\mathbf{B}). \qquad \mathbf{(4.8)}$$

All deviations from (4.6–4.8) are thus nonnegative, but when we use (−) instead (+) we find that nonnegativity is not true for all fuzzy numbers. The higher-order deviations δ, $n = 1, 2, 3, 4, \ldots$, that use the operators (+) in R or $(\cdot)$ in R^+ guarantee nonnegativity.

Let us now examine some more properties, namely,

$$\delta_\ell\,(n\mathbf{A}) = \int_{\alpha=0}^{1} [nx_m - nx_\ell\,(\alpha)]\,d\alpha$$
$$= n\int_{\alpha=0}^{1} [x_m - x_\ell\,(\alpha)]\,d\alpha$$
$$= n\delta_\ell\,(\mathbf{A}), \qquad n \in N_0.$$

In the same way we have

$$\delta_r(n\mathbf{A}) = n\delta_r(\mathbf{A}), \qquad n \in N_0,$$

and

$$\delta(n\mathbf{A}) = n\delta(\mathbf{A}), \qquad n \in N_0.$$

Also,

$$\delta_\ell\,(n\mathbf{A})/n = \delta_\ell\,(\mathbf{A}),$$
$$\delta_r(n\mathbf{A})/n = \delta_r(\mathbf{A}),$$

and

$$\delta(n\mathbf{A})/n = \delta(\mathbf{A}).$$

We now examine some special cases.

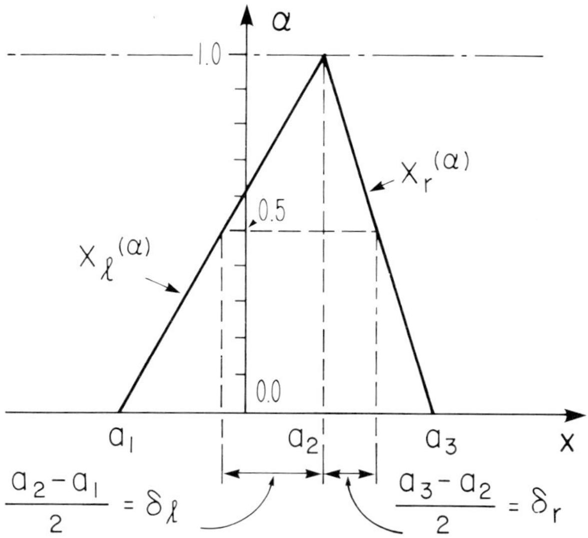

Figure 4.2 First-order deviation in a TFN **A**.

1. Let us take a TFN fuzzy number **A**, as is shown in Figure 4.2. Then,

$$\mathbf{A} = (a_1, a_2, a_3),$$

$$x_\ell(\alpha) = \alpha(a_2 - a_1) + a_1,$$

$$x_r(\alpha) = -\alpha(a_3 - a_2) + a_3,$$

and

$$A_\alpha = [(a_2 - a_1)\alpha + a_1, -(a_3 - a_2)\alpha + a_3].$$

Computing the deviations, we obtain

$$\delta_\ell(\mathbf{A}) = (a_2 - a_1)/2,$$

$$\delta_r(\mathbf{A}) = (a_3 - a_2)/2,$$

and

$$\delta(\mathbf{A}) = (a_3 - a_1)/2.$$

The ordinate α that corresponds to the left deviations and to the right deviations is $\alpha = 0.5$. Now let us consider $\delta_\ell = (a_2 - a_1)/2$ to the left of a_2. We then have

$$x = a_2 - (a_2 - a_1)/2 = (a_1 + a_2)/2$$

and

$$(a_1 + a_2)/2 = (a_2 - a_1)\alpha + a_1. \tag{4.9}$$

The solution of equation (4.9) is $\alpha = 0.5$. Of course, the same result is obtained on the right. Hence, for TFN's the mean deviations are given by the ordinate $\alpha = 0.5$; that is, half the base of the triangle.

2. Suppose instead that **A** is a fuzzy number of type L–R. (from Dubois and Prade[1]). Then,

$$x_\ell(\alpha) = x_m + u \cdot F_\ell^{-1}(\alpha),$$

$$x_r(\alpha) = x_m + v \cdot F_r^{-1}(\alpha),$$

$$\mathbf{A}_\alpha = [x_m + u \cdot F_\ell^{-1}(\alpha),\ x_m + v \cdot F_r^{-1}(\alpha)],$$

$$\delta_\ell(\mathbf{A}) = \int_{\alpha=0}^{1} [x_m - (x_m + uF_\ell^{-1}(\alpha)]\, d\alpha$$

$$= -u \int_{\alpha=0}^{1} F_\ell^{-1}(\alpha)\, d\alpha.$$

$$\delta_r(\mathbf{A}) = \int_{\alpha=0}^{1} \{[x_m + vF_r^{-1}(\alpha)] - x_m\}\, d\alpha$$

$$= v \int_{\alpha=0}^{1} F_r^{-1}(\alpha)\, d\alpha.$$

In addition,

$$\delta(\mathbf{A}) = \delta_\ell(\mathbf{A}) + \delta_r(\mathbf{A})$$

$$= -u \int_{\alpha=0}^{1} F_\ell^{-1}(\alpha)\, d\alpha + v \int_{\alpha=0}^{1} F_r^{-1}(\alpha)\, d\alpha.$$

EXAMPLE 4.1

In this example we consider two reference-like functions.

$$\begin{aligned} F_L(x') &= 0, & x' &\le -1, \\ &= \sqrt{1 + x'}, & -1 \le x' &\le 0. \\ F_r(x') &= 1 - (x')^2, & 0 \le x' &\le 1, \\ &= 0, & 1 &\le x'. \end{aligned}$$

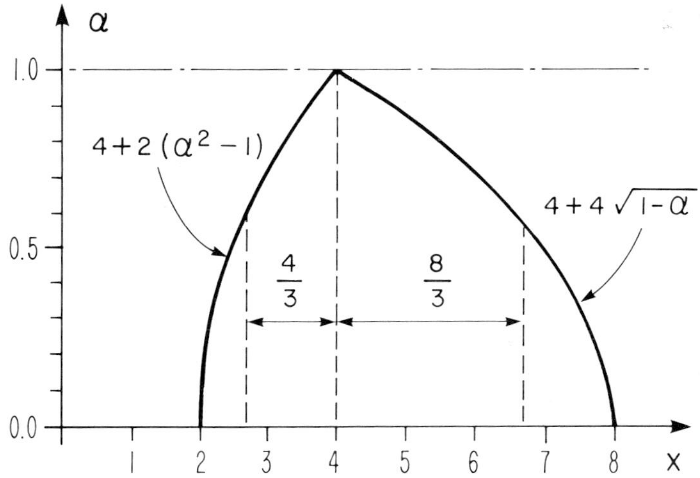

Figure 4.3 Membership function for Example 4.1.

Now, let $x_m = 4$, $u = 2$, and $v = 3$. From these data

$$F_L^{-1}(\alpha) = \alpha^2 - 1,$$
$$F_R(\alpha) = \sqrt{1-\alpha},$$

we find

$$\delta_\ell(\mathbf{A}) = -2 \int_{\alpha=0}^{1} (\alpha^2 - 1)\, d\alpha = 4/3, \quad \mathbf{(4.10)}$$

$$\delta_r(\mathbf{A}) = 4 \int_{\alpha=0}^{1} \sqrt{(1-\alpha)}\, d\alpha = 8/3, \quad \mathbf{(4.11)}$$

and

$$\delta(\mathbf{A}) = 4/3 + 8/3 = 4. \quad \mathbf{(4.12)}$$

Figure 4.3 shows the membership function given by

$$A_\alpha = [4 + 2(\alpha^2 - 1),\ 4 + 4\sqrt{1-\alpha}].$$

The Case of an L-R Fuzzy Number with a Flat

If a flat spot exists between x_{m_1} and x_{m_2}, as shown in Figure 4.4, an arbitrary point x_m between x_{m_1} and x_{m_2} is taken such that

$$x_{m_1} \le x_m \le x_{m_2}.$$

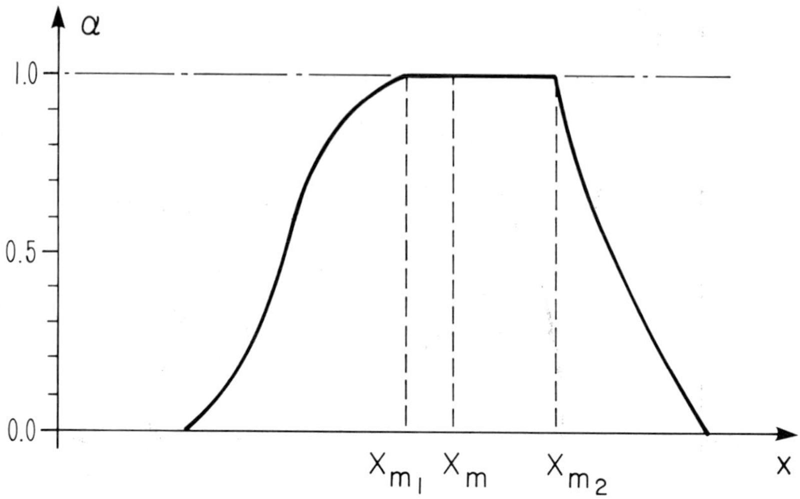

Figure 4.4 L–R fuzzy number with a flat.

Then δ_ℓ (**A**) and δ_r(**A**) depend on x_m, but δ(**A**) does not vary with x_m.

The Case of an Ordinary Number (nonfuzzy)

In an ordinary number, we have

$$\delta_\ell\,(A) = 0, \qquad \delta_r(A) = 0, \qquad \delta(A) = 0.$$

The Case of Intervals of Confidence Independent of α

In the case shown in Figure 4.5, we have

$$\delta(\mathbf{A}) = x_r - x_\ell\,,$$

where δ_ℓ (**A**) and δ_r(**A**) depend on the arbitrary point x_m.

Relative Mean Deviation

At times it is useful to work with the relative mean deviation. In such cases

$$\xi_\ell\,(\mathbf{A}) = \delta_\ell\,(\mathbf{A})/(x_m - x_\ell\,), \tag{4.13}$$

$$\xi_r(\mathbf{A}) = \delta_r(\mathbf{A})/(x_r - x_m), \tag{4.14}$$

$$\xi(\mathbf{A}) = \xi_\ell\,(\mathbf{A}) + \xi_r(\mathbf{A}), \tag{4.15}$$

or

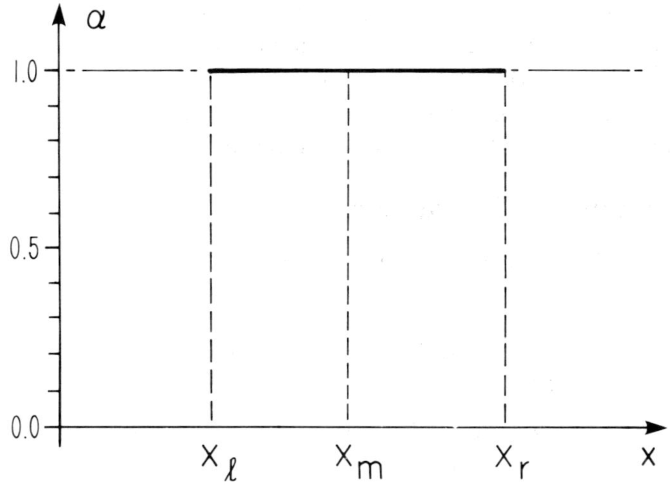

Figure 4.5 Interval of confidence independent of α.

$$\xi'(\mathbf{A}) = \delta(\mathbf{A})/(x_r - x_\ell). \tag{4.16}$$

We note that $\xi(\mathbf{A}) \neq \xi'(\mathbf{A})$ except in special cases.

As a simple numerical example we use Example 4.1 and equations (4.13) to (4.16). Thus

$$\xi_\ell(\mathbf{A}) = \frac{4/3}{4-2} = \frac{2}{3},$$

$$\xi_r(\mathbf{A}) = \frac{8/3}{8-4} = \frac{2}{3},$$

$$\xi(\mathbf{A}) = 2/3 + 2/3 = 4/3,$$

$$\xi'(\mathbf{A}) = 4/6 = 2/3.$$

Relative Divergence

Other ratios can also be useful. Consider, for example,

$$e_\ell(\mathbf{A}) = \delta_\ell(\mathbf{A})/\,|x_m|, \qquad \text{left relative divergence},$$

$$e_r(\mathbf{A}) = \delta_r(\mathbf{A})/\,|x_m|, \qquad \text{right relative divergence},$$

$$e_r(\mathbf{A}) = \delta(\mathbf{A})/\,|x_m|$$

$$= e_\ell(\mathbf{A}) + e_r(\mathbf{A}), \qquad \text{relative divergence}.$$

Again taking (4.10), (4.11), and (4.12) we obtain

$$e_\ell(\mathbf{A}) = \frac{4/3}{4} = \frac{1}{3}, \qquad e_r(\mathbf{A}) + \frac{8/3}{4} = \frac{2}{3}, \qquad e(\mathbf{A}) = 1.$$

The relative divergence can be considered like a relative index of fuzziness for a fuzzy number. Moreover, more e(**A**) is small the lesser is the fuzziness **A** compared to its modal value. If the fuzzy number has a flat, an arbitrary value x_m is chosen between the bounds of the flat. The relative divergence of an ordinary number is then zero, even for the case of the nonfuzzy number 0.

Mean Interval of Confidence

Let $\mathbf{A} \subset R$ be a fuzzy number such that

$$A^m = [x_m - \delta_l(\mathbf{A}), x_m + \delta_r(\mathbf{A})] \qquad \textbf{(4.17)}$$

is the mean interval of confidence of **A** (Fig. 4.6).

It is easy to prove that

$$(\mathbf{C} = \mathbf{A}\ (+)\ \mathbf{B}) \Rightarrow (C^m = A^m\ (+)\ B^m).$$

If a fuzzy number **A** has a maximum value (flat) $[x_{m_1}, x_{m_2}]$ then

$$x_m = (x_{m_1} + x_{m_2})/2.$$

EXAMPLE 4.2

If we reconsider Example 4.1 and Figure 4.3, we find that

$$x_m = 4,$$
$$\delta_l(\mathbf{A}) = 4/3,$$
$$\delta_r(\mathbf{A}) = 8/3,$$

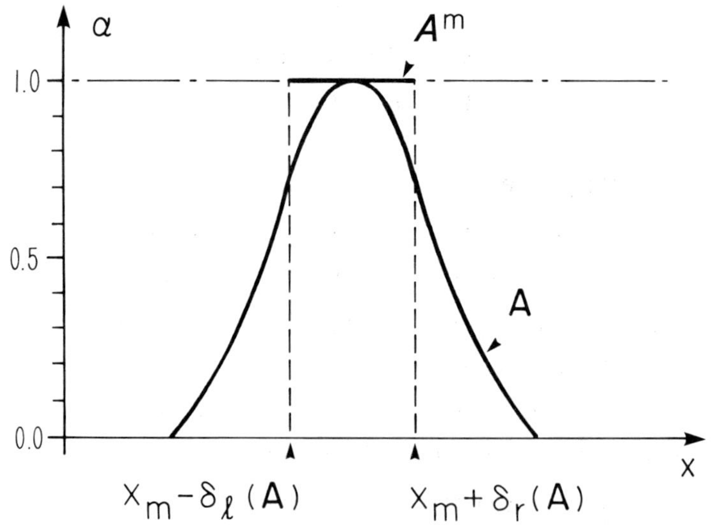

Figure 4.6 Mean interval of confidence.

$$A^m = [4 - 4/3,\ 4 + 8/3] = [8/3,\ 20/3] = [2.6666,\ 6.6666].$$

The concept of the mean interval of confidence allows the substitution of the notion of interval of confidence for the fuzzy numbers, of course, with a loss of subjective information.

Another Presentation for the Mean Interval of Confidence

Let us once again take (4.1) and (4.2) and write

$$\delta_l\,(\mathbf{A}) = x_m - \int_{\alpha=0}^{1} x_l\,(\alpha)\,d\alpha,$$

$$\delta_r(\mathbf{A}) = -x_m + \int_{\alpha=0}^{1} x_r(\alpha)\,d\alpha.$$

We can then rewrite (4.17) as follows:

$$A^m = [x_m - (x_m - \int_0^1 x\ (\alpha)\,d\alpha),\ x_m + (-x_m + \int_0^1 x_r(\alpha)\,d\alpha]$$

$$= [\int_0^1 x_l\,(\alpha)\,d\alpha - \int_0^1 x_r(\alpha)\,d\alpha].$$

Using this notation gives

$$A^m\,(+)\,B^m = [\int_0^1 ((x_{al}\,(\alpha) + x_{bl}\,(\alpha))\,d\alpha,\ \int_0^1 ((x_{ar}(\alpha) + x_{br}(\alpha))\,d\alpha]$$

and

$$A^m\,(-)\,B^m = [\int_0^1 ((x_{al}\,(\alpha) - x_{br}(\alpha))\,d\alpha,\ \int_0^1 ((x_{ar}(\alpha) - x_{bl}\,(\alpha))\,d\alpha].$$

It is interesting to associate the concept of the mean interval of confidence with the concept of the mean, or the mathematical expectation, of a random variable in the case of a hybrid number. Let $(\mathbf{A}, \mathrm{L})$ be a hybrid number, where

$$A_\alpha = [x_{al}\,(\alpha),\ x_{ar}(\alpha)],$$

and

$$\overline{\ell} = \int_{R} \ell \cdot f(\ell)\, d\ell ,$$

Then

$$(A^m, \overline{\ell}) = A^m[+]\,\overline{\ell} = [\overline{\ell} + \int_{\alpha=0}^{1} x_{a\ell}(\alpha)\, d\alpha,\ \overline{\ell} + \int_{\alpha=0}^{1} x_{ar}(\alpha)\, d\alpha].$$

Remark

Conversion of a fuzzy number to an interval of confidence is an operation that decreases the amount of subjective information available. (Note that several fuzzy numbers can have the same mean interval of confidence even though they are not alike.) This is similar to taking the mathematical expectation of a random variable that reduces the amount of objective information present. These mathematical operations are of course very useful, but it is always important to keep their limitations in mind, that is a loss of subjective information.

4.3 AGREEMENT INDEX FOR A FUZZY NUMBER

Let us define the concept of a *fuzzy upper bound*. Consider a function $h(x)$, $x \in R$ such that

$$\begin{aligned} h(x) &= 1, && x \le x_1, \\ &= \text{a monotonically decreasing function}, && x_1 \le x \le x_2, \\ &= 0, && x \ge x_2 . \end{aligned}$$

We represent this function by a fuzzy subset $\mathbf{H} \subset R$.

Let us now consider a fuzzy number $\mathbf{A} \subset R$, which we call the *agreement index* of $\mathbf{A}$ with regard to $\mathbf{H}$, the ratio being defined in the following way:[1]

$$\iota(\mathbf{A}, \mathbf{H}) = (\text{area } \mathbf{A} \cap \mathbf{H})/(\text{area } \mathbf{A}) \in [0, 1],$$

as shown in Figure 4.7.

When $\mathbf{A}$ is nonfuzzy, that is,

$$\begin{aligned} \mu(\mathbf{A}) &= 1, && x = n, \\ &= 0, && x \ne n, \end{aligned}$$

then

$$\iota(\mathbf{A}, \mathbf{H}) = h(n).$$

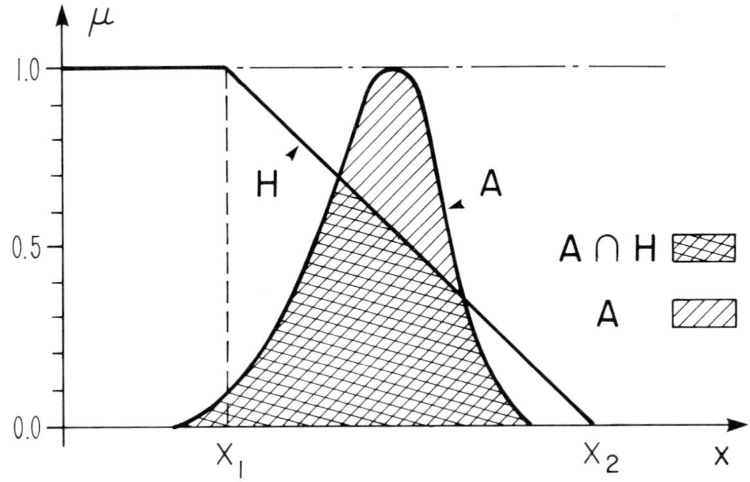

Figure 4.7 Agreement index for fuzzy numbers.

When **H** is nonfuzzy, that is,

$$
\begin{aligned}
h(x) &= 1, \qquad x \le x_1, \\
&= 0, \qquad x > x_1.
\end{aligned}
$$

In this case we compute the area of **A** to the left of x_1.

EXAMPLE 4.3

Consider the TFN $\mathbf{A} \subset R$ shown in Figure 4.8:

$\forall\ x \in R$:

$$
\begin{aligned}
\mu_{\mathbf{A}}(x) &= 0, & x &\le 3, \\
&= (x - 3)/7, & 3 \le x &\le 10, \\
&= (-x + 13)/3, & 10 \le x &\le 13, \\
&= 0, & x &\ge 13, \qquad \textbf{(4.18)}
\end{aligned}
$$

with the following function **H**:

$\forall\ x \in R$:

$$
\begin{aligned}
h(x) &= 1, & x &\le 8, \\
&= (-x + 14)/6, & 8 \le x &\le 14, \\
&= 0, & x &\ge 14. \qquad \textbf{(4.19)}
\end{aligned}
$$

We must now describe the surface $\mathbf{A} \cap \mathbf{H}$ and compute its area as follows.

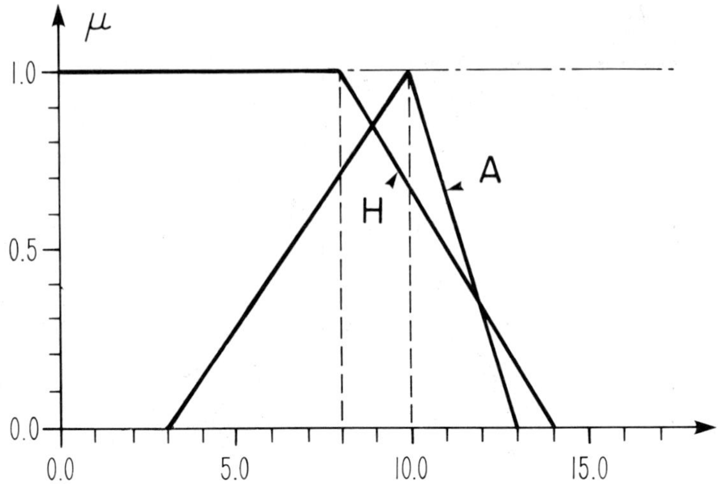

Figure 4.8 Agreement index (Example 4.3).

1. $(x-3)/7 = (-x+14)/6$ gives $x = 8.92307$, $\mu_{\mathbf{A}}(x) = 0.84615$.
2. $(-x+13)/3 = (-x+14)/6$ gives $x = 12$, $\mu_{\mathbf{A}}(x) = 0.33333$.

Let us now compute the area of $\mathbf{A} \cap \mathbf{H}$:

$$\text{area } \mathbf{A} \cap \mathbf{H} = \int_{x=3}^{8.92307} (x-3)/7 \, dx + \int_{x=8.92307}^{12} (-x+14)/6 \, dx$$

$$+ \int_{x=12}^{13} (-x+13)/3 \, dx$$

$$= 2.50530 + 1.81459 + 0.16666$$

$$= 4.48715.$$

$$\text{area } \mathbf{A} = (11 \times 1)/2 = 5.5.$$

Thus we find that

$$\iota(\mathbf{A}, \mathbf{H}) = (4.48715)/5.5 = 0.81854.$$

We consider two more examples.

EXAMPLE 4.4

Let us consider again the TFN **A**, which is given in Example 4.3, Figure 4.8, but with a nonfuzzy H defined as

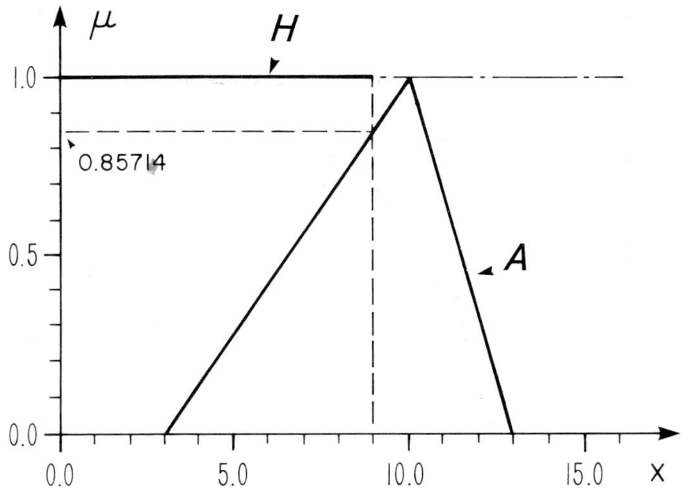

Figure 4.9 Agreement index (Example 4.4).

$$h(x) = 1, \quad x \leq 9,$$
$$= 0, \quad x > 9.$$

as shown in Figure 4.9.

Let us next define the surface $\mathbf{A} \cap H$ (here H is not fuzzy). Thus $(x-3)/7$ for $x=9$ gives $(9-3)/7 = 0.85714$, area $\mathbf{A} \cap H = (6 \times 0.85714)/2 = 2.57142$, and area $\mathbf{A} = 5.5$. Therefore $\iota(\mathbf{A}, H) = (2.57142/5.5) = 0.46753$.

EXAMPLE 4.5

The agreement index can be used in Z in the same way that it is being used in R. In this case, the sum obtained by integration is replaced by one that is discrete. For example, let us consider **A** and **H** given by

	0	1	2	3	4	5	6	7	8	≥9
A =	0	0.2	0.3	0.6	0.9	1	0.8	0.4	0.1	0

(4.20)

	0	1	2	3	4	5	6	7	8	≥9
H =	1	1	1	1	1	0.8	0.4	0.1	0	0

	0	1	2	3	4	5	6	7	8	≥9
A ∩ **H** =	0	0.2	0.3	0.6	0.9	0.8	0.4	0.1	0	0

(4.21)

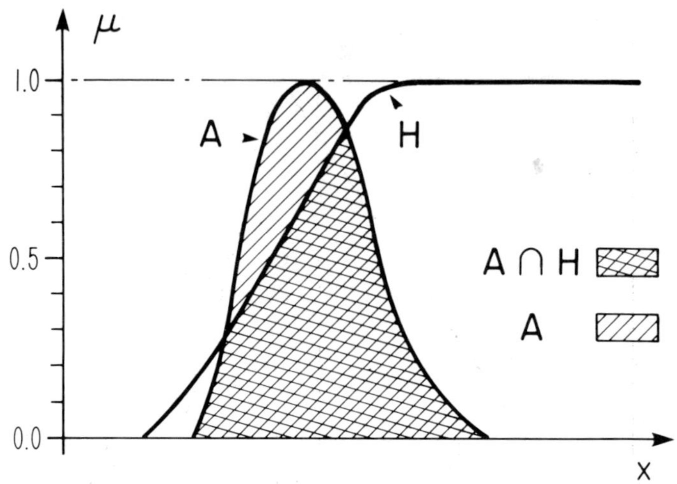

Figure 4.10 Fuzzy lower bound.

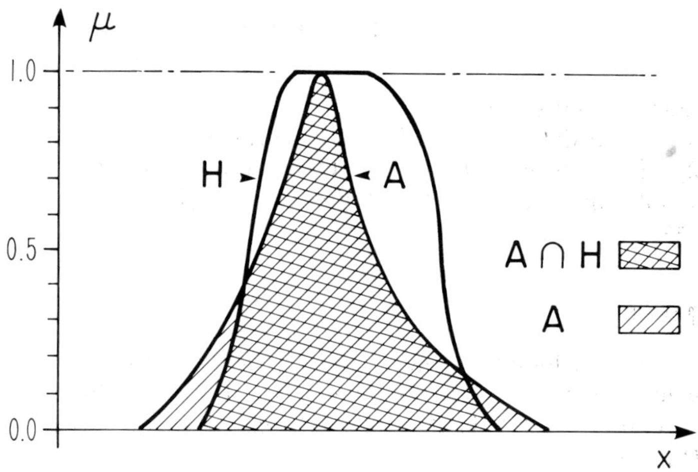

Figure 4.11 Fuzzy range.

$$\text{sum } \mathbf{A} \cap \mathbf{H} = 0.2 + 0.3 + 0.6 + 0.9 + 0.8 + 0.4 + 0.1 = 3.3,$$

$$\text{sum } \mathbf{A} = 0.2 + 0.3 + 0.6 + 0.9 + 1 + 0.8 + 0.4 + 0.1 = 4.3,$$

$$\iota(\mathbf{A}, \mathbf{H}) = 3.3/4.3 = 0.76744.$$

This process can also be used for determining the *fuzzy lower bound* as shown in Figure 4.10, or the *fuzzy range* as shown in Figure 4.11.

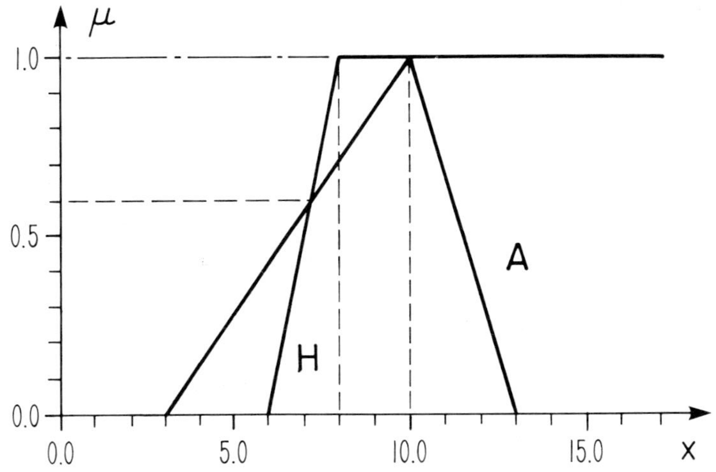

Figure 4.12 Agreement index (Example 4.6).

EXAMPLE 4.6

Using (4.18) again, but this time with a fuzzy lower bound, we have

$$\begin{aligned} h(x) &= 0, && x \le 6, \\ &= (x-6)/3, && 6 \le x \le 8, \\ &= 1, && 8 \le x. \end{aligned}$$

Let us now compute the surface and the area $\mathbf{A} \cap \mathbf{H}$ (Fig. 4.12):

$$(x-3)/7 = (x-6)/3,$$

which gives x = 7.2 and $\mu_{\mathbf{A}}(x) = 0.6$. Therefore, we have

$$\begin{aligned} \text{area } \mathbf{A} \cap \mathbf{H} &= \int_{x=6}^{7.2} [(x-6)/2]\, dx + \int_{x=7.2}^{10} [(x-3)/7]\, dx + \\ &\quad \int_{x=10}^{13} [(-x+13)/3]\, dx \\ &= 0.36 + 2.24 + 1.5 \\ &= 4.1, \\ \text{area } \mathbf{A} &= 5.5, \end{aligned}$$

therefore,

$$\begin{aligned} \iota(\mathbf{A}, \mathbf{H}) &= 4.1/5.5 \\ &= 0.74545. \end{aligned}$$

EXAMPLE 4.7

We now consider an example pertaining to the concept of the fuzzy range. Again we will use (4.18), the fuzzy range being shown in Figure 4.13.

$$
\begin{aligned}
h(x) &= 0, & x &\le 4,\\
&= x - 4, & 4 \le x &\le 5,\\
&= 1, & 5 \le x &\le 11,\\
&= -x + 12, & 11 \le x &\le 12,\\
&= 0, & 12 &\le x.
\end{aligned}
$$

For computing the area, we have

$$(x - 3)/7 = x - 4, \text{ which gives } x = 4.16666,\ \mu_{\mathbf{A}}(x) = 0.16666;$$

$$(-x + 13)/3 = x + 12, \text{ which gives, } x = 11.5,\ \mu_{\mathbf{A}}(x) = 0.5.$$

$$
\begin{aligned}
\text{area } \mathbf{A} \cap \mathbf{H} = &\int_{x=4}^{4.16666} [(x - 4)]\,dx + \int_{x=4.16666}^{10} [(x - 3)/7]\,dx\\
&+ \int_{x=10}^{11.5} [(-x + 13)/3]\,dx + \int_{x=11.5}^{12} [(-x + 12)]\,dx\\
&= 0.01389 + 3.40278 + 1.125 + 0.125 = 4.66666,
\end{aligned}
$$

$$\text{area } \mathbf{A} = 5.5.$$

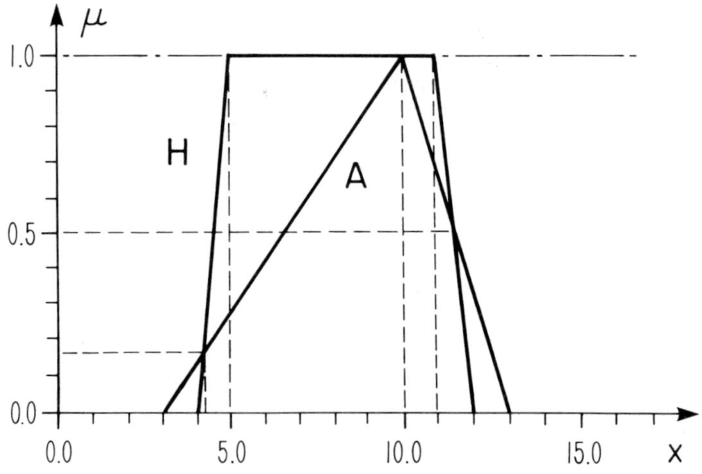

Figure 4.13 Agreement index (Example 4.7).

Therefore the fuzzy agreement index is

$$\iota(\mathbf{A} \cap \mathbf{H}) = (4.66666/5.5) = 0.84848.$$

Remark

Two fuzzy numbers **A** and **B** may not be alike but still may have the same agreement index with regard to the same function $h(x)$, which is a real-life phenomenon.

Utilization of a Law of Possibility

In the theory of fuzzy subsets the law of possibility plays a role similar to that played by the law of probability in measurement theory for ordinary sets.[2] Thus if

$\forall\ x \in R$:

$$h(x) \in [0, 1], \qquad \text{and} \qquad \bigvee_{x \in R} h(x) = 1, \tag{4.22}$$

then $h(x)$ is called the *possibility law* on R.

In addition, if **A** is a fuzzy subset in R, we call the *possibility of* **A** *for the law* $h(x)$ to be

$$\text{poss}_{\mathbf{H}}\, \mathbf{A} = \bigvee_{x \in R} (\mu_{\mathbf{A}}(x) \wedge h(x)). \tag{4.23}$$

A law like Equation (4.23) can be considered as a fuzzy subset of R and is denoted by **H**.

We now examine how this concept of the possibility law can be applied to fuzzy numbers and fuzzy bounds. We also make some comparisons by employing the agreement index.

EXAMPLE 4.8

Using (4.18) and (4.19), which are similar to the possibility law, we have $\text{poss}_{\mathbf{H}}\, \mathbf{A} = 0.84815$, which is shown in Figure 4.14.

EXAMPLE 4.9

Considering the data of Example 4.4 as shown in Figure 4.15, we have $\text{poss}_{\mathbf{H}}\, \mathbf{A} = 0.85714$.

EXAMPLE 4.10

Taking the data of the Example 4.5, Figure 4.10, and considering Equation (4.21), we have $\text{poss}_{\mathbf{H}}\, \mathbf{A} = 0.9$.

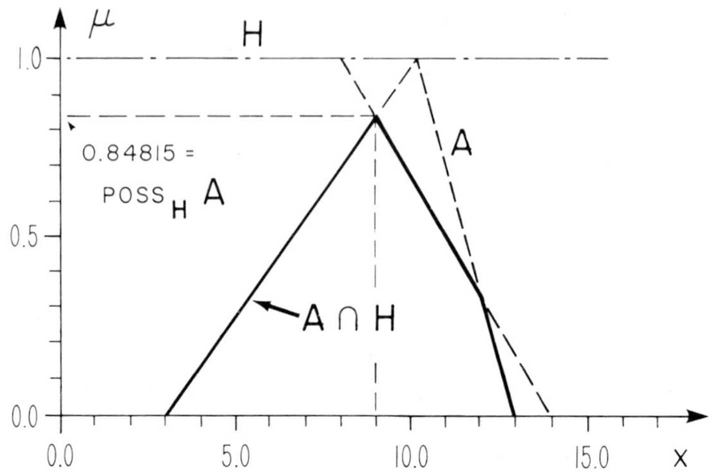

Figure 4.14 Possibility law: poss $_{\mathbf{H}}$ **A** (Example 4.8).

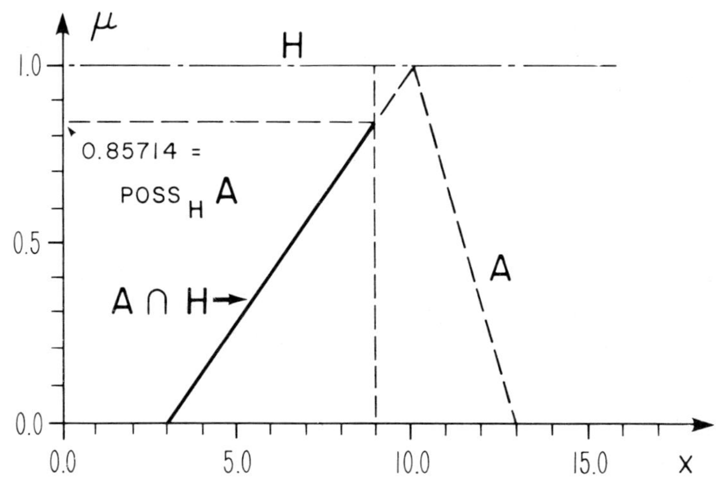

Figure 4.15 Possibility law: poss $_{\mathbf{H}}$ **A** (Example 4.9).

EXAMPLE 4.11

From the data of Example 4.6, Figure 4.12, and using $\iota = 0.74545$, we find that poss $_{\mathbf{H}}$ $\mathbf{A} = 1$. Results are shown in Figure 4.16.

EXAMPLE 4.12

Using the data from Example 4.7, Figure 4.13, with $\iota = 0.84848$, we find that poss $_{\mathbf{H}}$ $\mathbf{A} = 1$. Figure 4.17 shows the results.

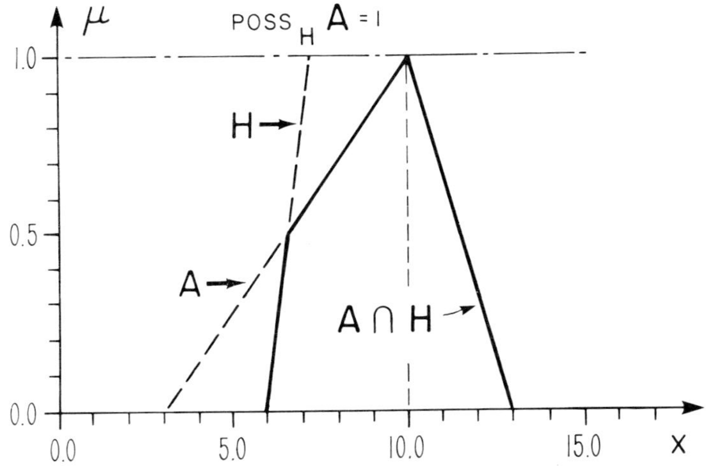

Figure 4.16 Possibility law: poss $_{\mathbf{H}}$ **A** = 1 (Example 4.11).

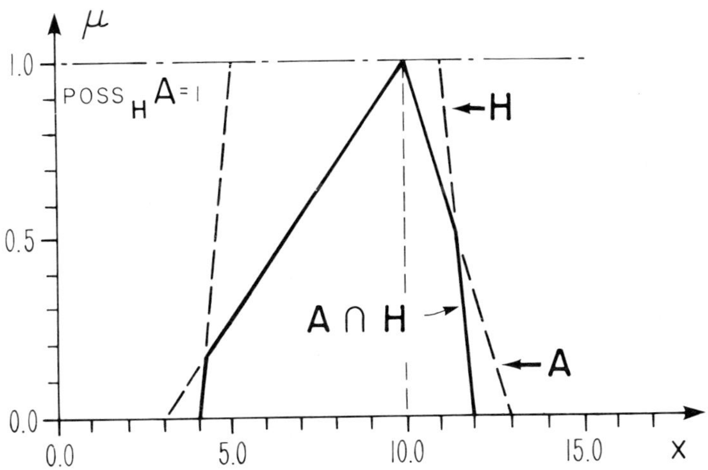

Figure 4.17 Possibility law: poss $_{\mathbf{H}}$ **A** = 1 (Example 4.12).

After examining so many examples we must note that the concept of possibility does not take into account sufficiently the shape of fuzzy bounds and fuzzy numbers. From this point of view, the agreement index appears to be better. It lacks, however, the interesting algebraic properties of the concept of possibility. For example, the agreement index does not have the property of valuation. Nevertheless, from a practical viewpoint, the agreement index is a very good indicator.

Another question we must examine is the comparison between the concept of possibility and the agreement index.

EXAMPLE 4.13

Again we use the $\mu_{\mathbf{A}}(x)$ given in Equation (4.18) and examine the values of μ for which we have poss **A** = 0.6, when we consider a nonfuzzy bound as shown in Figure 4.18. We see that two values of x with the same possibility exist. They are

1. $(x - 3)/7 = 0.6$, which gives $x = 7.2$
2. $(-x + 13)/3 = 0.6$, which gives $x = 11.2$.

Obviously such a result can be very disturbing.

Now let us use $\iota(\mathbf{A}, \mathbf{H})$. We then have

1. for

$$\begin{aligned} h(x) &= 1, \qquad x \le 7.2, \\ &= 0, \qquad x > 7.2, \\ \iota(\mathbf{A}, \mathbf{H}) &= 1.26/5.5 = 0.22909; \end{aligned}$$

2. for

$$\begin{aligned} h(x) &= 1, \qquad x \le 11.2, \\ &= 0, \qquad x > 11.2, \\ \iota(\mathbf{A}, \mathbf{H}) &= 4.46/5.5 = 0.81890. \end{aligned}$$

This result shows that the agreement index provides better information in such cases than does the concept of possibility, because the cummulative property of the agreement index removes the ambiguity.

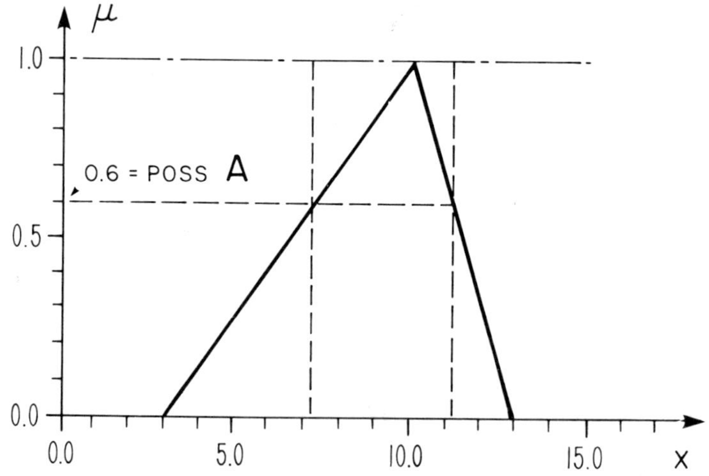

Figure 4.18 Possible values of x for poss **A** = 0.6 (Example 4.13).

Quadratic Agreement Index

It is possible, and useful, to define an index that favors values nearer to $x = h^{-1}(\alpha)$ by using a quadratic form. For this purpose consider

$$\iota_2(\mathbf{A}, \mathbf{H}) = \frac{\{\int_{\mathbf{A} \cap \mathbf{H}} [x^*(\alpha) - x_1(\alpha)]^2 \, d\alpha\}}{[\int_{\mathbf{A}} [x_2(\alpha) - x_1(\alpha)]^2 \, d\alpha]}, \qquad \alpha \in [0, 1], \qquad \textbf{(4.24)}$$

where $x^*(\alpha)$ is the curve that limits the surface $\mathbf{A} \cap \mathbf{H}$ at the right, $x_1(\alpha)$ is the curve that limits $\mathbf{A}$ at the left, and $x_2(\alpha)$ is the curve that limits $\mathbf{A}$ at the right. Figure 4.19 (on page **276**) shows how (4.24) is obtained by using double integrals.[3]

$$I_{\mathbf{A} \cap \mathbf{H}} = \int_{\alpha=0}^{\max_x \alpha^*(x)} \int_{x = x_1(\alpha)}^{x^*(\alpha)} [x^*(\alpha) - x] \, dx \, d\alpha$$

$$= \int_{\alpha=0}^{\max_x \alpha^* x} \left(\left| x \cdot x^*(\alpha) - (x^2)/2 \right|_{x = x_1(\alpha)}^{x = x^*(\alpha)} \right) d\alpha$$

$$= \int_{\alpha=0}^{\max_x \alpha^*(x)} \{[x^*(\alpha)]^2/2 - x^*(\alpha) \cdot x_1(\alpha) + [x_1(\alpha)]^2/2\} \, d\alpha$$

$$= \frac{1}{2} \int_{\alpha=0}^{\max_x \alpha^*(x)} [x^*(\alpha) - x_1(\alpha)]^2 \, d\alpha.$$

In the same way, we find

$$I_{\mathbf{A}} = \frac{1}{2} \int_{\alpha=0}^{1} [x_2(\alpha) - x_1(\alpha)]^2 \, d\alpha.$$

The ratio $(I_{\mathbf{A} \cap \mathbf{H}})/I_{\mathbf{A}}$ effectively gives (4.24), and we can easily check that this ratio belongs to [0, 1]. We now illustrate the concept of quadratic agreement index by means of a numerical example.

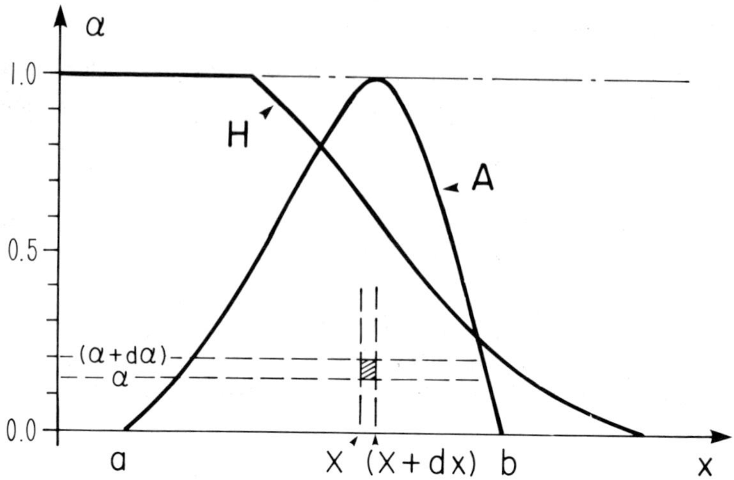

Figure 4.19 Quadratic agreement index.

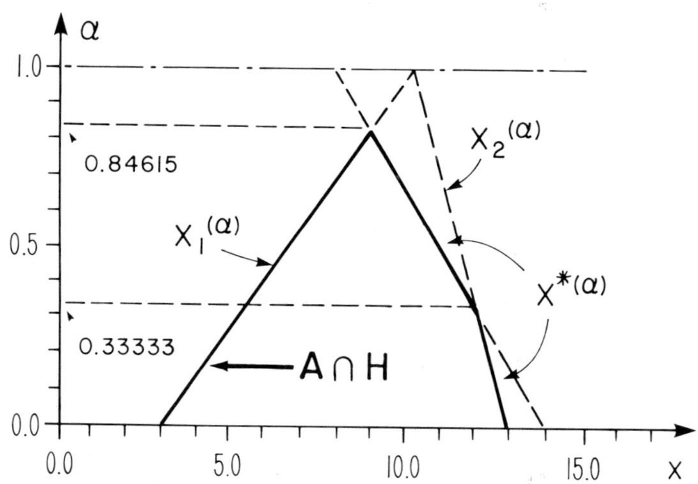

Figure 4.20 Quadratic agreement index (Example 4.14).

EXAMPLE 4.14

Once again let us use (4.18) and (4.19). Let us now compute the function $x(\alpha)$ from $\alpha(x)$ (Fig. 4.20).

1. $\alpha = (x - 3)/7$ gives $x_I = 3 + 7\alpha$,
2. $\alpha = (14 - x)/6$ gives $x_{II} = 14 - 6\alpha$,
3. $\alpha = (13 - x)/3$ gives $x_{III} = 13 - 3\alpha$.

We therefore have

$$x^* = x_{III} = 13 - 3\alpha, \qquad \alpha \in [0, 0.33333],$$

$$x^* = x_{II} = 14 - 6\alpha, \qquad \alpha \in [0.33333, 0.84615]$$

and

$$I_{\mathbf{A} \cap \mathbf{H}} = \frac{1}{2} \int_{\alpha=0}^{0.33333} [(13 - 3\alpha) - (3 + 7\alpha)]^2 \, d\alpha$$

$$+ 2 \int_{\alpha=0.33333}^{0.84615} [(14 - 6\alpha) - (3 + 7\alpha]^2 \, d\alpha$$

$$= \frac{1}{2} \int_{\alpha=0}^{0.33333} 10^2(1-\alpha)^2 \, d\alpha + 2 \int_{\alpha=0.33333}^{0.84615} (11 - 13\alpha)^2 \, d\alpha$$

$$= (1/2)\,(23.45753).$$

$$I_{\mathbf{A}} = \frac{1}{2} \int_{\alpha=0}^{1} [(13 - 3\alpha) - (3 + 7\alpha)]^2 \, d\alpha$$

$$= \frac{1}{2} \int_{\alpha=0}^{1} 10^2(1-\alpha)^2 \, d\alpha$$

$$= (1/2)\,(33.33333).$$

Therefore we have $\iota_2(\mathbf{A}, \mathbf{H}) = (23.45753)/(33.33333) = 0.93616$.

EXAMPLE 4.15

Taking again the data from Example 4.4, shown in Figure 4.9, where $x^* < c^{te} = 9$, we have

$$I_{\mathbf{A} \cap \mathbf{H}} = \frac{1}{2} \int_{\alpha=0}^{0.85714} [9 - (3 + 7\alpha)]^2 \, d\alpha$$

$$= \frac{1}{2} \int_{\alpha=0}^{0.85714} (6 - 7\alpha)^2 \, d\alpha$$

$$= (1/2)\,(10.28571).$$

$$I_{\mathbf{A}} = (1/2)\,(33.33333),$$

and

$$\iota(\mathbf{A}, \mathbf{H}) = (10.28571)/(33.33333) = 0.30857.$$

Now we study an example in Z.

EXAMPLE 4.16

Using the data of Example 4.5 and taking $\mathbf{A} \cap \mathbf{H}$ and $\mathbf{A}$ level by level for $\alpha = 0.1, 0.2, \ldots, 0.9, 1$ (level 0 is not significant here), we have the results shown in Table 4.1 for $\mathbf{A} \cap \mathbf{H}$, using Equation (4.21). For $\mathbf{A}$ from Equation (4.20) we have the results shown in Table 4.2. Hence $\iota(\mathbf{A}, \mathbf{H}) = 96/153 = 0.62745$.

4.4 THEORY OF CATASTROPHE AND FUZZY NUMBERS

Let us review catastrophic theory by means of an example.[4,5]

Let $f(x, b)$ be a function of two variables $x \in R$ and $b \in R$, where $f(x, b)$ takes its values in R. Let this function be given by

$$f(x, b) = -x^3 + x + b, \tag{4.25}$$

which represents a surface in the space 0-f-x-b. The intersection of this surface with the plane 0-x-b gives a line whose equation is

$$-x^3 + x + b = 0, \qquad \text{or} \qquad b = -x + x^3. \tag{4.26}$$

It is easy to draw this curve on the plane b-0-x, as shown in Figure 4.21 (on page **280**).

Now consider

$$b'_x = db/dx = -1 + 3x^2,$$
$$b''_x = d^2b/dx^2 = 6x.$$

Two points (b_2, x_2) and (b_1, x_1) appear where $b'_x = 0$ and one point (b_3, x_3) where $b''_x = 0$. It follows then that

$$(b_2, x_2) = (-(-\sqrt{3}/3) + (-\sqrt{3}/3)^3, -\sqrt{3}/3)$$
$$= (0.38490, -0.57735), \tag{4.27}$$
$$(b_1, x_1) = (-(\sqrt{3}/3) + (\sqrt{3}/3)^3, -(-\sqrt{3}/3)$$
$$= (-0.38490, 0.57735), \tag{4.28}$$

and

$$(b_3, x_3) = (0, 0). \tag{4.29}$$

Table 4.1

α	$(\mathbf{A} \cap \mathbf{H})\alpha$	Interval	$(\text{Interval})^2$
0.1	1 to 7	6	36
0.2	2 to 6	5	25
0.3	3 to 6	4	16
0.4	3 to 6	3	9
0.5	3 to 5	2	4
0.6	3 to 5	2	4
0.7	4 to 5	1	1
0.8	4 to 5	1	1
0.9	4 to 4	0	0
1	$\emptyset$	0	0
			Total = 96

Table 4.2

α	$\mathbf{A}_\alpha$	Interval	$(\text{Interval})^2$
0.1	1 to 8	7	49
0.2	1 to 7	6	36
0.3	2 to 7	5	25
0.4	3 to 7	4	16
0.5	3 to 6	3	9
0.6	3 to 6	3	9
0.7	4 to 6	2	4
0.8	4 to 6	2	4
0.9	4 to 5	1	1
1	5 to 5	0	0
			Total = 153

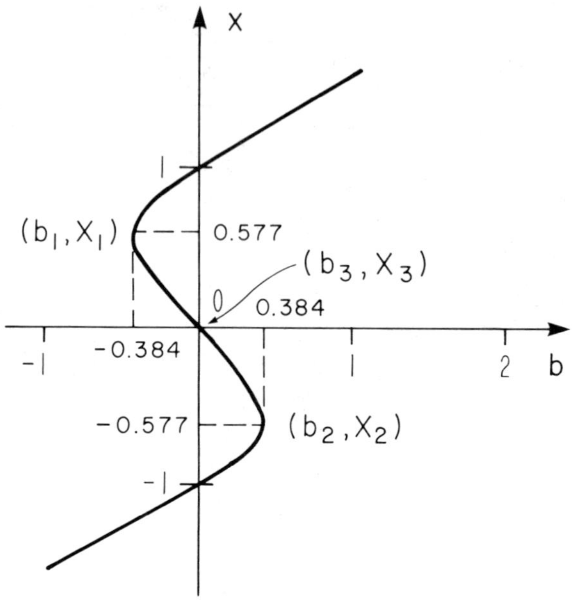

Figure 4.21 The curve $-x^3 + x + b = 0$.

This curve has the shape of an S in the *b*-0-x plane as shown in Figure 4.21. Figure 4.22 shows a surface that is generated by Equation (4.25) for various values of *f*. This surface is a cylinder generated by (4.26) and is called a *fold*. More generally, a fold does not have to be a cylinder; it can be any surface.

Let us now deform a fold in such a way so that it disappears and becomes a cusp, as shown in Figure 4.23. To obtain a cusp we consider the function

$$f(x, a, b) = -x^3 + ax + b, \qquad x, a, b \in R. \tag{4.30}$$

Another surface with topological peculiarities in more than three dimensions is studied later.

We now review some of the concepts that were introduced by Thom[4] and which are relevant to our topic. However, because this book is also concerned with applications rather than with pure theory, we simplify some of the points of interest. Because catastrophic theory is more easy to explain using the theory of systems, we use the latter concept in deterministic cases.

Let us represent a system by *S* and, at least initially, let us consider *m* state variables, where *m* is finite, $e_i(t)$, $i = 1, 2, 3, \ldots, m$, $t \in R^+$, and *t* is the time of observation. In addition, each variable $e_i(t)$ takes its values in *R*. We define an *input* to *S* at the time *t*, by a vector

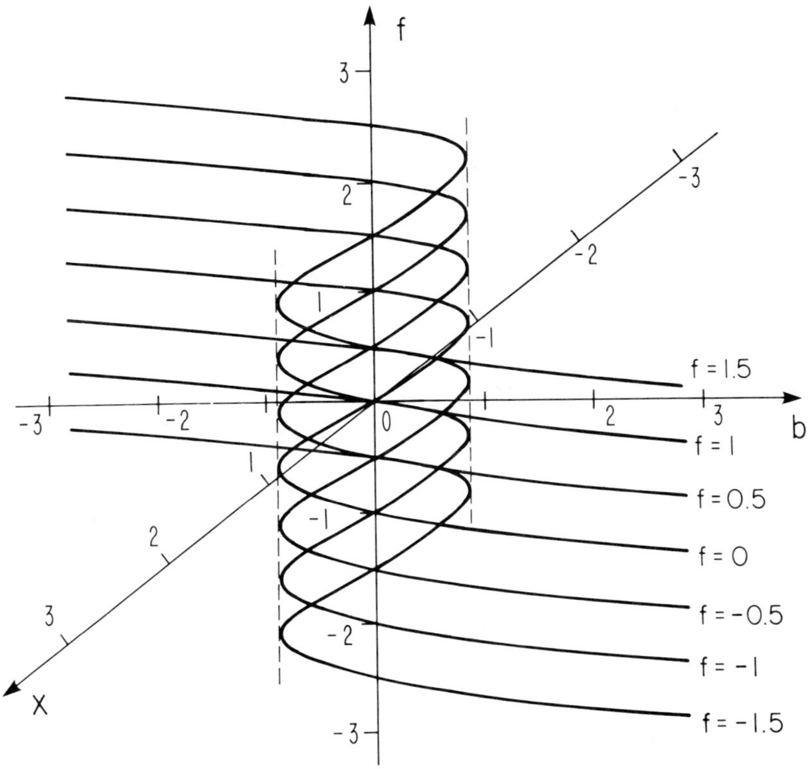

Figure 4.22 The surface $f(x, b) = -x^3 + x + b$.

$$u(t) = \{u_1(t), u_2(t), \ldots, u_r(t)\},$$

where r is finite.

Finally, in such systems several *output variables*, $y_j(t)$, $j = 1, 2, 3, \ldots, s$, where s is finite, exist which take their values in R or in subsets of R. Here we have considered continuous time, but there is no reason why it cannot be discrete.

For a given physical system, a mathematical model can be expressed as

$$\begin{aligned} d(e(t))/dt &= F[e(t), u(t)], \text{: state equation,} \\ y(t) &= G[e(t), u(t)], \text{: output equation.} \end{aligned} \qquad \textbf{(4.31)}$$

Or, in discrete time,

$$\begin{aligned} e(t+1) &= \mathcal{F}[e(t), u(t)], \\ y(t) &= \mathcal{G}[e(t), u(t)]. \end{aligned}$$

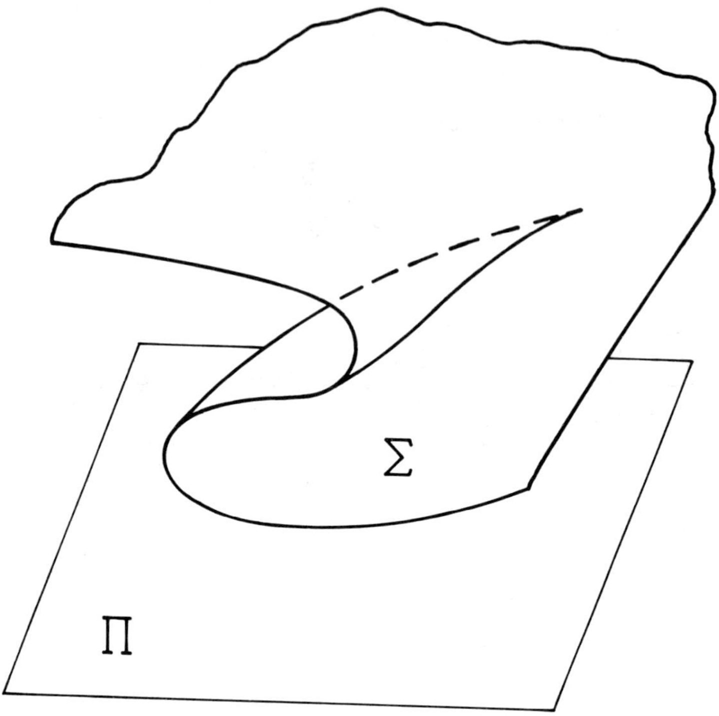

Figure 4.23 Cusp that is a deformation of a fold.

If we assume that the system S exists in a space with $(m + r + s)$ dimensions, a general vector can be considered as follows:

$$v(t) = (e_1(t), e_2(t), \ldots, e_m(t), u_1(t), u_2(t), \ldots, u_r(t), y_1(b), y_2(t), \ldots, y_s(t)).$$

Changing the notation, we can write

$$v(t) = [v_1(t), v_2(t), \ldots, v_m(t), v_{m+1}(t), v_{m+2}(t), \ldots, v_{m+r}(t), v_{m+r+1}(t), \ldots, v_{m+r+s}(t)]. \quad \textbf{(4.32)}$$

Each vector (4.32) defines a point in R^{m+r+s}. Now, let us suppose that it is possible to separate the $(m + r + s)$ variables into two different groups, some with fast evolution and others with slow evolution. We call those with fast evolution the *variables of catastrophic evolution*, and those with slow evolution the *inputs*, *outputs*, or *states*. Since only the catastrophic model is considered here, we assume S contains n fast variables $x^{(1)}(t), x^{(2)}(t), \ldots, x^{(n)}(t)$ and k only the slow variables $\pi^{(1)}(t), \pi^{(2)}(t) \ldots, \pi^{(k)}(t)$.

Further, let us now assume the existence of the function

$$f(x^{(1)}(t), x^{(2)}(t), \ldots, x^{(n)}(t), \pi^{(1)}(t)\ \pi^{(2)}(t), \ldots, \pi^{(k)}(t)).$$

Let us also define

$$dx(t)/dt = f(x(t), \pi(t). \qquad \textbf{(4.33)}$$

This equation corresponds to (4.31) with the specification given by (4.33).

In catastrophic theory we assume the existence of a potential function V for the system S:

$$V(x(t), \pi(t)),$$

such that when $\pi(t)$ is constant the system S tends to minimize the potential function V. We must note at this point that it is not necessary that V be a known function. This function V can be any scalar type function such as energy, cost, electric potential, temperature, or entropy function.

The minimization of V corresponds to setting $x(t)$ to an equilibrium state $x^*(t)$ that does not have to be unique. This equilibrium corresponds to

$$f_{x(\alpha)}(x(t), \pi(t)) = -\text{grad}_{x(\alpha)} V\{x(t), u(t)\}, \qquad \textbf{(4.34)}$$

for

$$f_{x(\alpha)}(x(t), \pi(t)) = 0, \qquad \alpha = 1, 2, \ldots, n.$$

The notation used in (4.34) means that gradient of V is obtained with respect to each variable $x^1(t), x^2(t), \ldots, x^n(t)$. We use the symbol M_f to represent the surface of a fold. To study the concept of catastrophe, we construct a set of M_f for the space of $\pi^{(1)}(t), \pi^{(2)}, \ldots, \pi^{(k)}(t)$, where π_f is the projection and is called the *branching layer*, or *bifurcation layer*. This projection gives singular points of the surface of the fold.

The simplest cases are those in which $n = 1$ and $k = 1$; that is, when there is only one fast variable $x(t)$ and one slow variable $\pi(\tau)$. In this case, let us assume that the surface of the fold of $x^*(\pi)$ is of the form shown in Figure 4.24. Here, where $n = k = 1$, this surface is a line π-0-x. In Figure 4.24 there is one and only one value of x for each value of π. From this figure we see that if $\pi = \pi_1$, the initial state x_1, the system will run with $\pi = \pi_1 =$ constant, reaching the stable point x_1^*. Similarly, for $\pi = \pi_2$, the system will proceed from x_2 to x_2^*. If in the system π increases from π_1 to π_2, the system S will travel from x_1^* to x_2^*, but the system tends to reach S and always remains on the curve $x^*(\pi)$.

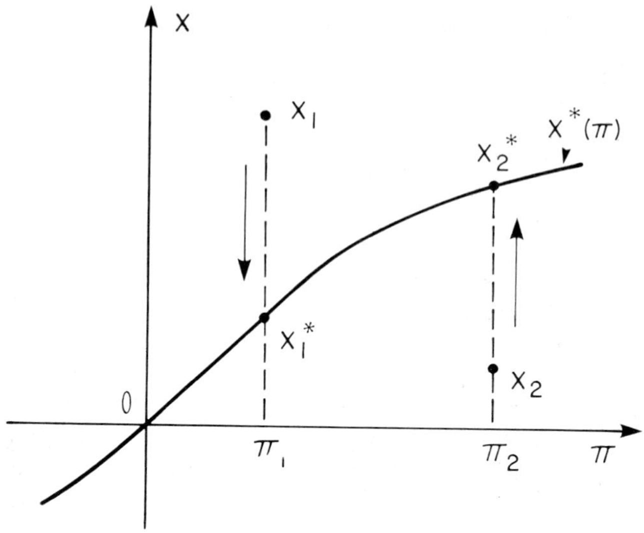

Figure 4.24 A stable surface of folding $x^*(\pi)$.

Catastrophic Phenomena

A different type of example is shown in Figure 4.25. Over a certain range for a single value of π, the function $x^*(\pi)$ may have multiple values. Let us assume that the system starts at point x_0^* on the left of the $x^*(\pi)$ curve. As π increases, the point moves along the lower branch of the curve, passing through 0, until it reaches the point A_1. At A_1 there are two possible solution points, A_1 and A_1'. Between A_1 and A_2 there are three possible solution points. The function, however, remains on the lower branch, moving along $x^*(\pi)$ from A_1 to A_2. Beyond A_2 the only possible solution is the point A_2' on the upper branch, and the system must therefore jump from A_2 to A_2', causing a discontinuity in its behavior. As π increases beyond the value A_2, $x^*(\pi)$ follows the upper branch in a uniform manner. Therefore we have found a discontinuity at $\pi = a_2$. If π is now decreased from the starting point at $x_0'^*$ (on the right) for $\pi > a_2$, $x^*(\pi)$ decreases uniformly along the upper branch through A_2' to A_1'. For values of π less than a_1, there is only one solution, and it is A_1. The function $x^*(\pi)$ must therefore jump catastrophically from A_1' to A_1. We see, then, that this function $x^*(\pi)$ is single-valued for π less than a_1 and greater than a_2. It is double-valued in the region A_1 to A_2. It takes on the values A_1 to A_2 as π increases and the values A_2' to A_1' as π decreases. The region A_1' to A_2 (or A_2 to A_1') is unstable. The phenomenon described here is called a jump *discontinuity*, or a *hysteresis*, which we refer to occasionally as a *catastrophic* effect. As explained earlier, this phenomenon exists for

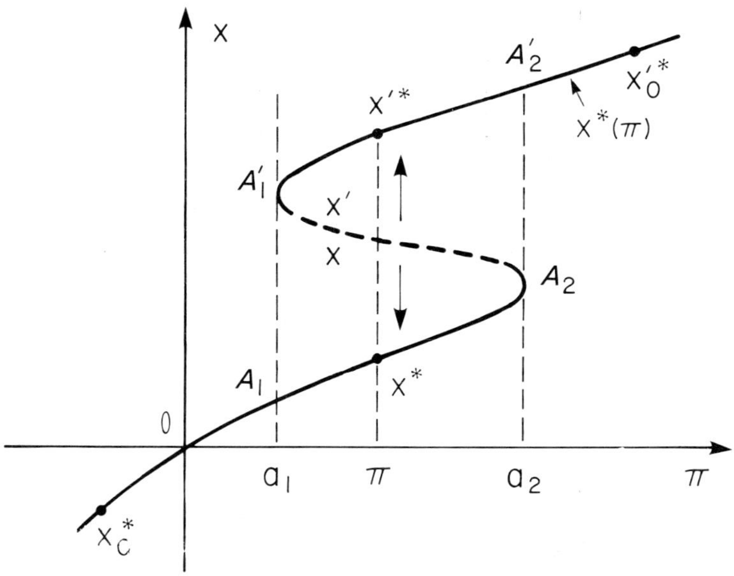

Figure 4.25 Catastrophe in the surface $x^*(\pi)$.

examples in polynomials of degree 3; see Equation (4.30). As discussed in this example, the region A_2A_1' is unstable and is called *repellor*, while the remaining parts of the curve are called *attractor*.

In catastrophe theory, we sometimes deal with more general cases such as multivalued functions having nonunique solutions, for example, hysteresis, and divergence for more than two variables. Thom[4] has given a list of elementary catastrophes, and we briefly describe them here. Let us assume that in the couplet (n, k), n is the number of fast variables and k is the number of slow variables. In Table 4.3 we will find cases corresponding to

$$\begin{aligned}(n, k) &= (1, 0),\\ (n, k) &= (1, 1),\\ (n, k) &= (1, 2),\\ (n, k) &= (1, 3),\\ (n, k) &= (1, 4),\end{aligned}$$

where the number of slow variables k is sometimes called *codimension*. Also note that the partial derivatives of the potential function V

$$f'_x = (\partial V)/(\partial x),$$

$$f'_y = (\partial V)/(\partial y) \qquad \text{(case when } n = 2\text{)},$$

characterize the catastrophes. We can see this more clearly in Table 4.3.

Table 4.3

n	k	*Potential function V (Thom's archetype)*	*Name*
1	0	$V = \frac{x^2}{3}$	single minimum
1	1	$V = \frac{x^4}{4} + p\,\frac{x^2}{2}$	fold
1	2	$V = \frac{x^4}{4} + p\,\frac{x^2}{2} + qx$	cusp
1	3	$V = \frac{x^5}{5} + p\,\frac{x^3}{3} + q\,\frac{x^2}{2} + rx$	swallow's tail
1	4	$V = \frac{x^6}{6} + s\,\frac{x^4}{4} + p\,\frac{x^3}{3} + q\,\frac{x^2}{2} + rx$	butterfly

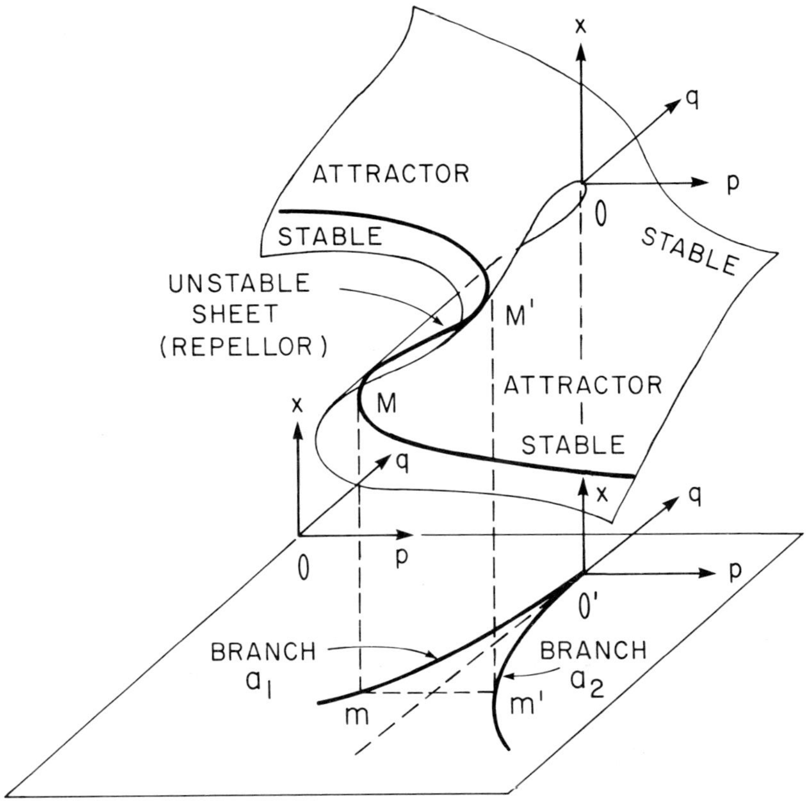

Figure 4.26 Nonunivocity, discontinuity, hysteresis, and divergence on a cusp-type catastrophe.

The cusp in Table 4.3, by way of example, displays the following four phenomena: nonunivocity, discontinuity, hysteresis, and divergence. These phenomena are shown in Figure 4.26. In the plane p-0-x we see a fold that progressively disappears as we move from $-q$ toward 0. A projection of this fold is shown on p-0-x plane. This projection appears in the form of a curve on which the phenomena of nonunivocity, discontinuity, and hysteresis are perceptible. In order that the divergence may be seen, the fold must be projected on to p-0-q. If the center is moved to ∞, the cusp becomes a fold. On the other hand, if it is moved to $-\infty$, the fold disappears. A surface like the cusp is also called a *Riemann-Hugonnot surface.*

The projection of the cusp onto p-0-q gives a curve called the *semicubic parabola*, or *Whitney's cusp*, whose equation is

$$4p^3 + 27q^2 = 0.$$

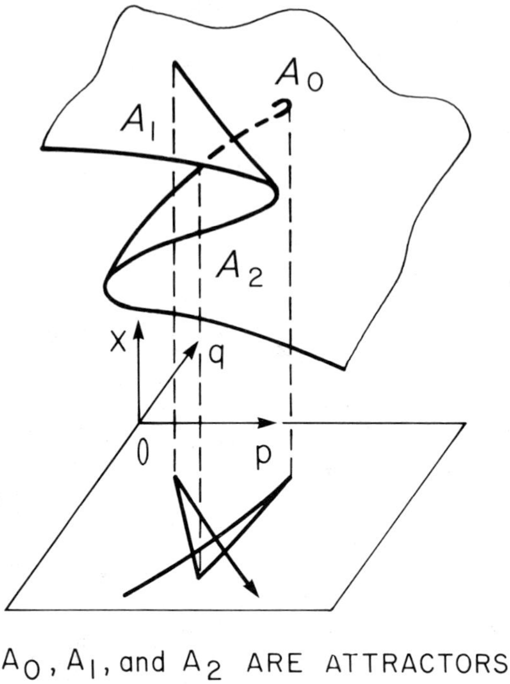

Figure 4.27(*A*) A butterfly-type catastrophe.

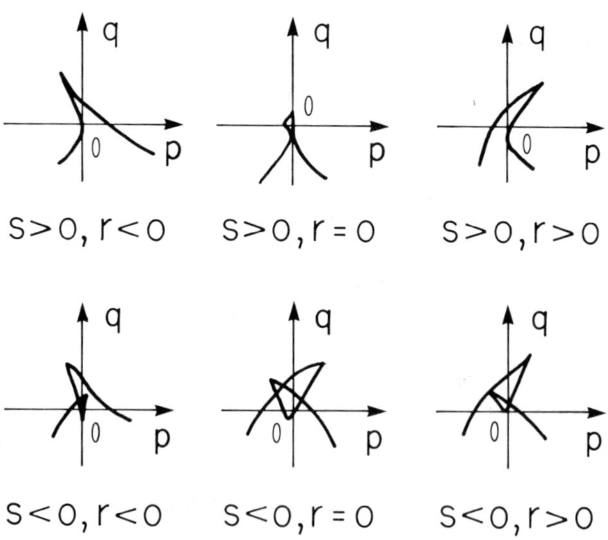

Figure 4.27(*B*) Influence of parameters *s* and *r* on catastrophe.

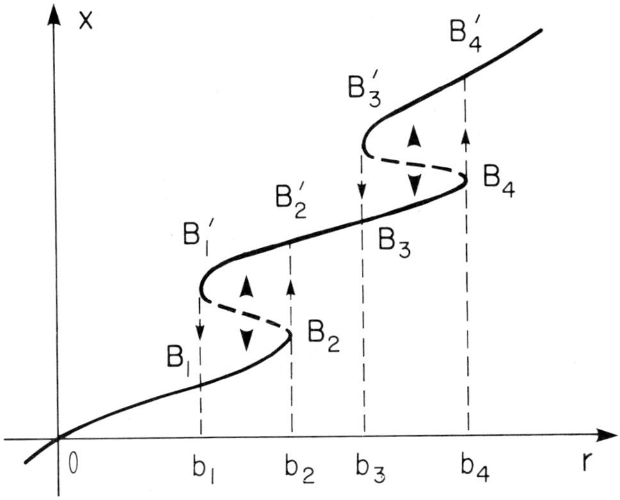

Figure 4.28 Projection of butterfly-type catastrophe on p-o-x plane.

Figure 4.27 shows a butterfly-type catastrophe (Table 4.3) projected on to the *p*-0-*q* plane. From this figure we see that there are three stable domains, or attractors, A_0, A_1, and A_2. This result shows that the buttterfly is really a double-cusp. The influence of the parameters *s* and *r* on the catastrophe are also shown in Figure 4.27(*B*).

If we correctly cut a butterfly-type surface for appropriately selected values of *q*, we obtain a projection on to the plane *p*-0-x that appears as double fold (Figure 4.28). In this figure the stable parts are shown by a solid line and the unstable parts by a dotted line.

The basic concepts of the theory of catastrophe that have just been described are next extended by applying the concept of fuzziness to them.

Introduction to the Theory of Fuzzy Catastrophes

Let us consider the case of the fold in Table 4.3.

$$V(x, p) = x^4/4 + p(x^2/2).$$

Its partial derivative is

$$\partial V/\partial x = x^3 + px.$$

We write $b = \partial V/\partial x$. For $p < 0$, we have

$$x^3 + px - b = 0.$$

Taking a particular numerical case, $p = -1$, we obtain

$$-x^3 + x + b = 0.$$

The curve of this equation in the b-0-x plane is S-shaped and is the projection of the cusp defined in Table 4.3 for $(n, k) = (1, 2)$. This S curve has a part that is attractive (disjoint) and a part that is repulsive, as is shown in Figure 4.29.

Let us now examine the effect of the segment $(\mathcal{B}_1, \mathcal{B}_2)$ on the intervals of confidence on the 0-x axis as it moves along the 0-b axis from $-\infty$ to $+\infty$. This is equivalent to first increasing b and then deceasing it. We have already used (4.27–4.30) to compute points of correction on this curve. Taking these values, we get

$$(b_2, x_2) = (0.38490, -0.57735),$$

$$(b_0, x_0) = (0, 0),$$

and

$$(b_1, x_1) = (-0.38490, 0.57735).$$

Let us write $b = f(x) = x^3 - x$ and $x = \phi(b) = f^{-1}(b)$. In addition, for the part of the curve from $(-\infty, -\infty)$ to (b_2, x_2) let $x = \phi_1(b)$. And for the part from (b_1, x_1) to (∞, ∞), $x = \phi_2(b)$. That is, if we put $x = \phi_1(b)$ for the low branch, and $x = \phi_2(b)$ for the high branch of the curve shown in Figure 4.30, we obtain

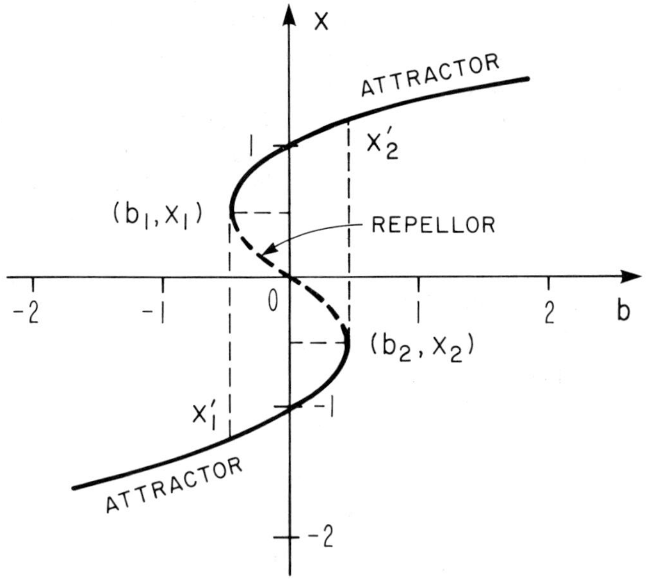

Figure 4.29 A catastrophe curve $-x^3 + x + b = 0$.

$$\phi_1^{-1}(x) = x^3 - x, \qquad x \in (-\infty, -0.57735],$$
$$\phi_2^{-1}(x) = x^3 - x, \qquad x \in [0.57735, \infty).$$

Now let us move the segment β along the 0-b axis, where $\beta = [\beta_1, \beta_2]$, such that $\beta_2 - \beta_1 = 1$ and $\beta = [\beta_1, \beta_1 + 1]$. Moving from $-\infty$ to $+\infty$ along 0-b gives

$$\mathcal{X} = \psi^+([\beta_1, \beta_1 + 1]).$$

We see, then, that this is the segment on the 0-x axis that constitutes the function ψ^+, which corresponds to $[\beta_1, \beta_1 + 1]$ on the 0-b axis.

From $-\infty$ to b_2, the function $\psi^+([\beta_1, \beta_1 + 1])$ increases regularly and monotonically. That is, if $\beta_1 + 1 \leq b_2$, then

$$\mathcal{X} = \psi^+([\beta_1, \beta_1 + 1]) = [\phi_1(\beta_1), \phi_2(\beta_2)].$$

If $\beta_1 \leq b_2 < \beta_1 + 1$, ψ^+ is not regular, and

$$\psi^+([\beta_1, \beta_2] = [\phi_1(\beta_1), x_2] \cup [x'_2, \phi_1(\beta_1 + 1)],$$

where $x'_2 = \phi_2(b_2)$ and the two segments $[\phi_1(\beta_1), x_2]$ and $[x_2', \phi_2(\beta_1 + 1)]$ are disjoint $(x_2 < x_2')$.

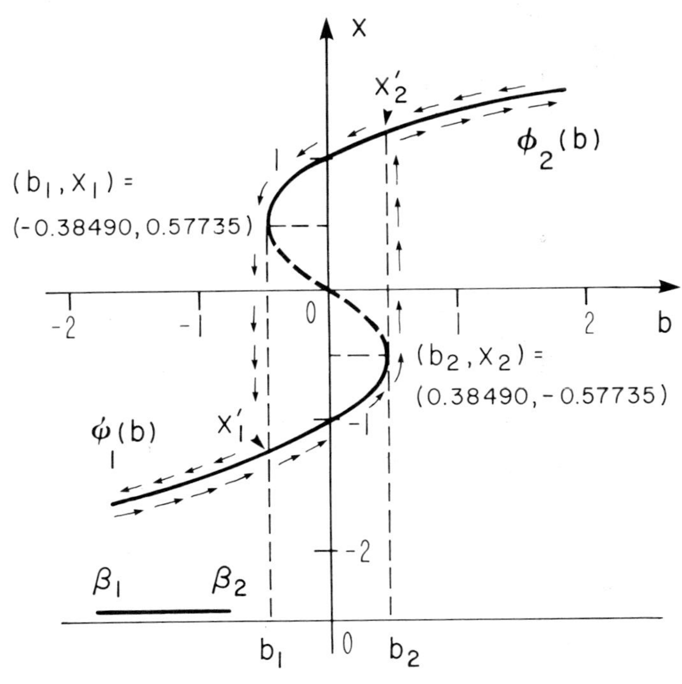

Figure 4.30 Stable and unstable segments in the interval of confidence x.

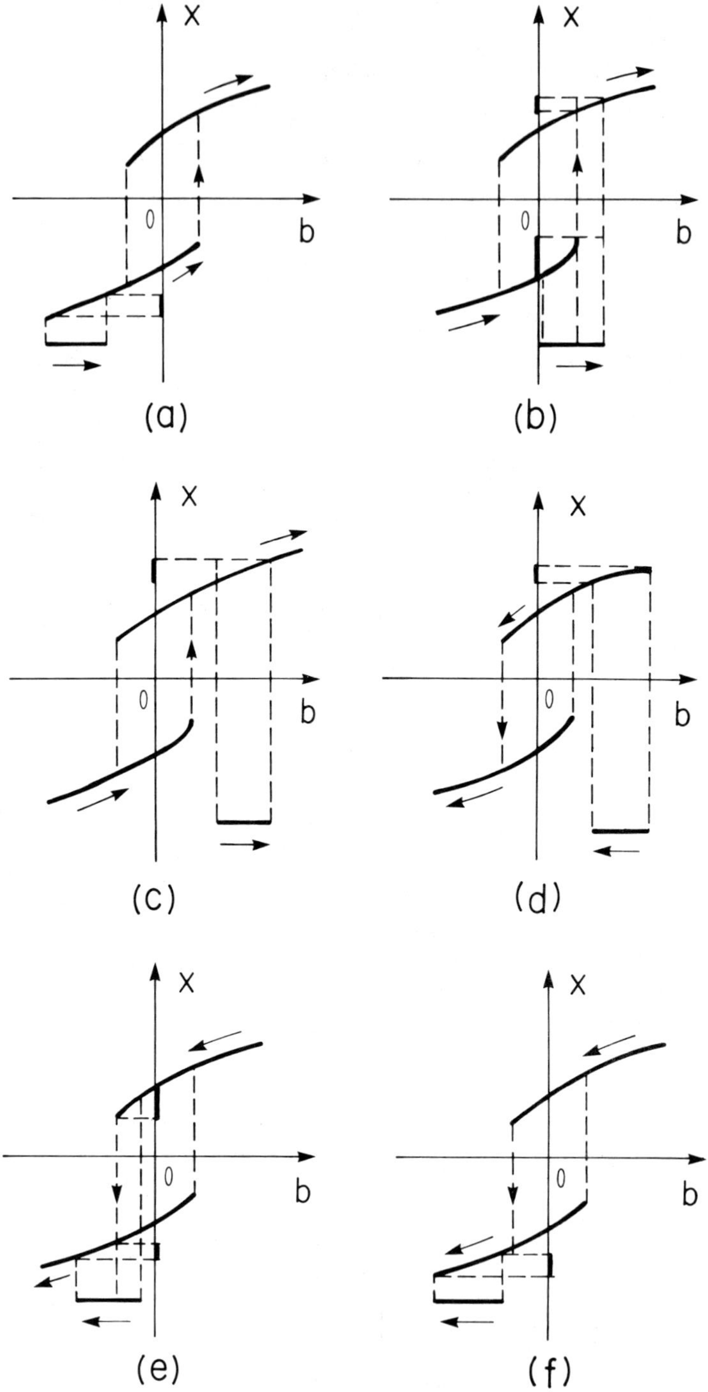

Figure 4.31 Catastrophe-induced disjunctions into the interval of confidence x.

Finally, if $\beta_1 \geq \beta_2$, $\psi^+([\beta_1, \beta_1 + 1])$ becomes regular again and increases monotonically. Thus

$$\psi^+([\beta_1, \beta_1 + 1]) = [\phi_2(\beta_1), \phi_2(\beta_1 + 1)].$$

Now let us shift $[\beta_1, \beta_1 + 1]$ in the opposite direction, from $+\infty$ to $-\infty$ along the 0-b axis. We see then that from $+\infty$ to b_1, $\psi^-([\beta_1, \beta_1 + 1])$ increases regularly and monotonically.[5] Thus, if $\beta_1 \geq b_1$, then

$$\mathcal{X} = \psi^-([\beta_1, \beta_1 + 1]) = [\phi_2(\beta_1), \phi_2(\beta_1 + 1)].$$

If $\beta_1 < b_1 < \beta_1 + 1$, ψ^- is not regular,

$$\psi^-([\beta_1, \beta_1 + 1]) = [\phi_1(\beta_1), (x_1')] \cup [x_1, \phi_2(\beta_1 + 1)],$$

where $x_1' = \phi_1(b_1)$ and the two segments $[\phi_1(\beta_1), x_1']$ and $[x_1, \phi_2(\beta_1 + 1)]$ are disjoint ($x_1' < x_1$).

Finally, if $\beta_1 + 1 \leq b_1$, $\psi^-(\beta_1, \beta_1 + 1])$ becomes regular again,

$$\psi^-([\beta_1, \beta_1 + 1]) = [\phi_1(\beta_1), \phi_1(\beta_1 + 1)].$$

As we have just seen, a catastrophe induces a disjunction into the interval of confidence. Figures 4.31(a) to 4.31(f) show such results.

We now extend these concepts of catastrophe theory to fuzzy numbers.

Catastrophe with Fuzzy Numbers

Let us consider a curve defined by $\phi(u, w) = 0$ in the u-0-w plane and assume a fuzzy number $\mathbf{W} \subset R$ such that $\mu_{\mathbf{W}}(w)$ maintains its exact shape when it is shifted to the left or to the right along 0-w.

From $\phi(u, w) = 0$, we can obtain $u = f(w)$, which may be a multi-valued function of w.

When a level α exists for $U_\alpha = f(W_\alpha)$ such that

$$U_\alpha = f(W_\alpha) = [u_1^{(\alpha)}, u_2^{(\alpha)}] \cup [u_3^{(\alpha)}, u_4^{(\alpha)}], \tag{4.35}$$

with

$$[u_1^{(\alpha)}, u_2^{(\alpha)}] \cap [u_3^{(\alpha)}, u_4^{(\alpha)}] = \emptyset, \tag{4.36}$$

then

$$u_2^{(\alpha)} < u_3^{(\alpha)}.$$

The curve $\phi(u, w) = 0$ is said to have a simple catastrophe of rank 1 for the fuzzy number $\mathbf{W}$ from $-\infty$ to $+\infty$ along 0-W.

When the number $\mathbf{W}$ is not fuzzy, condition (4.35) is replaced by the following: when N passes from w to $w + \Delta w$, we have $f(w) = u_1$, $f(w + \Delta w) = u_2$ and $(u_2 - u_1)$ does not tend to 0 during the time

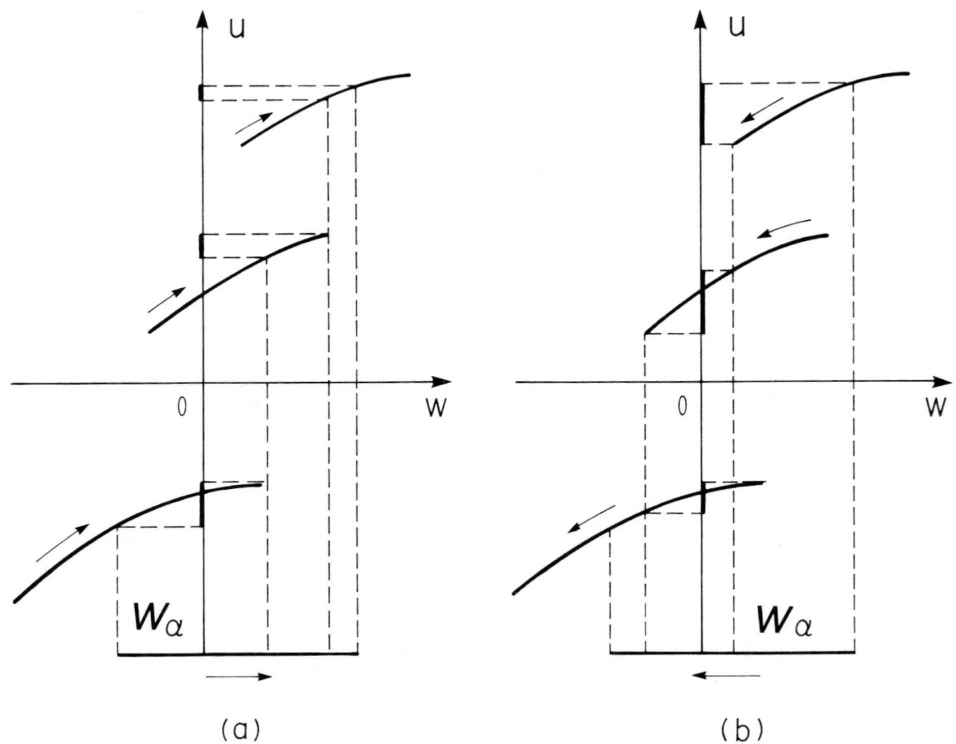

Figure 4.32 $\Phi(u, w) = 0$: catastrophe of rank 2.

$\Delta w \to 0$; that is, $f(w)$ jumps from u_1 to u_2 when W passes through different values of w.

The cut (4.36), or the corresponding jump in the case of fuzzy numbers, is called a *hiatus*.

When a curve $\phi(u, w) = 0$ has n hiatus jumps for a fuzzy number **W**, the simple catrastrophe is said to be of rank n, and n jumps will exist at a level α. Figure 4.32 shows a curve $\phi(u, w) = 0$ that has a simple catastrophe of rank 2 for the fuzzy number **W**. At the level α, there are two hiatus and it is possible that at a level $\alpha' > \alpha$ there is only one hiatus at **W**. At $\alpha = 1$ it is possible that the hiatus can become a jump discontinuity.

In our elementary exposition to the fuzzy catastrophic theory, we have limited ourselves to simple catastrophes of the fold type (*archetypes*, Table 4.3, $(n, k) = (1, 1)$ of Thom[41]).

EXAMPLE 4.17

Let us return to Figure 4.30. We have here

$$w = u^3 - u. \tag{4.37}$$

Now let us consider a fuzzy number like the one shown in Figure 4.33, where $w_0 = 0.5$. We then have

$$\begin{aligned} \mu_{\mathbf{W}}(w) &= 0, & w &\leq 0, \\ &= 2w, & 0 \leq w &\leq 0.5, \\ &= -4w + 3, & 0.5 \leq w &\leq 0.75, \\ &= 0, & 0.75 &\leq w. \end{aligned} \tag{4.38}$$

We now look for the fuzzy subset that is inducted on 0-u by the non-univocal curve in Figure 4.33 when **W** moves from $-\infty$ to $+\infty$, as shown in Figure 4.34.

To find what we are looking for we must solve the third-degree equation 4.37.

Suppose that $u^3 + pu + q = 0$. There are three possible cases that must be considered. They are

1. Three real and different roots if
$$4p^3 + 27q^2 < 0.$$
2. One real root and two complex conjugate roots if
$$4p^3 + 27q^2 > 0.$$
3. One real single root and two real double roots if
$$4p^3 + 27q^2 = 0.$$

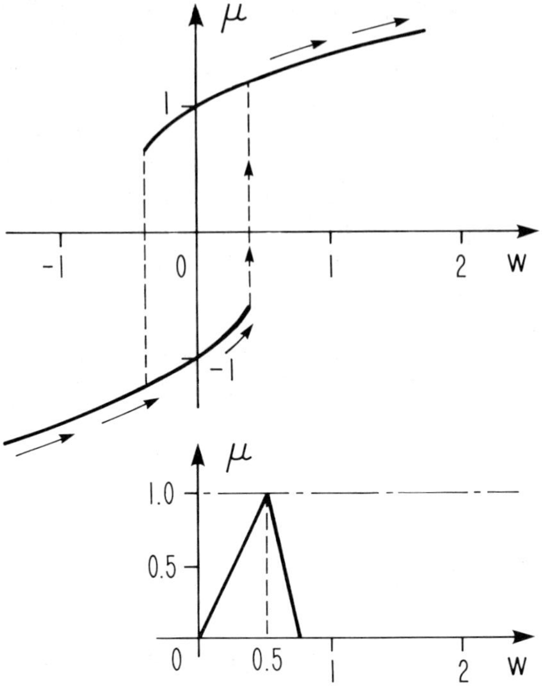

Figure 4.33 Catastrophe with fuzzy numbers (Example 4.17).

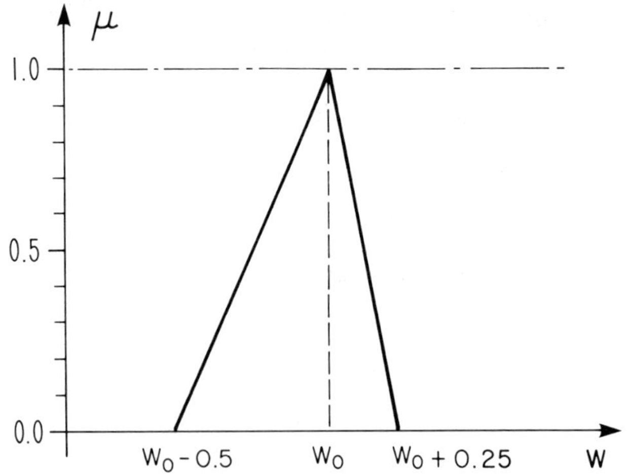

Figure 4.34 Shifting of **w** in the fuzzy number (Example 4.17).

Case 1

In this case $4p^3 + 27q^2 < 0$. Let

$$\cos \phi = -(3/-p)^{3/2}(q/2).$$

Then

$$u_1 = (2/\sqrt{3})\sqrt{-p}\cos \phi/3,$$

$$u_2 = (2/\sqrt{3})\sqrt{-p}\cos[(2\pi/3) + (\phi/3)],$$

and

$$u_3 = (2/\sqrt{3})\sqrt{-p}\cos[(4\pi/3) + (\phi/3)].$$

Case 2

Here $4p^3 + 27q^2 > 0$.

(a) $p < 0$ and $q < 0$. Let

$$\cosh \phi = -(3/-p)^{3/2}(q/2),$$

then

$$u_1 = (2/\sqrt{3})\sqrt{-p}\cosh(\phi/3).$$

(b) $p > 0$ and $q < 0$. Let

$$\sinh \phi = -(3/p)^{3/2}(q/2),$$

then

$$u_1 = (2/\sqrt{3})\sqrt{p}\sinh(\phi/3).$$

(c) $q > 0$. In this subcase the sign of q will change; the sign of the root will also change in the result.

Case 3

For this case, $4p^3 + 27q^2 = 0$. We then have

$$u_1 = 3p/q.$$

$$u_2 = u_3 = -3q/2p.$$

The procedure of Case 2 is used here, where the complex roots are obtained by dividing $(u^3 + pu + q)$ by $(u - u_1)$. When this is done, a second-degree equation is obtained whose roots are complex.

Let us now consider (4.37), where we have

$$u^3 - u - w = 0.$$

Using the equations given above, and taking into account the value of w for each level α, we write

Table 4.4

α	$[\frac{\alpha}{2}, \frac{3}{4} - \frac{\alpha}{4}]$	U_α
0	[0, 0.750]	[−1, −0.5773] ∪ [1.1547, 1.2625]
0.1	[0.050, 0.725]	[−0.9739, −0.5773] ∪ [1.1547, 1.2558]
0.2	[0.100, 0.700]	[−0.9456, −0.5773] ∪ [1.1547, 1.2491]
0.3	[0.150, 0.675]	[−0.9142, −0.5773] ∪ [1.1547, 1.2423]
0.4	[0.200, 0.650]	[−0.8788, −0.5773] ∪ [1.1547, 1.2353]
0.5	[0.250, 0.625]	[−0.8375, −0.5773] ∪ [1.1547, 1.2283]
0.6	[0.300, 0.600]	[−0.7864, −0.5773] ∪ [1.1547, 1.2211]
0.7	[0.350, 0.625]	[−0.7140, −0.5773] ∪ [1.1547, 1.2119]
0.7698	[0.38490, 0.55755]	−0.5773 ∪ [1.1547, 1.2088]
0.8	[0.400, 0.550]	[1.1594, 1.2065]
0.9	[0.450, 0.525]	[1.1758, 1.1990]
1	0.500	1.1914

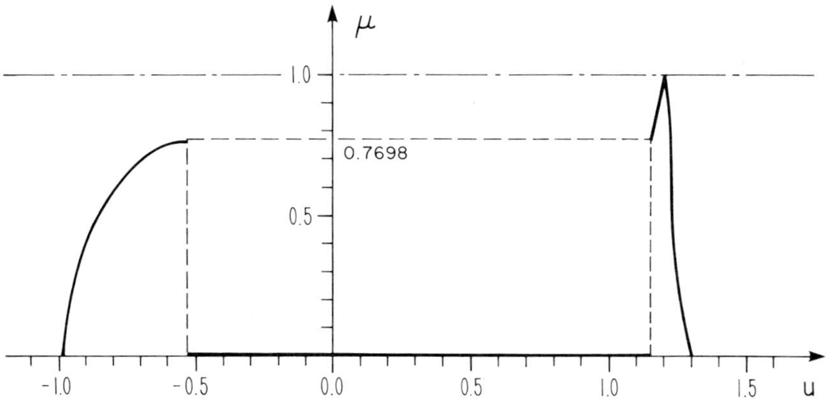

Figure 4.35 W_α for different level α (Example 4.17).

$$W_\alpha = [w_1^{(\alpha)}, w_2^{(\alpha)}] = [\alpha/3, 3/4 - \alpha/4].$$

We thus obtain, for increasing w, Table (4.4) and Figure 4.35.

Now, let us examine how the curve $x^3 - x = w$ is affected as w is decreased from $+\infty$ to $-\infty$, as shown in Figure 4.36. In this case, the fuzzy number (4.38) results in a table, derived from Table 4.5 for values of μ. Since the fuzzy number **W** is never less than $W = 0$, the interval of confidence of μ is evaluated on the upper side of the curve only.

These results are shown in Figure 4.37. If we compare Figures 4.35 and 4.37 we see that for the same fuzzy number the inducted fuzzy subset differs according to the direction in which the fuzzy number is moved. If the motion is to the left, there is no hiatus, but if the motion is to the right, there is a hiatus. This difference, which is the result of the direction in which the fuzzy number is moved, is due to the presence of a simple catastrophe, which is a fold according to Thom's archetype. As in the case of ordinary numbers, a non-univocity is the cause of a catastrophe.

Discontinuity and Catastrophe

A discontinuity in the derivative of a potential function V that is associated with nonunivocity causes a catastrophic phenomenon. A discontiniuty in $u = f(w)$ gives a hiatus in the fuzzy subset **u** such that $\mathbf{u} = f(\mathbf{W})$, where **W** is a fuzzy number. If the discontinuity appears at the point w, we have

$\exists\ \alpha \in [0, 1]$,

$$w^* = [w_1^{(\alpha)}, w_2^{(\alpha)}].$$

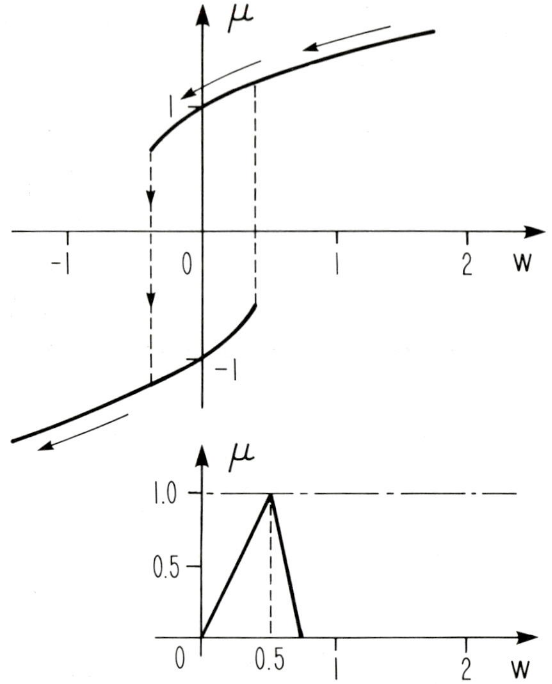

Figure 4.36 The curve $x^3 - x = w$ with interval of confidence (Example 4.17).

Table 4.5

α	$[\frac{\alpha}{2}, \frac{3}{4} - \frac{\alpha}{4}]$	U_α
0	[0, 0.750]	[1, 1.2625]
0.1	[0.050, 0.725]	[1.0241, 1.2558]
0.2	[0.100, 0.700]	[1.0466, 1.2431]
0.3	[0.150, 0.675]	[1.0679, 1.2423]
0.4	[0.200, 0.650]	[1.0880, 1.2353]
0.5	[0.250, 0.625]	[1.1071, 1.2283]
0.6	[0.300, 0.600]	[1.1254, 1.2211]
0.7	[0.350, 0.625]	[1.1429, 1.2119]
0.8	[0.400, 0.550]	[1.1597, 1.2065]
0.9	[0.450, 0.525]	[1.1758, 1.1990]
1	0.500	1.1914

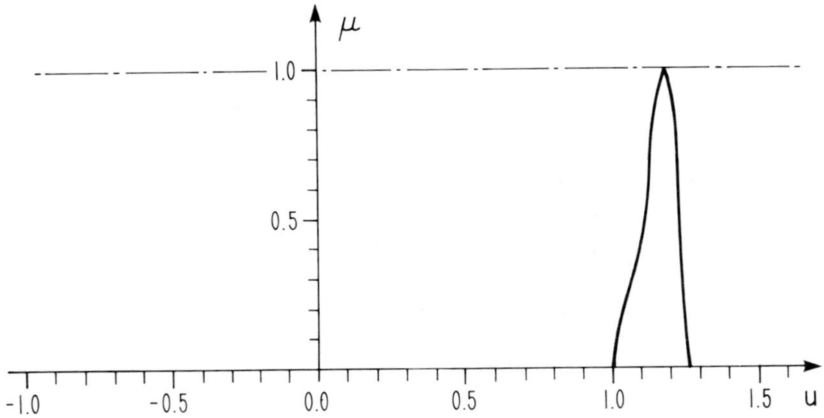

Figure 4.37 α-values for different confidence interval u (Example 4.17).

Of course, every discontinuity does not give a catastrophe. As far as fuzzy numbers are concerned, a catastrophe will appear when a fuzzy number **W** corresponds to at least a pair of fuzzy subsets that are normal and unlike ($\mathbf{U}_1 = f_1(\mathbf{W})$, $\mathbf{U}_2 = f_2(\mathbf{W})$), where at least one hiatus exists when going from $-\infty$ to $+\infty$, and another exists when going in the reverse direction.

We see, therefore, that catastrophes generate fuzzy numbers with hiati. Let us study in more detail this kind of fuzzy number.

Fuzzy Numbers with Hiatus

Let us inspect now the two kinds of discontinuity shown in Figures 4.38 and 4.39. In Figure 4.38 we see that a discontinuity at the point $w = a$ is such that any point between A and B (except A or B) belongs to the curve. In this case, we observe that a fuzzy number **W** will be transformed into a fuzzy number with a hiatus that corresponds to points A and B.

If we consider the curve $u = f(w)$ such that every point that lies between A and B belongs to it, the curve $w = f^{-1}(u)$ will have a flat spot on it between A and B. For this condition **W** will be transformed into a fuzzy number without a hiatus.

It is obvious that in the theory of catastrophes discontinuities occur according to Figure 4.38, with a discontinuous jump. Hence, a catastrophe will generate a fuzzy number with a hiatus. The parts of curves that are being considered here are called the attractors, while those parts that are not being considered are unstable.

Another aspect related to both catastrophes and fuzzy numbers is

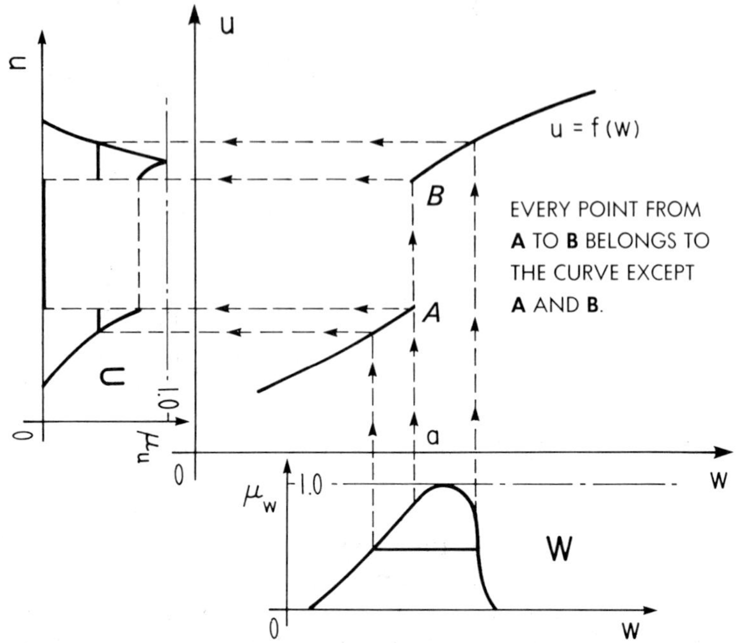

Figure 4.38 A fuzzy number with a hiatus corresponding to A and B.

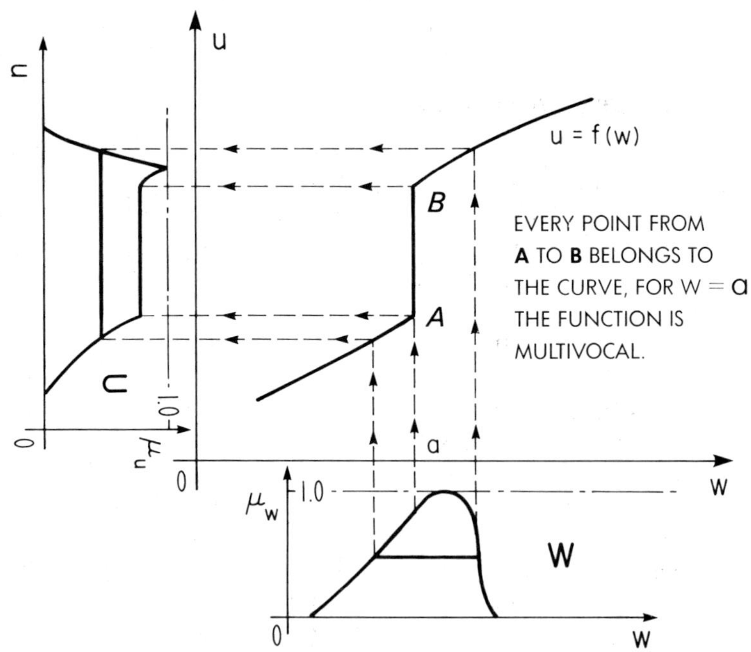

Figure 4.39 A fuzzy number without a hiatus.

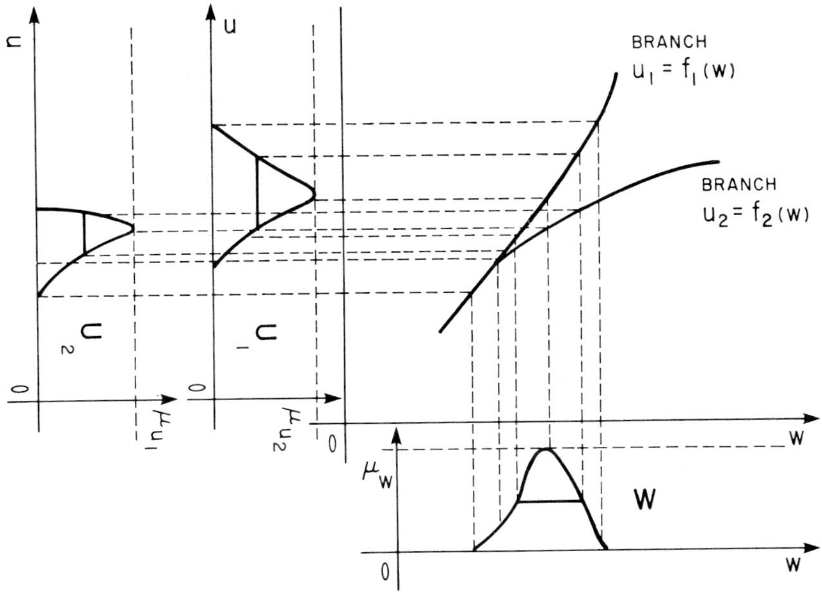

Figure 4.40 Divergence by branching in catastrophe.

divergence by branching, which is shown in Figure 4.40. We see from this figure that the effect of a branching point on a fuzzy number **W** is a doubling of the number of branches. Instead of there being a single **U**, two branches $\mathbf{U}_1$ and $\mathbf{U}_2$ are obtained.

In the following paragraphs, we briefly study fuzzy numbers with hiatus. We first consider the fuzzy number $\mathbf{A} \subset R$, and then the fuzzy subset, where

$\forall\, x \in R$:

$$\begin{aligned} \mu_{\mathbf{X}}(x) &= \mu_{\mathbf{A}}(x), && x \le x_1, \\ &= 0, && x_1 \le x \le x_2, \\ &= \mu_{\mathbf{A}}(x - (x_2 - x_1)), && x \le x_2. \end{aligned}$$

is the *fuzzy number with a hiatus*. The choice of the word *number* can be critical because the number **X** does not have the convexity possessed by the fuzzy number **A**. Nevertheless, we use this word for convenience. Figure 4.41 shows a fuzzy number with a hiatus.

The concept of a hiatus can be extended to Z as follows:

	−2	−1	0	1	2	3	4	5	6	7	8	9	10
X =	0	0.2	0.5	0.8	1	0.7	0	0	0	0.5	0.3	0.1	0
							*	*	*				

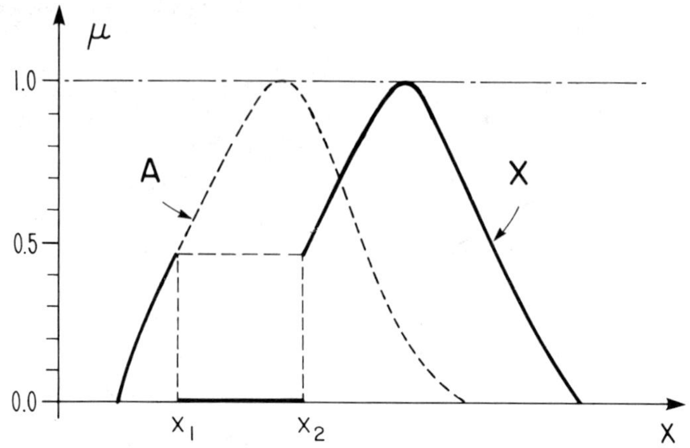

Figure 4.41 A fuzzy number **X** with a hiatus.

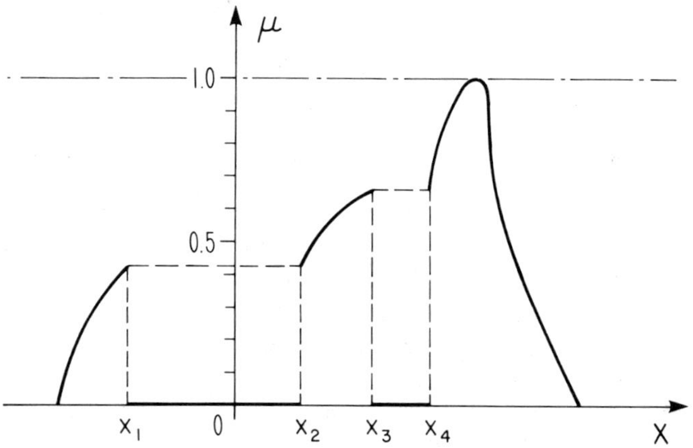

Figure 4.42 A fuzzy number **X** with two hiatus.

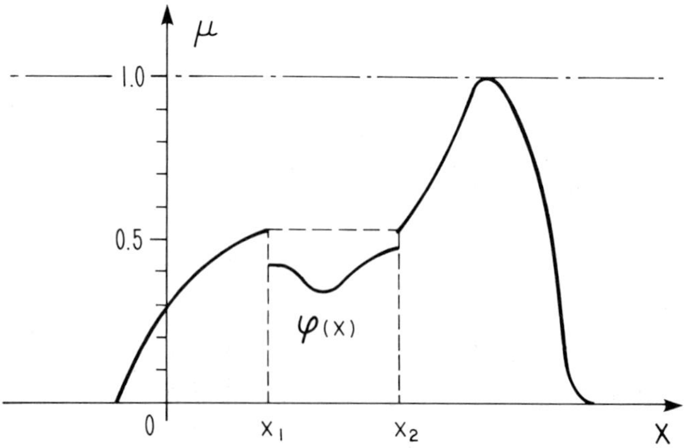

Figure 4.43 A fuzzy number **X** that is almost "convex."

A fuzzy number may have several hiati. Figure 4.42 shows a fuzzy number with two hiatus.

The concept of a hiatus can be generalized in many ways. One of these generalizations is a fuzzy number that is "almost" convex and which is defined as follows:

$\forall\ x \in R$:

$$\begin{aligned} \mu_{\mathbf{X}}(x) &= \mu_{\mathbf{A}}(x), && x \leq x_1, \\ &= \phi(x), && x_1 \leq x \leq x_2, \\ &= \mu_{\mathbf{A}}(x - (x - x_1)), && x_2 \leq x \end{aligned}$$

where $\mu_{\mathbf{A}}(x)$ is a given fuzzy number, $\phi(x) \leq \mu_{\mathbf{A}}(x_1)$, and is continuous in (x_1, x_2).

Figure 4.43 represents such a fuzzy number that is almost convex.

It is also possible for this type of hiatus to occur several times, in which case the fuzzy number is said to be a *fuzzy number n—almost convex.*

We now come to several theorems of considerable interest.

THEOREM 4.1

Suppose that $\mathbf{X} \subset R$ is a fuzzy number with a hiatus for $(a_2, a_3) \subset R$, and that $\mathbf{Y} \subset \mathbf{R}$ is another fuzzy number without a hiatus. A necessary and sufficient condition for $\mathbf{X}\ (+)\ \mathbf{Y}$ to be a fuzzy number without a hiatus is then

$$\forall\ \alpha \in \mu_{\mathbf{X}}(a_2), \qquad (a_3 - a_2) \leq (b_2^{(\alpha)} - b_1^{(\alpha)}), \tag{4.39}$$

where $[b_1^{(\alpha)}, b_2^{(\alpha)}]$ is the interval of confidence at the level α of $\mathbf{Y}$.

Proof:

The α-cut of $\mathbf{X}$ for $\alpha \leq \mu_{\mathbf{X}}(a_2)$ is

$$[a_1^{(\alpha)}, a_2] \cup [a_3, a_4^{(\alpha)}].$$

Let us now add this union to the segment $[b_1^{(\alpha)}, b_2^{(\alpha)}]$, a union that is distributive with respect to $(+)$.

$$\begin{aligned} &([a_1^{(\alpha)}, a_2] \cup [a_3, a_4^{(\alpha)}]\ (+)\ [b_1^{(\alpha)}, b_2^{(\alpha)}] \\ &= [a_1^{(\alpha)} + b_1^{(\alpha)}, a_2 + b_2^{(\alpha)}] \cup [a_3 + b_1^{(\alpha)}, a_4^{(\alpha)} + b_2^{(\alpha)}]. \end{aligned}$$

The corresponding intersection will not be empty iff

$$a_2 + b_2^{(\alpha)} \geq a_3 + b_1^{(\alpha)}$$

or

$$a_3 - a_2 \leq b_2^{(\alpha)} - b_1^{(\alpha)}. \tag{4.40}$$

Let us now study what happens in an example when condition (4.40) is not satisfied.

EXAMPLE 4.18

A fuzzy number $\mathbf{X} \subset R$ with a hiatus and a fuzzy number $\mathbf{Y} \subset R$ without a hiatus are given in Figure 4.44(*A*). We can show easily that condition (4.40) is not satisfied. Let us compute, from $\alpha = 0$ to $\alpha = 1$, the interval of confidence using a step of 0.1. The results of such a calculation are shown in Figure 4.44(*B*) and Table (4.6), from which we see that the sum $\mathbf{X}\ (+)\ \mathbf{Y}$ is not convex. There is no convexity between 33.5 and 35. If (a_2, a_3) were shorter, the gap would not exist and the sum would be convex.

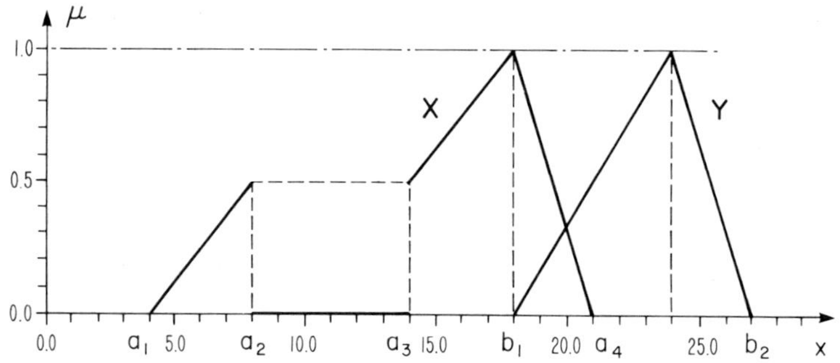

Figure 4.44(*A*) Fuzzy numbers **X** (with hiatus) and **Y** (without hiatus).

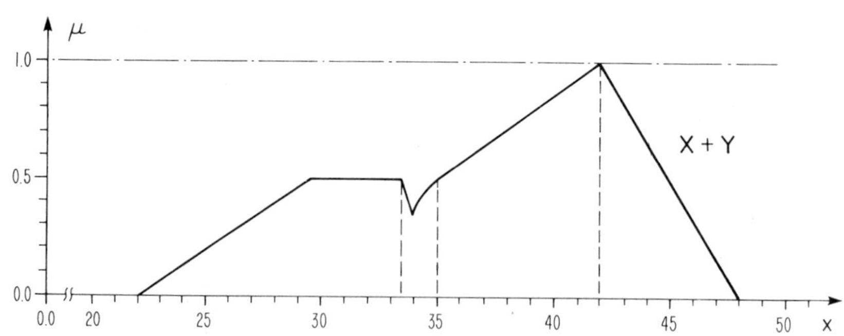

Figure 4.44(*B*) The sum **X** + **Y** that is not convex.

Table 4.6

α	**X**	**Y**	**X (+) Y**
0	$[4, 8] \cup [14, 21]$	$[18, 27]$	$[22, 48]$
0.1	$[4.8, 8] \cup [14, 20.7]$	$[18.6, 26.7]$	$[23.4, 47.4]$
0.2	$[5.6, 8] \cup [14, 20.4]$	$[19.2, 26.4]$	$[24.8, 46.8]$
0.3	$[6.4, 8] \cup [14, 20.1]$	$[19.8, 26.1]$	$[26.2, 46.1]$
0.4	$[7.2, 8] \cup [14, 19.8]$	$[20.4, 25.8]$	$[27.6, 33.8] \cup [34.4, 45.6]$
0.5	$[8, 8] \cup [14, 19.5]$	$[21, 25.5]$	$[29, 33.5] \cap [35, 45]$
0.6	$[14.8, 19.2]$	$[21.6, 25.2]$	$[36.4, 44.4]$
0.7	$[15.6, 18.9]$	$[22.2, 24.9]$	$[37.8, 43.8]$
0.8	$[16.4, 18.6]$	$[22.8, 24.6]$	$[39.2, 43.2]$
0.9	$[17.2, 18.3]$	$[23.4, 24.3]$	$[40.6, 42.6]$
1	18	24	42

THEOREM 4.2

Suppose that $\mathbf{X} \subset R$ and $\mathbf{Y} \subset R$ are fuzzy numbers with respective hiatus (a_2, a_3) and (b_2, b_3). The sum $\mathbf{X}\ (+)\ \mathbf{Y}$ does not have a hiatus iff

$$\left.\begin{aligned} a_2 - a_1^{(\alpha)} &\geq b_3 - b_2 \quad (4.41) \\ a_4^{(\alpha)} - a_3 &\geq b_3 - b_2 \quad (4.42) \end{aligned}\right\} \alpha \leq \mu_{\mathbf{X}}(b_2),$$

or

$$\left.\begin{aligned} b_2 - b_1^{(\alpha)} &\geq a_3 - a_2 \quad (4.43) \\ b_4^{(\alpha)} - b_2 &\geq a_3 - a_2 \quad (4.44) \end{aligned}\right\} \alpha \leq \mu_{\mathbf{Y}}(a_2).$$

Proof:

From the hiatus (a_2, a_3) of $\mathbf{X}$,

$$a_1^{(\alpha)} \leq a_2 \leq a_3 \leq a_4^{(\alpha)},$$

and from the hiatus (b_2, b_3) of $\mathbf{Y}$,

$$b_1^{(\alpha)} \leq b_2 \leq b_3 \leq b_4^{(\alpha)}.$$

Let us compute $\mathbf{X}\ (+)\ \mathbf{Y}$:

$$\begin{aligned} \mathbf{X}\ (+)\ \mathbf{Y} &= ([a_1^{(\alpha)}, a_2] \cup [a_3, a_4^{(\alpha)}])\ (+)\ ([b_1^{(\alpha)}, b_2] \cup [b_3, b_4^{(\alpha)}]) \\ &= [a_1^{(\alpha)} + b_1^{(\alpha)}, a_2 + b_2] \cup [a_1^{(\alpha)} + b_3, a_2 + b_4^{(\alpha)}] \\ &\quad \cup [a_3 + b_1^{(\alpha)}, a_4^{(\alpha)} + b_2] \cup [a_3 + b_3, a_4^{(\alpha)} + b_4^{(\alpha)}]. \end{aligned}$$

Now let us write

$$\begin{aligned} A_\alpha &= [a_1^{(\alpha)} + b_1^{(\alpha)}, a_2 + b_2], \\ B_\alpha &= [a_1^{(\alpha)} + b_3, a_2 + b_4^{(\alpha)}], \\ C_\alpha &= [a_3 + b_1^{(\alpha)}, a_4^{(\alpha)} + b_2], \\ D_\alpha &= [a_3 + b_3, a_4^{(\alpha)} + b_4^{(\alpha)}]. \end{aligned}$$

In addition, let us impose $B_\alpha \cap C_\alpha \neq \varnothing$, that is,

$$(a_2 + b_4^{(\alpha)}) \geq (a_3 + b_1^{(\alpha)}),$$

and

$$a_4^{(\alpha)} + b_2 \geq a_1^{(\alpha)} + b_3,$$

or

$$a_3 - b_2 \leq b_4^{(\alpha)} - b_1^{(\alpha)},$$

and

$$b_3 - b_2 \leq a_4^{(\alpha)} - a_1^{(\alpha)}. \qquad \mathbf{(4.45)}$$

Also, let $A_\alpha \cap B_\alpha \neq \varnothing$, or

$$a_2 + b_2 \geq a_1^{(\alpha)} + b_3, \qquad \mathbf{(4.46)}$$

and

$$a_2 + b_4^{(\alpha)} \geq a_1^{(\alpha)} + b_1^{(\alpha)}. \qquad \mathbf{(4.47)}$$

We can now eliminate equation (4.47) because $b_2 \leq b_4^{(\alpha)}$ and $b_3 \geq b_1^{(\alpha)}$. We then obtain from (4.46)

$$b_3 - b_2 \leq a_2 - a_1^{(\alpha)}. \qquad \mathbf{(4.48)}$$

Let us also impose $C_\alpha \cap D_\alpha = \varnothing$. Therefore

$$a_4^{(\alpha)} + b_2 \geq a_3 + b_3 \qquad \mathbf{(4.49)}$$

and

$$a_4^{(\alpha)} + b_4^{(\alpha)} \geq a_3 + b_1^{(\alpha)}.$$

Under these conditions we can eliminate (4.49), obtaining the result

$$b_3 - b_2 \leq a_4^{(\alpha)} - a_3. \qquad \mathbf{(4.50)}$$

Let us assemble (4.45), (4.48) and (4.50)

$$a_3 - a_2 \leq b_4^{(\alpha)} - b_1^{(\alpha)}, \qquad \mathbf{(4.51)}$$

$$b_3 - b_2 \leq a_2 - a_1^{(\alpha)}, \qquad \mathbf{(4.52)}$$

and

$$b_3 - b_2 \leq a_4^{(\alpha)} - a_3. \qquad \mathbf{(4.53)}$$

We see that (4.52) and (4.53) are more restrictive than (4.51). It follows then that

$$b_3 - b_2 \leq a_2 - a_1^{(\alpha)}$$

and

$$b_3 - b_2 \leq a_4^{(\alpha)} - a_3.$$

Last, by the symmetry that exists between *a* and *b*, we obtain[6]

$$a_3 - a_2 \leq b_2 - b_1^{(\alpha)} \qquad \mathbf{(4.54)}$$

and

$$a_3 - a_2 \leq b_4^{(\alpha)} - b_3. \tag{4.55}$$

Remark

Let us see what happens to Theorem 4.2 if **Y**, or **X**, does not result in a hiatus. We found that Theorem 4.2 is valid, and in this case we replace conditions (4.54) and (4.55) by

$$a_3 - a_2 \leq b_4^{(\alpha)} - b_1^{(\alpha)}.$$

Since the hiatus disappears, index 4 changes to index 2, and we have

$$a_3 - a_2 \leq b_2^{(\alpha)} - b_1^{(\alpha)},$$

which again gives (4.39).

EXAMPLE 4.19

Let us examine the two fuzzy numbers **X** and **Y** with the hiatus shown in Figure 4.45(*A*). From conditions (4.41) and (4.42), or (4.43) and (4.44), we have for $\alpha = 0.6$

$$a_1^{(\alpha)} = 7.6, \qquad a_2 = 8.8, \qquad a_3 = 10.8, \qquad a_4^{(\alpha)} = 14,$$

and

$$b_2 = 23.4, \qquad b_3 = 24.4.$$

We see that conditions (4.41) and (4.42) are satisfied. Thus

$$a_2 - a_1^{(\alpha)} = 8.8 - 7.6 = 1.2, \qquad b_3 - b_2 = 1,$$

and

$$a_4^{(\alpha)} - a_3 = 14 - 10.8 = 3.2.$$

The sum **X** + **Y** therefore does not have a hiatus. The result is shown in Figure 4.45(*B*).

THEOREM 4.3

Suppose that $\mathbf{X} \subset R$ is a fuzzy number with a hiatus. Then **X** (+) **X** and **X** (+) **X** (+) **X** $\cdots$ will be fuzzy numbers that are almost convex, but never fuzzy numbers; that is, they will never be convex fuzzy subsets if the number of the sum is finite.

Proof:

From (4.41) and (4.42), or (4.43) and (4.45), we find that the conditions

$$a_2 - a_1^{(\alpha)} \geq a_3 - a_2$$

and

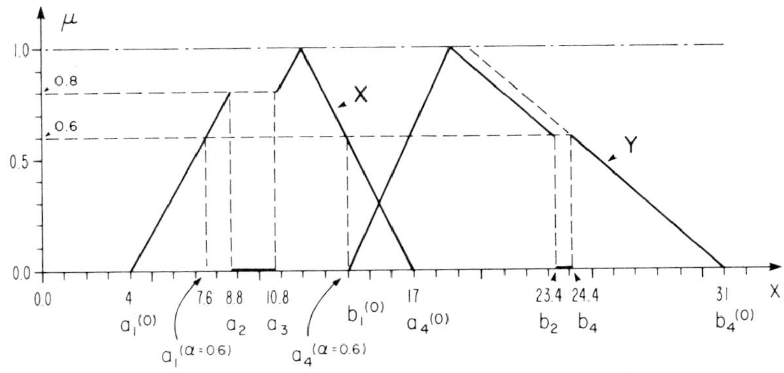

Figure 4.45(*A*) Fuzzy numbers **X** and **Y** with hiatus (Example 4.19).

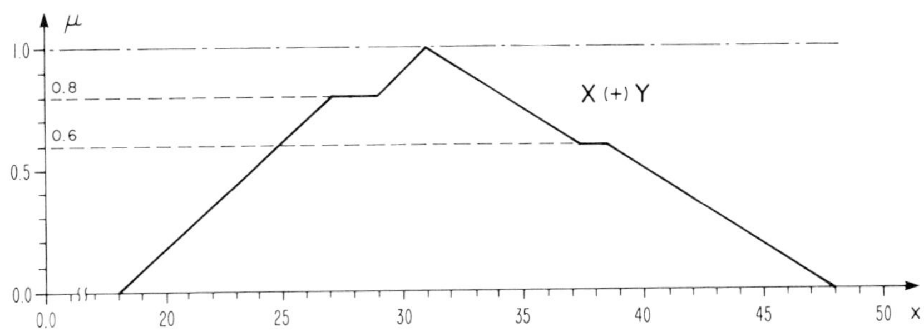

Figure 4.45(*B*) Sum of **X** and **Y** without a hiatus (Example 4.19).

$$a_4^{(\alpha)} - a_3 \geq a_3 - a_2$$

cannot be satisfied because there will be always an $\alpha \leq u_{\mathbf{X}}(a_2)$ from which one of these two conditions will not be satisfied.

If we compare Figures 4.46(*A*) and 4.46(*B*), it is readily apparent what happens in this situation. The same is true if we compare Figures 4.46(*C*) and 4.46(*D*).

If the hiatus is sufficiently small, the gap in **X** (+) **X** is reduced, but is never disappears. Similarly, if we compute the following sums from a fuzzy number **X** that has a hiatus that is small enough,

$$2 \cdot \mathbf{X} = \mathbf{X}\ (+)\ \mathbf{X},$$

and

$$3 \cdot \mathbf{X} = \mathbf{X}\ (+)\ \mathbf{X}\ (+)\ \mathbf{X},$$

$$n \cdot \mathbf{X} = (n-1)\ \mathbf{X}\ (+)\ \mathbf{X},$$

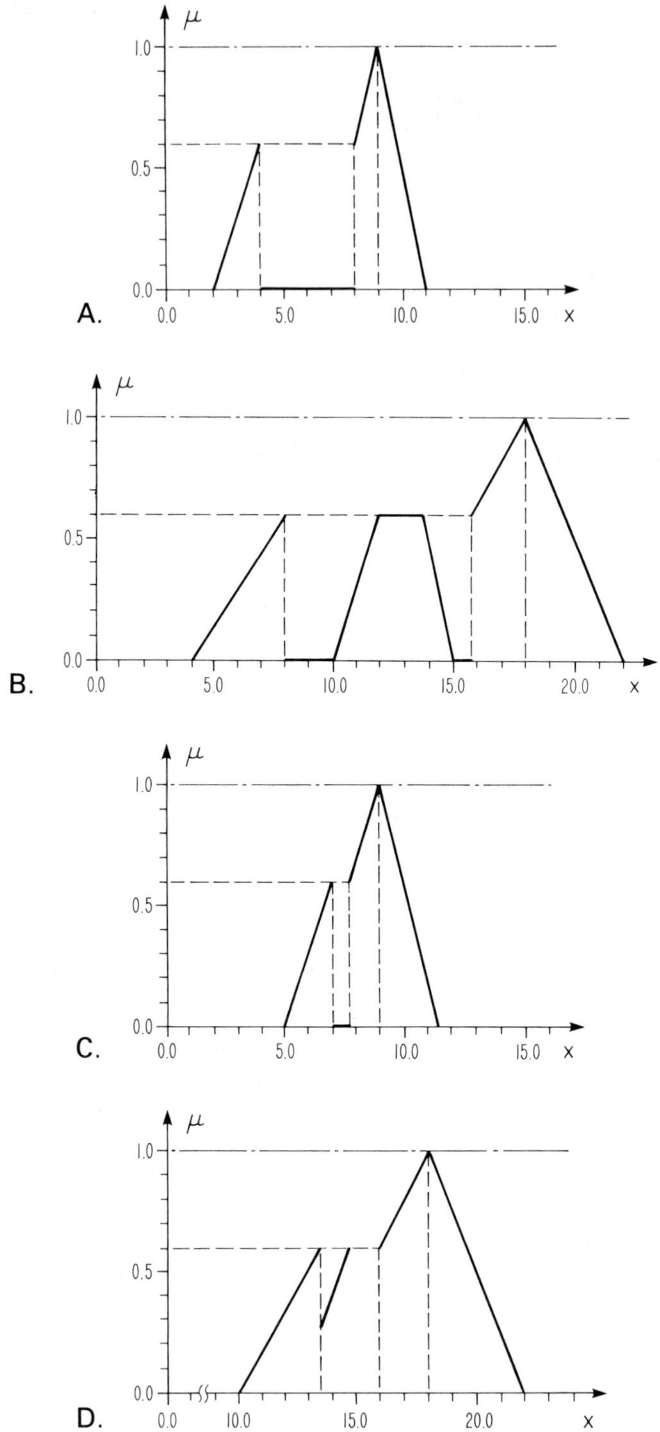

Figure 4.46 Fuzzy numbers with different hiatus.

we can prove that the sum tends to a fuzzy number, when $n \rightarrow \infty$, without a gap.

If n is finite, a gap will always remain present, no matter how small.

Bidirected Fuzzy Numbers

We now discuss a concept called the *bidirected fuzzy number.* Suppose an observer defines a fuzzy number in R, and in a particular situation the observer thinks of increasing the value of $x \in R$ and finds another fuzzy number in R when the value of $x \in R$ is decreasing. In this situation let us denote the bifuzzy number when x is increasing by $\mathbf{A}_{\rightarrow}$ and the fuzzy number when x is decreasing by $\mathbf{A}_{\leftarrow}$. Then the couplet

$$(\mathbf{A}_{\rightarrow}, \mathbf{A}_{\leftarrow})$$

will represent a bidirected fuzzy number. This situation is shown in Figure 4.47. We note that elements of the couplet may or may not have the same shape. However, both are convex and normal fuzzy numbers.

This concept can be generalized in the following way. Assume two fuzzy numbers $\mathbf{A} \subset R$ and $\mathbf{B} \subset R$. Starting with $\mu_{\mathbf{A}}(x)$ when $\mu_{\mathbf{A}}(x) = \alpha_1$, the observer passes into $\mu_{\mathbf{B}}(x)$ and remains there until he reaches $\mu_{\mathbf{B}}(x) = \alpha_2$, when he gets into $\mu_{\mathbf{A}}(x)$, where he remains on the curve. Figure 4.48 gives an example of this situation. However, this is

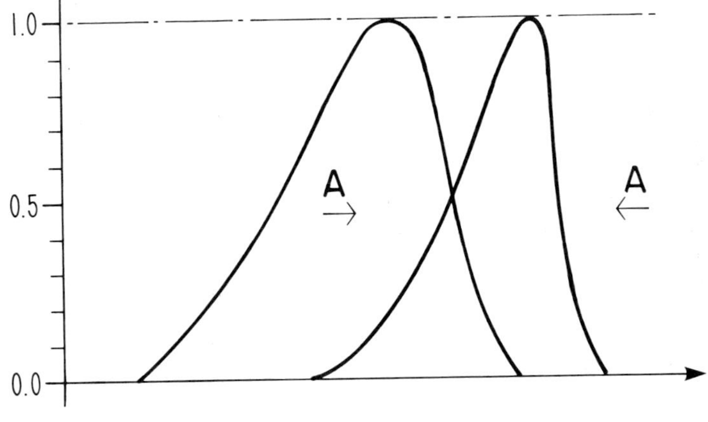

Figure 4.47 A bidirected fuzzy number.

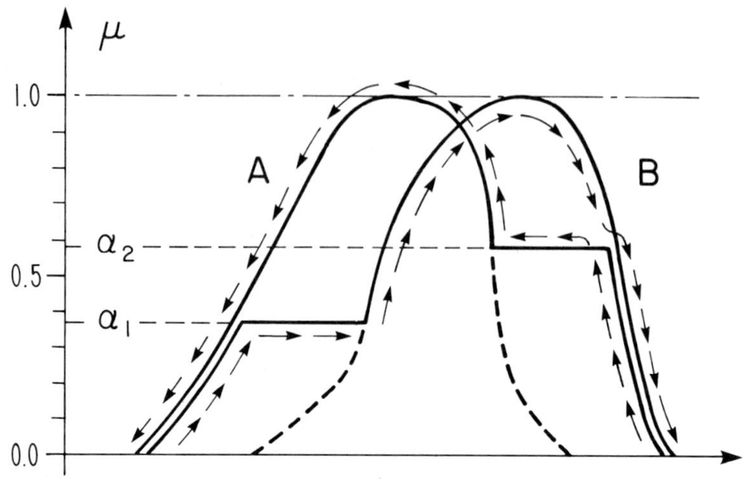

Figure 4.48 Two fuzzy numbers $\mu_{\mathbf{A}}(x)$ and $\mu_{\mathbf{B}}(x)$ with bidirectional motion.

one of many such motions. For example, if we take $\alpha_1 = \alpha_2 = 0$ we find again the concept of a bidirected fuzzy number.

We now conclude this chapter but the concepts introduced here may have several generalizations that may have some important practical applications. One of these is threshold, which must be considered in many applications and can be very useful. We do not, however, give the complete theory of such concepts here since this book is an introduction to this field. We encourage you to extend these concepts and theories further.

NOTES

1. ι is the Greek letter iota, corresponding to the *i* of the Latin alphabet.
2. The concept of possibility was introduced by L. A. Zadeh[2]; it is an important and powerful concept for the valuation of fuzzy subsets. A. Kaufmann[3] introduced the agreement index, which appears to be useful for fuzzy numbers.
3. $\alpha^*(x)$ is the inverse function of $x^*(\alpha)$.
4. This theory is the work of the very well known French mathematician Thom.
5. The use of the symbols + and − on ψ is clear. ψ^+ corresponds to positive motion from $-\infty$ to $+\infty$, and ψ^- to negative motion from $+\infty$ to $-\infty$.
6. To check $A_\alpha \cap D_\alpha \neq \varnothing$, $C_\alpha \cap A_\alpha = \varnothing$ and $B_\alpha \cap D_\alpha \neq \varnothing$.

REFERENCES

[1] D. Dubois and H. Prade, 1980, *Fuzzy Sets and Systems: Theory and Applications*, Academic Press, New York.

[2] L. A. Zadeh, 1968, Probability Measures of Fuzzy Events, *Journal of Mathematical Analysis and Applications*, **10**:421–427.

[3] A. Kaufmann, 1975, *Introduction to the Theory of Fuzzy Subsets Vol. 1*, Academic Press, New York.

[4] R. Thom, 1975, *Structural Stability and Morphogenesis*, W. A. Benjamin, Massachusetts.

[5] H. Noguchi, 1973, *Catastrophe Theory*, Kodansha Publisher, Tokyo.

APPENDIX A

THE MULTIPLICATION OF INTERVALS OF CONFIDENCE IN *R*

In Section 1.2 the multiplication of the interval of confidence was omitted when these segments belong to R instead of R^+. We present a more general case now.

The Table A.1 gives the outcome that is related to

$$A\,(\cdot)\,B = [a_1, a_2]\,(\cdot)\,[b_1, b_2], \qquad A, B, \subset R$$

Table A.1

	A	B	$A(\cdot)B = [a_1, a_2](\cdot)[b_1, b_2]$
1.	$0 \le a_1 \le a_2$	$0 \le b_1 \le b_2$	$[a_1 \cdot b_1, a_2 \cdot b_2]$
2.	$0 \le a_1 \le a_2$	$b_1 \le 0 \le b_2$	$[a_2 \cdot b_1, a_2 \cdot b_2]$
3.	$0 \le a_1 \le a_2$	$b_1 \le b_2 \le 0$	$[a_2 \cdot b_1, a_1 \cdot b_2]$
4.	$a_1 \le 0 \le a_2$	$0 \le b_1 \le b_2$	$[a_1 \cdot b_2, a_2 \cdot b_2]$
5.	$a_1 \le 0 \le a_2$	$b_1 \le 0 \le b_2$	$[\min(a_1 \cdot b_2, a_2 \cdot b_1), \max(a_1 \cdot b_1, a_2 \cdot b_2)$
6.	$a_1 \le 0 \le a_2$	$b_1 \le b_2 \le 0$	$[a_2 \cdot b_1, a_1 \cdot b_1]$
7.	$a_1 \le a_2 \le 0$	$0 \le b_1 \le b_2$	$[a_1 \cdot b_2, a_2 \cdot b_1]$
8.	$a_1 \le a_2 \le 0$	$b_1 \le 0 \le b_2$	$[a_1 \cdot b_2, a_1 \cdot b_1]$
9.	$a_1 \le a_2 \le 0$	$b_1 \le b_2 \le 0$	$[a_2 \cdot b_2, a_1 \cdot b_1]$

and the proof follows.

Proof:

1. $0 \le a_1 \le a_2,\ 0 \le b_1 \le b_2,$

$$A = [a_1, a_2], \qquad B = [b_1, b_2],$$
$$A\ (\cdot)\ B = [a_1, a_2]\ (\cdot)\ [b_1, b_2],$$
$$= [a_1 \cdot b_1, a_2 \cdot b_2].$$

2. $0 \le a_1 \le a_2,\ b_1 \le 0 \le b_2,$

$$A = [a_1, a_2], \qquad B = [b_1, 0] \cup [0, b_2],$$
$$A\ (\cdot)\ B = [a_1, a_2]\ (\cdot)\ ([b_1, 0] \cup [0, b_2]),$$
$$= [a_2 \cdot b_1, 0] \cup [0, a_2 \cdot b_2],$$
$$= [a_2 \cdot b_1, a_2 \cdot b_2].$$

3. $0 \le a_1 \le a_2,\ b_1 \le b_2 \le 0,$

$$A = [a_1, a_2], \qquad B = [b_1, b_2],$$
$$A\ (\cdot)\ B = [a_2 \cdot b_1, a_1 \cdot b_2].$$

4. $a_1 \le 0 \le a_2,\ 0 \le b_1 \le b_2,$

$$A = [a_1, 0] \cup [0, a_2], \qquad B = [b_1, b_2],$$
$$A\ (\cdot)\ B = ([a_1, 0] \cup [0, a_2])\ (\cdot)\ [b_1, b_2],$$
$$= ([a_1, 0]\ (\cdot)\ [b_1, b_2] \cup ([0, a_2]\ (\cdot)\ [b_1, b_2]),$$
$$= [a_1 \cdot b_2, 0] \cup [0, a_2 \cdot b_2],$$
$$= [a_1 \cdot b_2, a_2 \cdot b_2].$$

5. $a_1 \le 0 \le a_2,\ b_1 \le 0 \le b_2,$

$$A = [a_1, 0] \cup [0, a_2], \qquad B = [b_1, 0] \cup [0, b_2],$$
$$A\ (\cdot)\ B = ([a_1, 0]\ (\cdot)\ [b_1, 0]) \cup ([a_1, 0]\ (\cdot)\ [0, b_2]),$$
$$\cup\ ([0, a_2]\ (\cdot)\ [b_1, 0]) \cup ([0, a_2]\ (\cdot)\ [0, b_2]),$$
$$= [0, a_1 \cdot b_1] \cup [a_1 \cdot b_2, 0] \cup [a_2 \cdot b_1, 0] \cup [0, a_2 \cdot b_2],$$
$$= [\min\,(a_1 \cdot b_2, a_2 \cdot b_1), \max\,(a_1 \cdot b_1, a_2 \cdot b_2)].$$

6. $a_1 \le 0 \le a_2,\ b_1 \le b_2 \le 0,$

$$A = [a_1, 0] \cup [0, a_2], \qquad B = [b_1, b_2],$$
$$\begin{aligned} A\,(\cdot)\,B &= ([a_1, 0] \cup [0, a_2])\,(\cdot)\,[b_1, b_2], \\ &= [0, a_1 \cdot b_1] \cup [a_2 \cdot b_1, 0], \\ &= [a_2 \cdot b_1, a_1 \cdot b_1]. \end{aligned}$$

7. $a_1 \le a_2 \le 0,\ 0 \le b_1 \le b_2,$

$$A = [a_1, a_2], \qquad B = [b_1, b_2],$$
$$A\,(\cdot)\,B = [a_1 \cdot b_2, a_2 \cdot b_1].$$

8. $a_1 \le a_2 \le 0,\ b_1 \le 0 \le b_2,$

$$A = [a_1, a_2],\ B = [b_1, 0] \cup [0, b_2],$$
$$\begin{aligned} A\,(\cdot)\,B &= [0, a_1 \cdot b_1] \cup [a_1 \cdot b_2, 0], \\ &= [a_1 \cdot b_2, a_1 \cdot b_1]. \end{aligned}$$

9. $a_1 \le a_2 \le 0,\ b_1 \le b_2 \le 0,$

$$A = [a_1, a_2], \qquad B = [b_1, b_2],$$
$$A\,(\cdot)\,B = [a_2 \cdot b_2, a_1 \cdot b_1].$$

The above results are summarized in the following equation:

A, **B** $\subset R$:

$$A\,(\cdot)\,B = [\min(a_1 \cdot b_1, a_1 \cdot b_2, a_2 \cdot b_1, a_2 \cdot b_2), \max(a_1 \cdot b_1, a_1 \cdot b_2, a_2 \cdot b_1, a_2 \cdot b_2)]. \qquad \textbf{(A.1)}$$

Equation (A.1) can be generalized for three intervals of confidence.

$$\begin{aligned} (A\,(\cdot)\,B)\,(\cdot)\,C = [&\min(a_1 \cdot b_1 \cdot c_1, a_1 \cdot b_1 \cdot c_2, a_1 \cdot b_2 \cdot c_1, a_1 \cdot b_2 \cdot c_2, \\ &a_2 \cdot b_1 \cdot c_1, a_2 \cdot b_1 \cdot c_2, a_2 \cdot b_2 \cdot c_1, a_2 \cdot b_2 \cdot c_2), \max(a_1 \cdot \\ &b_1 \cdot c_1, a_1 \cdot b_1 \cdot c_2, a_1 \cdot b_2 \cdot c_1, a_1 \cdot b_2 \cdot c_2, a_2 \cdot b_1 \cdot c_1, \\ &a_2 \cdot b_1 \cdot c_2, a_2 \cdot b_2 \cdot c_1, a_2 \cdot b_2 \cdot c_2)]. \end{aligned}$$

We will obtain exactly the same result if we compute $A\,(\cdot)\,(B\,(\cdot)\,C)$, which proves associativity for $(\cdot)$ in R.[1]

The use of the intervals of confidence in R^+ leads to very simple computations, but the computations are more complicated when the intervals of confidence are in R.

The preceding remarks are also valid for fuzzy numbers in R.

In Table A.2, the results are given for the triple product.

INVERSE IN *R*

1. $0 < a_1 \leq a_2$: $[a_1, a_2]^{-1} = [1/a_2, 1/a_1]$,

2. $0 = a_1 < a_2$: $[a_1, a_2]^{-1} = [1/a_2, \infty]$,

3. $0 = a_1 = a_2$: only one point which moves to ∞,

4. $a_1 \leq a_2 < 0$: $[a_1, a_2]^{-1} = [1/a_2, 1/a_1]$,

5. $a_1 < a_2 = 0$: $[a_1, a_2]^{-1} = (-\infty, 1/a_1]$,

6. $a_1 \leq 0 \leq a_2$: $[a_1, a_2]^{-1} = [-\infty, 1/a_1] \cup [1/a_2, \infty]$. **(A.2)**

We do not have an interval of confidence in (A.2). This is due to the rejection of the inverse and the division in *R*.

MULTIPLICATION BY $k \in R$

$$k \geq 0: k \cdot [a_1, a_2] = [ka_1, ka_2],$$
$$k \leq 0: k \cdot [a_1, a_2] = [ka_2, ka_1].$$

NONDISTRIBUTIVITY IN *R*

Using a counter example we see that

$$(A\ (+)\ B)\ (\cdot)\ C \neq (A\ (\cdot)\ C\ (+)\ (B\ (\cdot)\ C).$$

Let

$$([3, 5]\ (+)\ [-2, 4]\ (\cdot)\ [2, 6] = [1, 9]\ (\cdot)\ [2, 6]$$
$$= [2, 54].$$

In addition, let

$$([3, 5]\ (\cdot)\ [2, 6])\ (+)\ ([-2, 4]\ (\cdot)\ [2, 6]) = [6, 30]\ (+)\ [-12, 24]$$
$$= [-6, 54].$$

The other type of distributivity is also not applicable.

$$(A\ (\cdot)\ B)\ (+)\ C \neq (A\ (+)\ C)\ (\cdot)\ (B\ (+)\ C).$$

This can be seen from Equation (1.54) the nondistributivity in R^+.

Table A.2

	$A=[a_1, a_2]$	$B=[b_1, b_2]$	$C=[c_1, c_2]$	$A(\cdot)B(\cdot)C$
1.	$0 \le a_1 \le a_2$	$0 \le b_1 \le b_2$	$0 \le c_1 \le c_2$	$[a_1b_1c_1, a_2b_2c_2]$
2.	$0 \le a_1 \le a_2$	$0 \le b_1 \le b_2$	$c_1 \le 0 \le c_2$	$[a_2b_2c_1, a_2b_2c_2]$
3.	$0 \le a_1 \le a_2$	$0 \le b_1 \le b_2$	$c_1 \le c_2 \le 0$	$[a_2b_2c_1, a_1b_1c_2]$
4.	$0 \le a_1 \le a_2$	$b_1 \le 0 \le b_2$	$0 \le c_1 \le c_2$	$[a_2b_1c_2, a_2b_2c_2]$
5.	$0 \le a_1 \le a_2$	$b_1 \le 0 \le b_2$	$c_1 \le 0 \le c_2$	$[\min(a_2b_1c_2, a_2b_2c_1), \max(a_2b_1c_1, a_2b_2c_2)]$
6.	$0 \le a_1 \le a_2$	$b_1 \le 0 \le b_2$	$c_1 \le c_2 \le 0$	$[a_2b_2c_1, a_2b_1c_1]$
7.	$0 \le a_1 \le a_2$	$b_1 \le b_2 \le 0$	$0 \le c_1 \le c_2$	$[a_2b_1c_2, a_1b_2c_1]$
8.	$0 \le a_1 \le a_2$	$b_1 \le b_2 \le 0$	$c_1 \le 0 \le c_2$	$[a_2b_1c_2, a_2b_1c_1]$
9.	$0 \le a_1 \le a_2$	$b_1 \le b_2 \le 0$	$c_1 \le c_2 \le 0$	$[a_1b_2c_2, a_2b_1c_1]$
10.	$a_1 \le 0 \le a_2$	$0 \le b_1 \le b_2$	$0 \le c_1 \le c_2$	$[a_1b_2c_2, a_2b_2c_2]$
11.	$a_1 \le 0 \le a_2$	$0 \le b_1 \le b_2$	$c_1 \le 0 \le c_2$	$[\min(a_1b_2c_2, a_2b_2c_1), \max(a_1b_2c_1, a_2b_2c_2)]$
12.	$a_1 \le 0 \le a_2$	$0 \le b_1 \le b_2$	$c_1 \le c_2 \le 0$	$[a_2b_2c_1, a_1b_2c_1]$
13.	$a_1 \le 0 \le a_2$	$b_1 \le 0 \le b_2$	$0 \le c_1 \le c_2$	$[\min(a_1b_2c_2, a_2b_1c_2), \max(a_1b_1c_2, a_2b_2c_2)]$
14.	$a_1 \le 0 \le a_2$	$b_1 \le 0 \le b_2$	$c_1 \le 0 \le c_2$	$[\min(a_1b_2c_2, a_2b_1c_2, a_1b_1c_1, a_2b_2c_1),$ $\max(a_1b_2c_1, a_2b_1c_1, a_1b_1c_2, a_2b_2c_2)]$

(continued)

Table A.2 *(continued)*

	$A = [a_1, a_2]$	$B = [b_1, b_2]$	$C = [c_1, c_2]$	$A (\cdot) B (\cdot) C$
15.	$a_1 \le 0 \le a_2$	$b_1 \le 0 \le b_2$	$c_1 \le c_2 \le 0$	$[\min(a_1b_1c_1, a_2b_2c_1), \max(a_1b_2c_1, a_2b_1c_1)]$
16.	$a_1 \le 0 \le a_2$	$b_1 \le b_2 \le 0$	$0 \le c_1 \le c_2$	$[a_2b_1c_2, a_1b_1c_2]$
17.	$a_1 \le 0 \le a_2$	$b_1 \le b_2 \le 0$	$c_1 \le 0 \le c_2$	$[\min(a_2b_1c_2, a_1b_1c_1), \max(a_2b_1c_1, a_1b_1c_2)]$
18.	$a_1 \le 0 \le a_2$	$b_1 \le b_2 \le 0$	$c_1 \le c_2 \le 0$	$[a_1b_1c_1, a_2b_1c_1]$
19.	$a_1 \le a_2 \le 0$	$0 \le b_1 \le b_2$	$0 \le c_1 \le c_2$	$[a_1b_2c_2, a_2b_1c_1]$
20.	$a_1 \le a_2 \le 0$	$0 \le b_1 \le b_2$	$c_1 \le 0 \le c_2$	$[a_1b_2c_2, a_1b_2c_1]$
21.	$a_1 \le a_2 \le 0$	$0 \le b_1 \le b_2$	$c_1 \le c_2 \le 0$	$[a_2b_1c_2, a_1b_2c_1]$
22.	$a_1 \le a_2 \le 0$	$b_1 \le 0 \le b_2$	$0 \le c_1 \le c_2$	$[a_1b_2c_2, a_1b_1c_2]$
23.	$a_1 \le a_2 \le 0$	$b_1 \le 0 \le b_2$	$c_1 \le 0 \le c_2$	$[\min(a_1b_2c_2, a_1b_1c_1), \max(a_1b_2c_1, a_1b_1c_2)]$
24.	$a_1 \le a_2 \le 0$	$b_1 \le 0 \le b_2$	$c_1 \le c_2 \le 0$	$[a_1b_1c_1, a_1b_2c_1]$
25.	$a_1 \le a_2 \le 0$	$b_1 \le b_2 \le 0$	$0 \le c_1 \le c_2$	$[a_2b_2c_1, a_1b_1c_2]$
26.	$a_1 \le a_2 \le 0$	$b_1 \le b_2 \le 0$	$c_1 \le 0 \le c_2$	$[a_1b_1c_1, a_1b_1c_2]$
27.	$a_1 \le a_2 \le 0$	$b_1 \le b_2 \le 0$	$c_1 \le c_2 \le 0$	$[a_1b_1c_1, a_2b_2c_2]$

NOTE

1. To prove associativity, we also use Table (A.2). We have:

   $([a_1, a_2]\,(\cdot)\,[b_1, b_2])\,(\cdot)\,[c_1, c_2]$. We then permute c and a
   $([c_1, c_2]\,(\cdot)\,[b_1, b_2])\,(\cdot)\,[a_1, a_2] = [a_1, a_2]\,(\cdot)$
   $([c_1, c_2]\,(\cdot)\,[b_1, b_2] = [a_1, a_2]\,(\cdot)\,([b_1, b_2]\,(\cdot)$
   $[c_1, c_2])$. Then, by permutation between a and c, we achieve the same results as in Table (A.2).

APPENDIX B

MULTIPLICATION OF FUZZY NUMBERS IN *R*

In Section 1.5 only the multiplication of fuzzy numbers in R^+ was considered. We now prove that a product of two fuzzy numbers in R gives a fuzzy number also in R. It is a trivial matter to prove the normality of the product, but this is not the case for convexity. To prove convexity, we need to examine several combinational cases.

In order to prove the convexity of **A** (·) **B** in R we write

$$A_\alpha = [a_1^{(\alpha)}, a_2^{(\alpha)}],$$

$$B_\alpha = [b_1^{(\alpha)}, b_2^{(\alpha)}],$$

$$A_{\alpha'} = [a_1^{(\alpha')}, a_2^{(\alpha')}],$$

and

$$B_{a'} = [b_1^{(\alpha')}, b_2^{(\alpha')}].$$

We now prove that

$\forall\ \alpha, \alpha' \in [0, 1]$, and **A**, **B** $\subset R$:

$$(\alpha' \geq \alpha) \Rightarrow (A_\alpha\ (\cdot)\ B_\alpha \supset A_\alpha{}'\ (\cdot)\ B_\alpha{}'.$$

The condition of monotonicity implies convexity, because the product of the two segments in R gives a segment in R (see Table A.1). Let us now check this monotonicity for the operator (·).

For purposes of simplicity, we write

$$[a_1, a_2] \text{ instead of } [a_1^{(\alpha)}, a_2^{(\alpha)}], \qquad \textbf{(B.1)}$$

$$[a'_1, a'_2] \text{ instead of } [a_1^{(\alpha')}, a_2^{(\alpha')}], \qquad \textbf{(B.2)}$$

$$[b_1, b_2] \text{ instead of } [b_1^{(\alpha)}, b_2^{(\alpha)}], \qquad \textbf{(B.3)}$$

and

$$[b'_1, b'_2] \text{ instead of } [b_1^{(\alpha')}, b_2^{(\alpha')}]. \qquad \textbf{(B.4)}$$

We also inspect the $5 \times 5 = 25$ different cases where segments (B.1–B.4) can be positioned. We use Table (A.1) for purposes of computation:

The property that we sought is proved in Table B.1. Thus,

$\forall$ **A**, **B** $\subset R$:

$$(\mathbf{A} \text{ and } \mathbf{B} \text{ convex}) \Rightarrow (\mathbf{A}\ (\cdot)\ \mathbf{B} \text{ convex}),$$

and also

$$(\mathbf{A} \text{ and } \mathbf{B} \text{ normal}) \Rightarrow (\mathbf{A}\ (\cdot)\ \mathbf{B} \text{ normal}).$$

Now we can write that

$$(\mathbf{A} \text{ and } \mathbf{B} \text{ are fuzzy numbers in } R) \Rightarrow (\mathbf{A}\ (\cdot)\ \mathbf{B} \text{ is a fuzzy number in } R).$$

It is also important to recall that in R all distributivities are not available except in some particular cases. For example,

$$([-3, 2]\ (+)\ [-4, 5])\ (\cdot)\ [-6, 1] = [1, 7]\ (\cdot)\ [-6, 1] = [-42, 7],$$

$$([-3, 2]\ (\cdot)\ [-6, 1])\ (+)\ ([4, 5]\ (\cdot)\ [-6, 1] = [-12, 18]\ (+)\ [-30, 5]$$
$$= [-42, 23].$$

Of course, the other distributivity is not true in R because it is not in R^+ (see Eq. 1.52a).

We have proven that fuzzy numbers in R are associative for $(\cdot)$ and the netural is the ordinary number 1. Therefore the structure is a commutative monoid.

We now study an example of the multiplication of two fuzzy numbers in R.

EXAMPLE B.1

In Figure B.1 two fuzzy numbers are defined by

$$A_\alpha = [a_1^{(\alpha)}, a_2^{(\alpha)}] = [\alpha - 5, -7\alpha + 3], \qquad \textbf{(B.5)}$$

and

$$B_\alpha = [b_1^{(\alpha)}, b_2^{(\alpha)}] = [5\alpha - 3, -2\alpha + 4]. \qquad \textbf{(B.6)}$$

Table B.1

	Conditions	Result
1.	$0 \le a_1 \le a'_1 \le a'_2 \le a_2$ $0 \le b_1 \le b'_1 \le b'_2 \le b_2$	$[a_1 b_1, a_2 b_2] \supset [a'_1 b'_1, a'_2 b'_2]$ indeed, $a_1 b_1 \le a'_1 b'_1$ $\quad$ $a_2 b_2 \ge a'_2 b'_2$
2.	$0 \le a_1 \le a'_1 \le a'_2 \le a_2$ $b_1 \le 0 \le b'_1 \le b'_2 \le b_2$	$[a_2 b_1, a_2 b_2] \supset [a'_1 b'_1, a'_2 b'_2]$ indeed, $a_2 b_1 \le a'_1 b'_1$ $\quad$ $a_2 b_2 \ge a'_2 b'_2$ (neg.) (pos.) $\quad$ (by hypothesis)
3.	$0 \le a_1 \le a'_1 \le a'_2 \le a_2$ $b_1 \le b'_1 \le 0 \le b'_2 \le b_2$	$[a_2 b_1, a_2 b_2] \supset [a'_2 b'_1, a'_2 b'_2]$ indeed, $a_2 b_1 \le a'_2 b'_1$ $\quad$ $a_2 b_2 \ge a'_2 b'_2$ (by hypothesis) $\quad$ (by hypothesis) (b_1 and b'_1 are negative)
4.	$0 \le a_1 \le a'_1 \le a'_2 \le a_2$ $b_1 \le b'_1 \le b'_2 \le 0 \le b_2$	$[a_2 b_1, a_2 b_2] \supset [a'_2 b'_1, a'_1 b'_2]$ indeed, $a_2 b_1 \le a'_2 b'_1$ $\quad$ $a_2 b_2 \ge a'_1 b'_2$ because $0 \le a'_2 \le a_2$ $\quad$ (pos.) (neg.) and $b_1 \le b'_1 \le 0$

(continued)

Table B.1 (*continued*)

	Condition	Result
5.	$0 \le a_1 \le a'_1 \le a'_2 \le a_2$	$[a_2 b_2, a_1 b_2] \supset [a'_2 b'_1, a'_1 b'_2]$
	$b_1 \le b'_1 \le b'_2 \le b_2 \le 0$	$a_2 b_1 \le a'_2 b_1 \le a'_2 b'_1$: by hypothesis
		$a_1 b_2 \le a'_1 b_2 \ge a'_1 b'_2$: by hypothesis
6.	$a_1 \le 0 \le a'_1 \le a'_2 \le a_2$	$[a_1 b_2, a_2 b_2] \supset [a'_1 b'_1, a'_2 b'_2]$
	$0 \le b_1 \le b'_1 \le b'_2 \le b_2$	$a_1 b_2 \ge a'_1 b'_1$ (neg.) (pos.) $\quad a_2 b_2 \ge a'_2 b'_2$ (by hypothesis)
7.	$a_1 \le 0 \le a'_1 \le a'_2 \le a_2$	$[\min(a_1 b_2, a_2 b_1), \max(a_1 b_1, a_2 b_2)] \supset [a'_1 b'_1, a'_2 b'_2]$
	$b_1 \le 0 \le b'_1 \le b'_2 \le b_2$	If $\min(a_1 b_2, a_2 b_1) = a_1 b_2$ then $a_1 b_2 \le a'_1 b'_1$ (neg.) (pos.)
		If $\min(a_1 b_2, a_2 b_1) = a_2 b_1$ then $a_2 b_1 \le a'_1 b'_1$ (neg.) (pos.)
		If $\max(a_1 b_1, a_2 b_2) = a_1 b_1$ then $a_1 b_1 \ge a_2 b_2 \ge a'_2 b'_2$ (by hypothesis)
		If $\max(a_1 b_1, a_2 b_2) = a_2 b_2$ then $a_2 b_2 \ge a'_2 b'_2$ (by hypothesis)

8.	$a_1 \le 0 \le a'_1 \le a'_2 \le a_2$ $b_1 \le b'_1 \le 0 \le b'_2 \le b_2$	$[\min(a_1 b_2, a_2 b_1), \max(a_1 b_1, a_2 b_2)]$ $\supset [a'_2 b'_1, a'_2 b'_2]$ If $\min(a_1 b_2, a_2 b_1) = a_1 b_2$, then $a_1 b_2 \le a_2 b_1 \le a'_2 b'_1$ because inequality $a_2 \ge a'_2$ is reinforced by $b_1 \le b'_1$ If $\min(a_1 b_2, a_2 b_1) = a_2 b_1$, then $a_2 b_1 \le a'_2 b'_1$ because $b_1 \le b'_1$ is reinforced by $a_2 \ge a'_2$ If $\max(a_1 b_1, a_2 b_2) = a_1 b_1$, then $a_1 b_1 \ge a_2 b_2 \ge a'_2 b'_2$ (by hypothesis) If $\max(a_1 b_1, a_2 b_2) = a_2 b_2$, then $a_2 b_2 \ge a'_2 b'_2$ (by hypothesis)
9.	$a_1 \le 0 \le a'_1 \le a'_2 \le a_2$ $b_1 \le b'_1 \le b'_2 \le 0 \le b_2$	$[\min(a_1 b_2, a_2 b_1), \max(a_1 b_1, a_2 b_2)]$ $\supset [a'_2 b'_1, a'_1 b'_2]$ If $\min(a_1 b_2, a_2 b_1) = a_1 b_2$, then $a_1 b_2 \le a_2 b_1 \le a'_2 b'_1$ If $\min(a_1 b_2, a_2 b_1) = a_2 b_1$, then $a_2 b_1 \le a'_2 b'_1$ If $\max(a_1 b_1, a_2 b_2) = a_1 b_1$, then $a_1 b_1 \ge a_2 b_2 \ge a'_1 b'_2$ (pos.) (pos.) (neg.)

(continued)

Table B.1 (*continued*)

	If $\max(a_1 b_1, a_2 b_2) = a_2 b_2$, then $a_2 b_2 \geq a_1' b_2'$ (pos.) (neg.)
10. $a_1 \leq 0 \leq a_1' \leq a_2' \leq a_2$ $b_1 \leq b_1' \leq b_2' \leq b_2 \leq 0$	$[a_2 b_1, a_1 b_1] \supset [a_2' b_1', a_1' b_2']$ $a_2 b_1 \leq a_2' b_1'$ because $b_1 \leq b_1'$ is reinforced by $a_2 \geq a_2'$ $a_1 b_1 \geq a_1' b_2'$ (pos.) (neg.)
11. $a_1 \leq a_1' \leq 0 \leq a_2' \leq a_2$ $0 \leq b_1 \leq b_1' \leq b_2' \leq b_2$	$[a_1 b_2, a_2 b_2] \supset [a_1' b_2', a_2' b_2']$ $a_1 b_2 \leq a_1' b_2'$ because $a_1 \leq a_1' \leq 0$ is reinforced by $0 \leq b_2' \leq b_2$ $a_2 b_2 \geq a_2' b_2'$ because $0 \leq a_2' \leq a_2$ is reinforced by $0 \leq b_2' \leq b_2$
12. $a_1 \leq a_1' \leq 0 \leq a_2' \leq a_2$ $b_1 \leq 0 \leq b_1' \leq b_2' \leq b_2$	$[\min(a_1 b_2, a_2 b_1), \max(a_1 b_1, a_2 b_2)]$ $\supset [a_1' b_2', a_2' b_2']$ If $\min(a_1 b_2, a_2 b_1) = a_1 b_2$, then $a_1 b_2 \leq a_1' b_2'$ If $\min(a_1 b_2, a_2 b_1) = a_2 b_1$, then $a_2 b_1 \leq a_1 b_2 \leq a_1' b_2'$ If $\max(a_1 b_1, a_2 b_2) = a_1 b_1$, then $a_1 b_1 \geq a_2 b_2 \geq a_2' b_2'$ If $\max(a_1 b_1, a_2 b_2) = a_2 b_2$, then $a_2 b_2 \geq a_2' b_2'$

13. $a_1 \le a'_1 \le 0 \le a'_2 \le a_2$

$b_1 \le b'_1 \le 0 \le b'_2 \le b_2$

$[\min(a_1 b_2, a_2 b_1), \max(a_1 b_1, a_2 b_2)]$

$\supset [\min(a'_1 b'_2, a'_2 b'_1), \max(a'_1 b'_1, a'_2 b'_2)]$

$\min(a_1 b_2, a_2 b_1) = a_1 b_2$

$\min(a'_1 b'_2, a'_2 b'_1) = a'_1 b'_2$

$a_1 b_2 \le a'_1 b'_2$ because $a_1 \le a'_1$ is reinforced by $b_2 \ge b'_2$

$\min(a_1 b_2, a_2 b_1) = a_1 b_2$

$\min(a'_1 b'_2, a'_2 b'_1) = a'_2 b'_1$

$a_1 b_2 \le a_2 b_1 \le a'_2 b'_1$ because $b_1 \le b'_1$ is reinforced by $a_2 \ge a'_2$

$\min(a_1 b_2, a_2 b_1) = a_2 b_1$

$\min(a'_1 b'_2, a'_2 b'_1) = a'_1 b'_2$

$a_2 b_1 \le a_1 b_2 \le a'_1 b'_2$ because $a_1 \le a'_1$ is reinforced by $b_2 \ge b'_2$

$\min(a_1 b_2, a_2 b_1) = a_2 b_1$

$\min(a'_1 b'_2, a'_2 b'_1) = a'_2 b'_1$

$a_2 b_1 \le a'_2 b'_1$ because $b_1 \le b'_1$ is reinforced by $a_2 \ge a'_2$

$\max(a_1 b_1, a_2 b_2) = a_1 b_1$

$\max(a'_1 b'_1, a'_2 b'_2) = a'_1 b'_1$

$a_1 b_1 \ge a'_1 b'_1$ because $a_1 \le a'_1 \le 0$ and $b_1 \le b'_1 \le 0$

(continued)

Table B.1 *(continued)*

$$\max(a_1 b_1, a_2 b_2) = a_1 b_1$$
$$\max(a'_1 b'_1, a'_2 b'_2) = a'_2 b'_2$$
$$a_1 b_1 \geq a_2 b_2 \geq a'_2 b'_2 \quad \text{because} \quad 0 \leq a'_2 \leq a_2 \quad \text{and} \quad 0 \leq b'_2 \leq b_2$$

$$\max(a_1 b_1, a_2 b_2) = a_2 b_2$$
$$\max(a'_1 b'_1, a'_2 b'_2) = a'_1 b'_1$$
$$a_2 b_2 \geq a_1 b_1 \geq a'_1 b'_1 \quad \text{because} \quad a_1 \leq a'_1 \leq 0 \quad \text{and} \quad b_1 \leq b'_1 \leq 0$$

$$\max(a_1 b_1, a_2 b_2) = a_2 b_2$$
$$\max(a'_1 b'_1, a'_2 b'_2) = a'_2 b'_2$$
$$a_2 b_2 \geq a'_2 b'_2 \quad \text{because} \quad 0 \leq a'_2 \leq a_2 \quad \text{and} \quad 0 \leq b'_2 \leq b_2$$

We do not have to go any further because by permutation between the *a* and the *b* we shall get similar results.

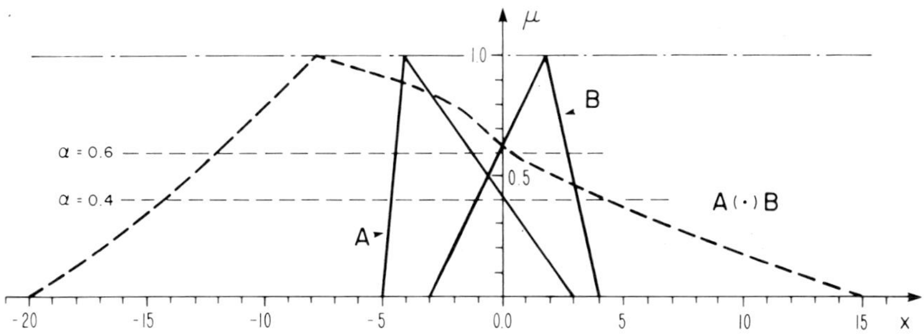

Figure B.1 Multiplication of two fuzzy numbers **A** and **B** in R.

From (B.5) and (B.6), and also from Figure B.1, we find three sections in the segment [0, 1] where α takes its various values in [0, 1]:

$$0 \le \alpha \le 0.4, \quad a_1^{(\alpha)} \le 0 \le a_2^{(\alpha)}, \quad b_1^{(\alpha)} \le 0 \le b_2^{(\alpha)},$$

$$0.4 \le \alpha \le 0.6, \quad a_1^{(\alpha)} \le a_2^{(\alpha)} \le 0, \quad b_1^{(\alpha)} \le 0 \le b_2^{(\alpha)},$$

and

$$0.6 \le \alpha \le 1, \quad a_1^{(\alpha)} \le a_2^{(\alpha)} \le 0, \quad 0 \le b_1^{(\alpha)} \le b_2^{(\alpha)}.$$

Using the results given in Table (A.1) we find that, for (1), $0 \le \alpha \le 0.4$

$$\begin{aligned} A_\alpha (\cdot) B_\alpha &= [\min (a_1^{(\alpha)} \cdot b_2^{(\alpha)}, a_2^{(\alpha)} \cdot b_1^{(\alpha)}), \\ &\quad \max (a_1^{(\alpha)} \cdot b_1^{(\alpha)}, a_2^{(\alpha)} \cdot b_2^{(\alpha)})] \\ &= [\min ((\alpha - 5)(-2\alpha + 4), (-7\alpha + 3)(5\alpha - 3)), \\ &\quad \max ((\alpha - 5)(5\alpha - 3), (-7\alpha + 3)(-2\alpha + 4))]. \end{aligned}$$

A computation based on second-degree equations gives

$$A_\alpha (\cdot) B_\alpha = [(\alpha - 5)(-2\alpha + 4), (\alpha - 5)(5\alpha - 3)]. \tag{B.7}$$

From (B.7) we have

$$\begin{aligned} A_0 (\cdot) B_0 &= [-20, 15], \\ A_{0.1} (\cdot) B_{0.1} &= [-18.62, 12.25], \\ A_{0.2} (\cdot) B_{0.2} &= [-17.28, 9.60], \\ A_{0.3} (\cdot) B_{0.3} &= [-16.16, 7.05], \\ A_{0.4} (\cdot) B_{0.4} &= [-14.72, 4.60]. \end{aligned}$$

For (2), $0.4 \leq \alpha \leq 0.6$

$$A_\alpha (\cdot) B_\alpha = [a_1^{(\alpha)} \cdot b_2^{(\alpha)}, a_1^{(\alpha)} \cdot b_1^{(\alpha)}]$$

$$= [(\alpha - 5)(-2\alpha + 4)\ (\alpha - 5)(5\alpha - 3)].$$

$$A_{0.4} (\cdot) B_{0.4} = [-14.72, 4.60],$$

$$A_{0.5} (\cdot) B_{0.5} = [-13.50, 2.25],$$

$$A_{0.6} (\cdot) B_{0.6} = [-12.32, 0].$$

For (3), $0.6 \leq \alpha \leq 1$

$$A_\alpha (\cdot) B_\alpha = [a_1^{(\alpha)} \cdot b_2^{(\alpha)}, a_2^{(\alpha)} \cdot b_1^{(\alpha)}]$$

$$= [(\alpha - 5)(-2\alpha + 4), (-7\alpha + 3)(5\alpha - 3)].$$

$$A_{0.6} (\cdot) B_{0.6} = [-12.32, 0],$$

$$A_{0.7} (\cdot) B_{0.7} = [-11.18, -0.95],$$

$$A_{0.8} (\cdot) B_{0.8} = [-10.08, -2.60],$$

$$A_{0.9} (\cdot) B_{0.9} = [-9.02, -4.95],$$

$$A_1 (\cdot) B_1 = [-8, -8] = -8.$$

The results are shown in Figure B.1.

APPENDIX C

CHANGE OF COORDINATES FOR A DOMAIN OF CONFIDENCE OF LEVEL α

We now consider changes of coordinates in R^2. It is useful to recall those equations that allow us to pass from a Cartesian system to another system by means of a simple rotation. In Figure C.1 we represent a point P_1 in R^2 and its coordinates (x_1, y_1) with regard to $x{-}0{-}y$, and its coordinates (x'_1, y'_1) with regard to $x'{-}0{-}y'$ after a rotation by β. This change of coordinates is shown in Figure C.1, and is given by

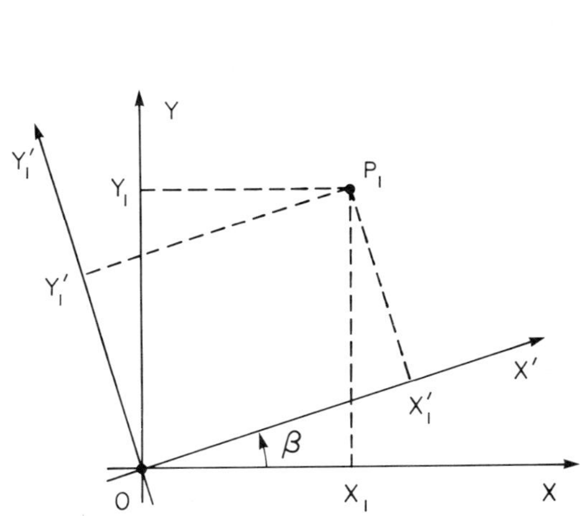

Figure C.1 Rotation of Cartesian coordinates in R^2.

$$x_1' = x_1 \cos\beta + y_1 \sin\beta,$$
$$y_1' = -x_1 \sin\beta + y_1 \cos\beta.$$

Conversely, we may write

$$x_1 = x_1' \cos\beta - y_1' \sin\beta,$$
$$y_1 = x_1' \sin\beta + y_1' \cos\beta.$$

Using matrices to represent these linear transformations, we have

$$\begin{bmatrix} \cos\beta & \sin\beta \\ -\sin\beta & \cos\beta \end{bmatrix} \cdot \begin{bmatrix} x_1 \\ y_1 \end{bmatrix} = \begin{bmatrix} x_1' \\ y_1' \end{bmatrix}.$$

$$\begin{bmatrix} \cos\beta & -\sin\beta \\ \sin\beta & \cos\beta \end{bmatrix} \cdot \begin{bmatrix} x_1' \\ y_1' \end{bmatrix} = \begin{bmatrix} x_1 \\ y_1 \end{bmatrix}.$$

Transformation in polar coordinates is not a linear transformation (Fig. C.2), and as we may see from

$$x_1 = \rho_1 \cos\theta_1,$$
$$y_1 = \rho_1 \sin\theta_1.$$

or

$$\rho_1 = \sqrt{x_1^2 + y_1^2},$$
$$\theta = \arctan y_1/x_1.$$

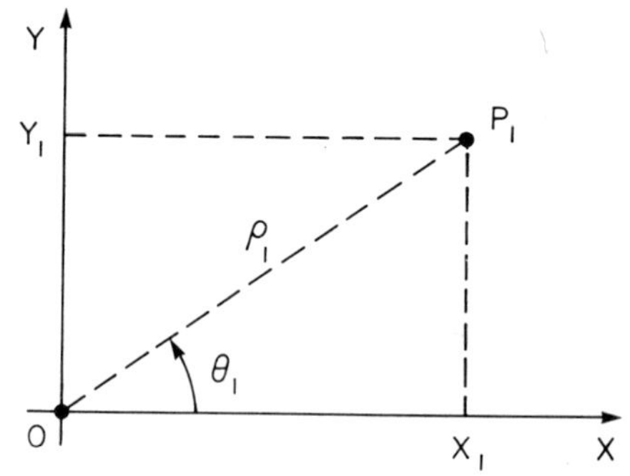

Figure C.2 Transformation in polar coordinates.

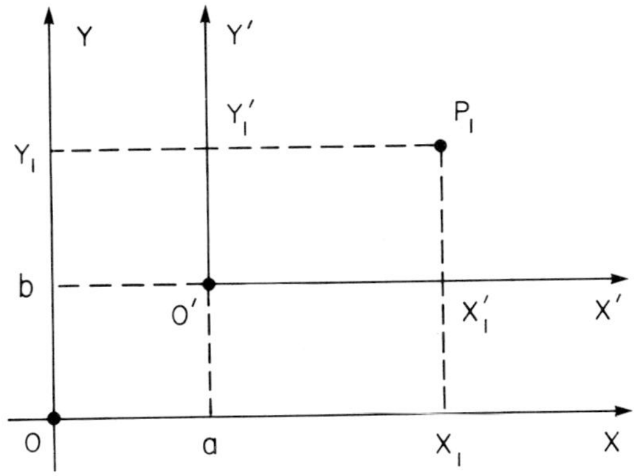

Figure C.3 Shifting of Cartesian coordinate in R^2.

We recall another typical and trivial change of coordinates. It is the linear transformation, shown in Figure C.3. We have

$$x'_1 = x_1 - a,$$

$$y'_1 = y_1 - b,$$

or

$$x_1 = x'_1 + a,$$

$$y_1 = y'_1 + b.$$

Using matrices we obtain

$$\begin{bmatrix} x_1 \\ \\ y_1 \end{bmatrix} + \begin{bmatrix} a \\ \\ b \end{bmatrix} = \begin{bmatrix} x'_1 \\ \\ y'_1 \end{bmatrix},$$

$$\begin{bmatrix} x'_1 \\ \\ y'_1 \end{bmatrix} - \begin{bmatrix} a \\ \\ b \end{bmatrix} = \begin{bmatrix} x_1 \\ \\ y_1 \end{bmatrix}.$$

In a general case, a change of coordinates is given by

$$x'_1 = f(x_1, y_1) \qquad \text{or} \qquad x_1 = \varphi(x'_1, y'_1),$$

$$y'_1 = g(x_1, y_1) \qquad \text{or} \qquad y_1 = \psi(x'_1, y'_1),$$

where the correspondence is biunivocal.

Now we show how the coordinates of a convex domain in R^2 are transformed.

EXAMPLE C.1

We examine the change of coordinates of a rectangle from the referential x–0–y to the referential x′–0–y′ by simple rotation, as shown in Figure C.4.

$$x_1 \leq x \leq x_2,$$

$$y_1 \leq y \leq y_2. \tag{C.1}$$

A rotation β around 0 gives the referential x′–0–y′. The coordinates (x, y) become (x′, y′) by the linear transformation

$$x = x' \cos\beta - y' \sin\beta,$$

$$y = x' \sin\beta + y' \cos\beta. \tag{C.2}$$

Keeping (C.1) in (C.2) we have

$$x_1 \leq x' \cos\beta - y' \sin\beta \leq x_2,$$

$$y_1 \leq x' \sin\beta + y' \sin\beta \leq x_2.$$

Of course, the rectangular domain of confidence is not given in a simple way after a rotation β, except for a multiple of $\pi/2$.

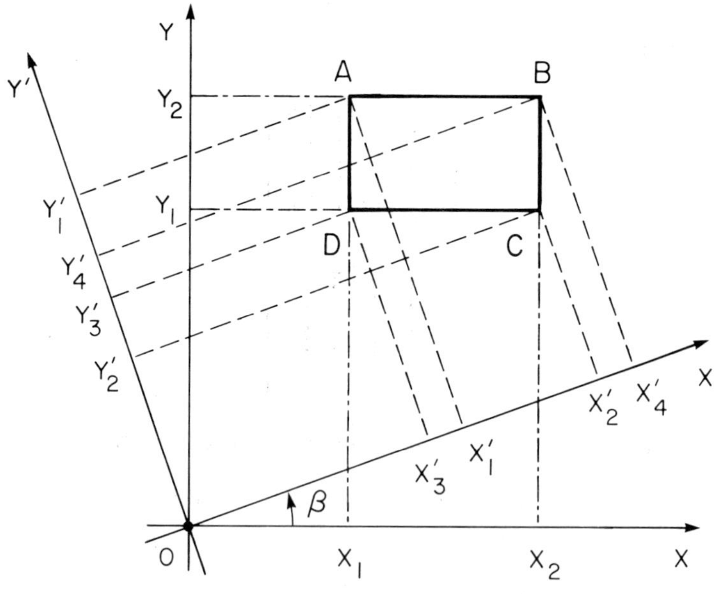

Figure C.4 Rotation of Cartesian coordinate for a convex domain in R^2 (Example C.1).

EXAMPLE C.2

To pass from a rectangle defined in $x-0-y$ with coordinates (x,y), as shown in Figure C.5, to the same rectangle defined in polar coordinates (ρ, θ) we have

$$x_1 \leq \rho \cos\theta \leq x_2,$$

$$y_1 \leq \rho \sin\theta \leq y_2.$$

This example can be used also to define a rectangle in the complex plane

$$z = x + jy \qquad j = \sqrt{-1}.$$

For this purpose we write

$$x_1 + jy_1 \preceq \rho(\cos\theta + j\sin\theta) \preceq x_2 + jy_2, \tag{C.3}$$

where the symbol $\preceq$ (partial order) must replace $\leq$ (total order). Equation (C.3) can also be written as

$$x_1 + jy_1 \preceq \rho\, e^{j\theta} \preceq x_2 + jy_2,$$

or, if we write

$$z_1 = x_1 + jy_1,$$

$$z_2 = x_2 + jy_2,$$

$$z = x + jy,$$

$$z_1 \preceq z \preceq z_2.$$

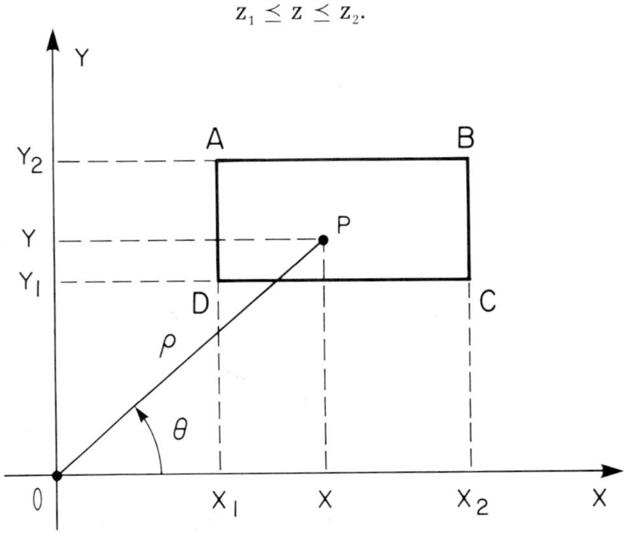

Figure C.5 Conversion of Cartesian coordinate to polar coordinate in R^2 for a convex domain in R^2 (Example C.2).

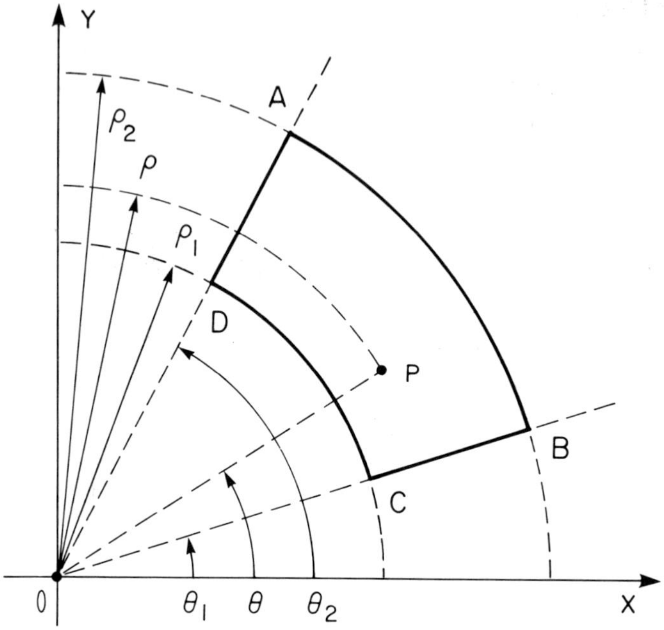

Figure C.6 Polar convexity (Example C.3).

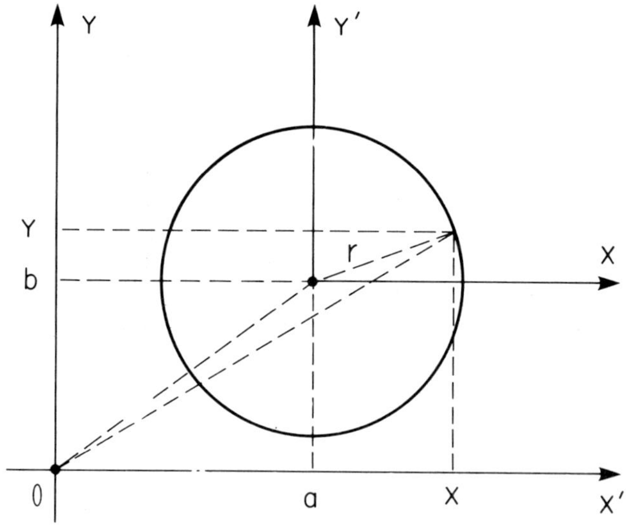

Figure C.7 A circular domain of confidence.

EXAMPLE C.3

As we explained in section 3.8, a fuzzy subset defined by

$$\rho_1 \leq \rho \leq \rho_2,$$
$$\theta_1 \leq \theta \leq \theta_2,$$

is not convex for the rectangular coordinates $x{-}0{-}y$, as we can see from Figure C.6. The lack of convexity is visible between C and D.

It is also possible to define another kind of convexity called *polar convexity* for a domain such as the one shown in Figure C.6.

In section 3.8 fuzzy complex numbers were defined by a couplet (ρ, θ) without the imaginary number

$$j = \sqrt{-1}.$$

The connection is not so simple, however, even though it can be defined.

EXAMPLE C.4

A circular domain of confidence with a center which is not $(x, y) = (0, 0)$ is shown in Figure C.7 and is defined by

$$0 \leq (x - a)^2 + (y - b)^2 \leq r^2.$$

Passing to coordinates $x'{-}0{-}y'$, we have

$$0 \leq x'^2 + y'^2 \leq r^2,$$

or

$$0 \leq \rho' \leq r.$$

In this appendix we recalled some elementary properties in the change of coordinates in R^2. They are well known, and can be easily extended to R^n, $n > 2$.

We have discussed domains of confidence without consideration to the presumption level α. To pass to fuzzy numbers of dimension n, it is necessary to check for convexity by nesting. Monotonicity is valid for the case when α moves from 0 to 1.

Note that changes in coordinates are made according to the type of operations that are being carried out on fuzzy numbers.

BIBLIOGRAPHY

ARTICLES SPECIALLY DEVOTED TO FUZZY NUMBERS AND VARIOUS ASSOCIATED CONCEPTS

E. P. Bonisson, 1982, A fuzzy sets based linquistic approach: Theory and Applications, in *Approximate Reasoning in Decision Analysis*, M. M. Gupta and E. Sanchez, eds., North Holland, Amsterdam, pp. 329–339.

S. Chanas and J. Kamburowski, 1981, The use of fuzzy variables in P.E.R.T., *Fuzzy Sets Syst.* **5**(1):11–20.

C. L. Chang, 1975, Interpretation and execution of fuzzy programs, in *Fuzzy Sets and Their Applications to Cognitive and Decision processes*, L. A. Zadeh, K. S. Fu, K. Tanaka, and M. Shimura, eds., Academic Press, New York, pp. 151–170.

A. DeLuca and S. Termini, 1979, Entropy and energy measures of a fuzzy set, in *Advances in Fuzzy Set Theory and Applications*, M. M. Gupta, R. K. Ragade, and R. R. Yager, eds., North-Holland, Amsterdam, pp. 321–338.

D. Dubois and H. Prade, 1978, Operations on fuzzy numbers, *Int. Jour. Syst. Sci.* **9**(6):613–626.

D. Dubois and H. Prade, 1979, Outline of fuzzy set theory, in *Advances in Fuzzy Set Theory and Applications*, M. M. Gupta, R. K. Ragade, and R. R. Yager, eds., North Holland, Amsterdam, pp. 27–48.

D. Dubois and H. Prade, 1979, Various kinds of interactives additions of fuzzy numbers, Applications to decision analysis in the presence of linguistic probabilities, *IEEE Conference on Decision and Control, 19th, Proceedings*, vol. 2, Fort Lauderdale, Fl., pp. 783–787.

D. Dubois and H. Prade, 1981, Addition of interactive fuzzy numbers, *IEEE Trans. Auto. Control* **26**(4):626–936.

D. Dubois and H. Prade, 1982, A class of fuzzy measures based on triangular norms—A general framework for the combination of uncertain information, *Int. Jour. Gen. Syst.* **8**(1):43–61.

D. Dubois and H. Prade, 1982, On several representations of an uncertain body of evidence, in *Approximate Reasoning in Decision Analysis*, M. M. Gupta and E. Sanchez, eds., North-Holland, Amsterdam, pp. 167–181.

I.R. Goodman, 1982, Characterizations of N-ary fuzzy set operations which induce homomorphic random set operations, in *Approximate Reasoning in Decision Analysis*, M. M. Gupta and E. Sanchez, eds., North-Holland, Amsterdam, pp. 203–211.

R. Jain, 1976, Tolerance analysis using fuzzy sets, *Int. Jour. Syst. Sci.* **12:**1393–1401.

A. Kaufmann, 1981, La Prevision a Long Terme par la Méthode Fuzzy Delphi, *Revue Busefal*, L.S.I., Univ. Paul Sabatier, Toulouse, pp. 60–71.

A. Kaufmann, 1981, La Theorie des Nombres Hybrides, *Revue Busefal*, 8, L.S.I., Univ. Paul Sabatier, Toulouse, pp. 105–113.

A. Kaufmann, 1981, Hybrid convolution, A way to combine fuzzy numbers and random numbers, *Fuzzy Math. (Wuhan, P.R. China)* **1**(2):13–20.

M. Kokawa, K. Nakamura and M. Oda, 1979, Fuzziness and catastrophe in estimation and decision processes, in *Advances in Fuzzy Set Theory And Applications*, M. M. Gupta, R. K. Ragade, and R. R. Yager, eds., North-Holland, Amsterdam, pp. 463–480.

H. Kwakernaak, 1979, Fuzzy random variables, algorithms and examples for the discrete case, *Inform. Sci.* **17:**615–616.

M. Mizumoto and K. Tanaka, 1976, Some properties of fuzzy sets of type 2, *Inform. Control* **31:**312–340.

N. Mizumoto and K. Tanaka, 1979, Some properties of fuzzy numbers, in *Advances in Fuzzy Sets Theory and Applications*, M. M. Gupta, R. K. Ragade, and R. R. Yager, eds., North Holland, Amsterdam, pp. 153–164.

S. Nahmias, 1977, Fuzzy Variables, *Fuzzy Sets Syst.* **1**(2):97–110.

S. Nahmias, 1979, Fuzzy variables in a random environment, in *Advances in Fuzzy Set Theory and Applications*, M. M. Gupta, R. K. Ragade, and R. R. Yager, eds., North Holland, Amsterdam, pp. 165–180.

C. V. Negoita, P. Flonder, and M. Sularia, 1977, On fuzzy environment in optimization problems, in *Modern Trends in Cybernetics and Systems*, J. Rose and C. Bilciu, eds., Springer Verlag, Berlin.

H. Prade, 1979, using fuzzy set theory in a scheduling problem, A case study, *Fuzzy Sets Syst.* **2:**153–166.

D. A. Ralescu, 1979, A survey of the Representation of fuzzy sets and its applications, in *Advances in Fuzzy Sets Theory and Applications*, M. M. Gupta, R. K. Ragade and R. R. Yager, eds., North Holland, Amsterdam, pp. 77–92.

N. B. Rao and A. Rashed, 1981, Some comments on fuzzy random variables, *Fuzzy Sets Syst.* **6**(3):285–292.

N. E. Stein and K. Talaki, 1981, Convex fuzzy random variables, *Fuzzy Sets Syst.* **6**(3):271–284.

G. Tamburrini and S. Termini, 1982, Some foundational problems in the formalization of vagueness, in *Fuzzy Information and Decision Processes*, M. M. Gupta and E. Sanchez, eds., North-Holland, Amsterdam, pp. 161–166.

E. Trillas and T. Riera, 1978, Entropies in finite fuzzy sets, *Inform. Sci.* **15**:159–168.

L. A. Zadeh, 1965, Fuzzy sets, *Inform. Control* **8**:338–353.

L. A. Zadeh, 1968, Probability measures of fuzzy events, *Jour. Math. Analysis Appl.* **10**:421–427.

L. A. Zadeh, 1971, Similarity relations and fuzzy orderings, *Inform. Sci.* **3**:177–200.

L. A. Zadeh, 1975, The concept of a linguistic variable and its application to approximate reasoning (I), (II), (III), *Inform. Sci.* **8**:199–249; **8**:301–357; **9**:43–80.

L. A. Zadeh, 1978, Fuzzy sets as a basis for a theory of possibility, *Fuzzy Sets Syst.* **1**(1):3–28.

[Note: From 1965 to 1984, over 4,500 papers and research reports have appeared on fuzzy set theory and its applications. A partial bibliography may be found in *Advances in Fuzzy Set Theory and Applications*, M. M. Gupta, R. K. Ragade and R. R. Yager, eds., North Holland, Amsterdam, 1979. The reader is also referred to the *International Journal of Fuzzy Sets and Systems* (North-Holland) and the *Journal of Fuzzy Mathematics* (Huazhong—Central China—University of Science and Technology, Wuhan, Hubei, People's Republic of China).]

BOOKS ON FUZZY SUBSETS THEORY AND APPLICATIONS

J. C. Bezdek, 1981, *Pattern Recognition with Fuzzy Cbjective Function Algorithms*, Plenum Press, New York.

D. Dubois and H. Prade, 1980, *Fuzzy Sets and Systems: Theory and Applications*, Academic Press, New York.

M. M. Gupta and E. Sanchez (eds.), 1982, *Approximate Reasoning in Decision Analysis*, North-Holland, Amsterdam.

M. M. Gupta and E. Sanchez (eds.), 1982, *Fuzzy Information and Decision Process*, North-Holland, Amsterdam.

M. M. Gupta (ed.), R. K. Ragade, and R. R. Yager (assoc. eds.), 1979, *Advances in Fuzzy Set Theory and Applications*, North-Holland, Amsterdam.

M. M. Gupta (ed.), G. Saridis, and B. R. Gaines (assoc. eds.), 1977, *Fuzzy Automata and Decision Processes*, North Holland, New York.

M. M. Gupta, A. Kandel, W. Bandler, and J. B. Kiszka (eds.), 1985, *Approximate Reasoning in Expert Systems*, North-Holland, Amsterdam.

A. Kandel, 1982, *Fuzzy Techniques in Pattern Recognition*, John Wiley and Sons, New York.

A. Kandel and S. C. Lee, 1979, *Fuzzy Switching and Automata*, Crane and Russak, New York.

A. Kaufmann, 1973, *Introduction a la Theorie des Sous-Ensembles Flous*, vol. 1, *Elements Théoriques de Base*, Masson et cie, Paris, France.

A. Kaufmann, 1975, *Introduction a la Theorie des Sous-Ensembles Flous*, vol. 2, *Applications a la Linguistique et la Semantique*, Masson et cie, Paris, France.

A. Kaufmann, 1975, *Introduction a la Theorie des Sous-Ensembles Flous*, vol. 3, *Applications a la Classification et la Reconnaissance des Formes, aux Automates et aux Systems, aux Choix des Critères*, Masson et cie, Paris, France.

A. Kaufmann, 1975, *Introduction to the Theory of Fuzzy Subsets*, vol. 1, Academic Press, New York.

A. Kaufmann, 1976, *Introduction a la Theorie des Sous-Ensembles Flous*, vol. 4, Ed. Masson, Paris.

A. Kaufmann, 1979, *Modèles Mathematiques pour la Stimulation Inventive*, Ed. Albin Michel, Paris.

A. Kaufmann, 1979, *Introduction to the Theory of Imprecise Sets* (in Russian), Radio, Moscow.

A. Kaufmann, 1982, *Introduccion a la teoria de los subconjuntos borrosos, Elementos teorieos de base*, C.E.C.S.A., Mexico.

A. Kaufmann, T. Dubois, and M. Cools, 1975, *Exercices avec Solutions sur la Theorie des Sous, Ensembles Flous*, Masson, Paris.

W. J. M. Kickert, 1979, *Fuzzy Theories on Decision Making*, Hortinus Nijhoff, The Hague, Boston, London.

E. H. Mamdani and B. R. Gaines, 1981, *Fuzzy Reasoning and Its Applications*, Academic Press, New York.

C. V. Negoita, 1979, *Managements Applications of System Theory*, Birkhauser, Bale and Stuttgart.

C. V. Negoita and D. A. Ralescu, 1975, *Applications of Fuzzy Sets to Systems Analysis*, Birkhauser, Bale and Stuttgart.

M. Prevot, 1980, *Sous ensembles Flous, une Approche Theorique*, Ed. Sirey, E. Trillas, 1977, Conjuntos Borrosos, Ed. Vincenso Universitad.

P. P. Wang and S. K. Chang (eds.), 1980, *Fuzzy Sets—Theory and Applications to Policy Analysis and Information Systems*, Plenum Press, New York.

L. A. Zadeh, K. S. Fu, K. Tanaka, and M. Shimura (eds.), 1974, *Fuzzy Sets and Their Applications to Cognitive and Decision Processes*, Academic Press, New York.

Index